AF601552

THE OIL PALM

A Potential Oil Yielding Palm for India

The Authors

Dr P. Rethinam obtained his graduation and post-graduation in agriculture from Tamil Nadu Agricultural University, Coimbatore. He started his career as Research Assistant in 1963, Assistant Lecturer and Assistant Professor up to 1976 at TNAU, Coimbatore and then joined Agricultural Research Service in ICAR in 1976. He became Project Coordinator, (Palms) in 1982 and was responsible for research coordination and management of palms. He was selected as Assistant Director General (Plantation Crops), ICAR, New Delhi in 1988 and coordinated the research on Plantation Crops, Medicinal and Aromatic Crops and Betel vine. He introduced the concept of irrigated Oil Palm as small holders crop. He was founder Director of the National Research Centre for Oil Palm at Pedavegi, A.P. and developed infrastructural facilities in a short time. Dr Rethinam was selected as the Chairman, Coconut Development Board, Kochi, Ministry of Agriculture, and Government of India in the year 2001. He has published 215 articles, co-edited two volumes of books on Recent Advances in Plantation Crops, edited 37 books and many technical bulletins, APCC Proceedings, COCOTECH Proceedings, Noni Search Proceedings and reports, DVDs, CDs of APCC intellectual database, etc.

Dr K. L. Chadha obtained Ph.D (Hort) from IARI, New Delhi. Occupying key positions e.g. Horticulture Commissioner (GoI) and DDG, Hort (ICAR) he steered R&D of horticulture for five decades. During this period, he established a number of Institutes and National Research Centres. He has made significant contributions to the commercialisation of oilpalm in India. He chaired the first working group on potentialities of oilpalm cultivation constituted by Ministry of Agriculture in 1986 and a national committee to review the progress in 2007. These reports suggested potential areas of oilpalm cultivation in 13 states. Most of the recommendations of his group were accepted, and resulted in setting up of large scale demonstrations to motivate farmers with he being the Secretary of the committee constituted by PMO to overview these. He led teams to 8 important oilpalm growing countries in the world to select the best source of planting material. His proposal to set up a NRC on oilpalm was approved by Govt. of India, which was later established at Pedavegi by ICAR. He has written/edited/contributed several books, research papers and popular articles. He received 24 awards, including Padma Shri. He is currently President of the Horticultural Society of India, New Delhi.

THE OIL PALM

A Potential Oil Yielding Palm for India

P. Rethinam

K. L. Chadha

2020

Daya Publishing House®

A Division of

Astral International Pvt. Ltd.

New Delhi – 110 002

ISBN: 9789388173858 (HB)
Reprinted 2020

Publisher's Note:

Every possible effort has been made to ensure that the information contained in this book is accurate at the time of going to press, and the publisher and author cannot accept responsibility for any errors or omissions, however caused. No responsibility for loss or damage occasioned to any person acting, or refraining from action, as a result of the material in this publication can be accepted by the editor, the publisher or the author. The Publisher is not associated with any product or vendor mentioned in the book. The contents of this work are intended to further general scientific research, understanding and discussion only. Readers should consult with a specialist where appropriate.

Every effort has been made to trace the owners of copyright material used in this book, if any. The author and the publisher will be grateful for any omission brought to their notice for acknowledgement in the future editions of the book.

Published by : **Daya Publishing House®**
A Division of
Astral International Pvt. Ltd.
– ISO 9001:2015 Certified Company –
4736/23, Ansari Road, Darya Ganj
New Delhi-110 002
Ph. 011-43549197, 23278134
E-mail: info@astralint.com
Website: www.astralint.com

Digitally Printed at : **Replika Press Pvt. Ltd.**

" WITH 27 MILLION HA UNDER OILSEEDS, INDIA PRODUCES 7 MILLION TONS OF OIL........

WITH 2 MILLION HA UNDER OIL PALM, INDIA CAN PRODUCE 7 MILLION TONS OIL."

Dr. S.K. PATTANAYAK
SECRETARY

भारत सरकार
कृषि एवं किसान कल्याण मंत्रालय
कृषि, सहकारिता एवं किसान कल्याण विभाग
Government of India
Ministry of Agriculture & Farmers Welfare
Department of Agriculture, Cooperation
& Farmers Welfare

Foreword

India is the largest producer of oilseeds in the world and the oilseed sector occupies an important position in the country's economy. The country accounts for 12-15 per cent of global oilseeds area, 6-7 per cent of vegetable oils production, and 9-10 per cent of the total edible oils consumption. In terms of acreage, production and economic value, oilseeds are second only to food grains. With the population increase on one side, and income increase on the other, increase in per capita consumption has considerably boosted the demand for vegetable oil, which in turn has necessitated the import. In order to reduce the dependence on imports, India launched the Technology Mission on Oilseeds with a view to increase the production and productivity of nine annual oilseeds. Even though this has increased the area and production, it still does not meet the demand of the country. Horizontal area expansion being a limiting factor, the productivity increase alone may not be adequate to meet the growing demand of palm oil. Considering this fact, the oil palm which is the highest oil yielding crop (4 to 8 tons of oil/ha/year) was introduced in many Asian countries. Malaysia and Indonesia have emerged as the largest palm producing oil in the world, pushing soybean to the second position. India also had introduced oil palm in the 1980s as a rainfed crop in the forest lands but is now promoting it as irrigated small holders' crop which shows promise in terms of yield and economics.

Oil palm, with its high oil yield potential, will not only help to narrow the gap between the demand and supply of vegetable oils but also provide employment opportunity, nutrition, livelihood and social security besides contributing substantially to bio-fuel energy.

Dr. P. Rethinam the former Director for National Research Centre for Oil Palm and Dr. K.L. Chadha, Former Horticulture Commissioner, Government of India and Deputy Director General, Horticulture, ICAR, New Delhi, with their vast experience in launching oil palm cultivation under irrigated conditions in the small and marginal holdings of the country have brought out this book on Oil palm. They have attempted to gather lot of information on all aspects of Oil palm cultivation, processing and by-products utilization. It is hoped that this book will provide guidance to the farmers, students, teachers and entrepreneurs, all of whom are interested to know about oil palm production and post-harvest management. I congratulate the author for bringing out this book on oil palm which is expected to provide guidance to R & D workers and farmers engaged in promotion of oil palm.

(S. K. Pattanayak)

Office : Krishi Bhawan, New Delhi-110001, दूरभाष / Phone : 23382651, 23388444 फैक्स सं० / Fax No. : 23386004
E-mail : secy-agri@gov.in

Preface

Oil Palm (*Elaeis guineensis*, Jacq), originated from West Africa is known for its nutritional and health effects as well as livelihood and social security to millions in the world. It travelled from its place of origin to different countries, reached Asian countries as ornamental plants in Bogor Botanical Garden in Indonesia in 1856 and Botanical garden in Kolkata and became a commercial highest oil yielding crop of the world. The four palms in Bogor Gardens formed the basic planting materials for Asia. Indonesia and Malaysia which are now the largest producers in the world. Palm oil has reached the highest contribution of 35 per cent of the total global vegetable oil production. A crop which has the highest yield potential of 18 t oil/ha/year, is able to produce 8.0-12.0 tons/ha of yield in larger plantations in Malaysia through continuous research and development. Oil Palm with two types of oil, palm oil and palm kernel oil, is now the major source of global vegetable oil economy for food, nutrition, industrial and biofuel purposes. While the commercial plantations in Malaysia and Indonesia have moved fast from 1960 and made further progress during eighties emerging as the global leaders in palm oil production. India started first commercial plantation in Thodupuzha, Kerala, in 1965 and expanded it to Kollam in Kerala. Little Andamans could not move forward due to ban on conversion of forest as well as improper management of plantations which resulted in poor performance and negative impact about oil palm cultivation. However, a historical change of irrigated oil palm concept introduced in 1986 with oil palm cultivation as small holders crop took a turn in Oil Palm Cultivation in India. Chadha Committee in 1986 and 2006 and Rethinam Committee 1985, 2000, and 2011 reports have identified 1.93 million ha in 18 states in the country including North East and implementation of Oil Palm Development by Ministry of Agriculture and Cooperation, Department of Agriculture, Government of India through State Governments have resulted in bringing about 300,000 ha under oil palm so for. The book on Oil Palm brought out now has all details of oil palm cultivation which the authors felt are essential to those who are involved in Oil Palm cultivation, processing and product development as well as by-product utilisation.

We the authors who are the pioneers in promoting irrigated oil palm successfully as small holders' crop in the country dedicate this book to the oil palm growers who are continuously making history in promoting of Oil Palm in India. It is hope that the publication will be useful to teachers, students, development agencies, processing industries and farmers involved in oil palm cultivation in India.

P. Rethinam

K. L. Chadha

Contents

Foreword vii

Preface ix

1. **The Oil Palm** 1
   - *1.1. Importance of Oil Palm*
   - *1.2. Importance of Palm Oil*

2. **Global Vegetable Oil Scenario** 7
   - *2.1. Global Scenario*
   - *2.2. Indian Scenario of Vegetable Oils*

3. **Global Palm Oil Scenario** 17
   - *3.1. Harvested Area, Production and Productivity of Palm Oil*
   - *3.2. World Major Producers of Palm Oil*
   - *3.3. Major Countries Exporting and Importing Palm Oil*
   - *3.4. Palm Oil Production in the Asia-Pacific Region*
   - *3.5. Future Contribution from Palm Oil*
   - *3.6. Global Oil Palm Research*
   - *3.7. Oil Palm Research in India*

4. **Introduction of Oil Palm in India** 45
   - *4.1. Why Oil Palm in India?*
   - *4.2. Early Introduction of Oil Palm*

4.3. *Commercial Plantations of Oil Palm*
4.4. *Reports on Oil Palm Cultivation in India*
4.5. *Efforts made for Development of Irrigated Oil Palm*
4.6. *Reservations in Oil Palm Cultivation in India*
4.7. *Problems that have Affected the Area Development in Oil Palm*
4.8. *Steps Suggested to Sustain Oil Palm Development*
4.9. *Way Forward*
4.10. *To Sum Up*

5. Climate and Soil Requirements 73
5.1. *Climate*
5.2. *Soils*

6. Origin, Habitat, Distribution and Botany of Oil Palm 93
6.1. *Origin*
6.2. *Habitat*
6.3. *Distribution*
6.4. *Botany of Oil Palm*

7. Crop Improvement 105
7.1. *Oil Palm Breeding*
7.2. *Improvement Programme*
7.3. *Breeding Methods*

8. Planting Material Production 115
8.1. *Hybrid Seed Production*
8.2. *Clonal Propagation*
8.3. *Planting Materials*

9. Establishment of Nurseries 133
9.1. *Nursery Establishment*
9.2. *Types of Polybag Nurseries*
9.3. *Planting and After Care*
9.4. *Fertilizer Application*
9.5. *Seed and Seedling Handling*
9.6. *Advanced Nursery Techniques for Primary Nursery*
9.7. *Management of Flood Affected Oil Palm Nursery*

10. Plantation Establishment and Management 151
10.1. *Plantation Establishment*

10.2. After Cultivation
10.3. Weed Management
10.4. Disaster Management
10.5. Oil Palm Cultivation in Wastelands

11. Water Management **179**
11.1. Water Requirement and Water Use Efficiency
11.2. Quantity of Water
11.3. Water Deficiency and its Effects
11.4. Methods of Irrigation
11.5. Drainage Management
11.6. Moisture Conservation Methods
11.7. Water Harvesting

12. Nutrient Management **189**
12.1. Nutrient Removal
12.2. Importance of Nutrients
12.3. Fertilizer Recommendations
12.4. Sources of Fertilizers
12.5. Methods of Application
12.6. Frequency of Fertilizer Application
12.7. Fertilizer Application Time
12.8. Soil and Leaf Analysis
12.9. Nutrient Deficiencies
12.10. Disorders
12.11. Yield Gap Analysis

13. Oil Palm Based Cropping/Farming Systems **215**
13.1. Multiple Cropping
13.2. Intercropping in Oil Palm
13.3. Mixed Cropping
13.4. Mixed Farming
13.5. Cover Crops

14. Plant Protection **225**
14.1. Insect Management
14.2. Disease Management

15. Harvesting, Yield and Handling **241**
15.1. Degree of Ripeness

15.2. *Bunch Ripeness*
15.3. *Pre-harvest and Harvest*
15.4. *The Ripening Process*
15.5. *Grading Procedures*
15.6. *Bunch Classifications*
15.7. *Grading Methods*
15.8. *Yield*
15.9. *Cost of Cultivation*
15.10. *Labour Productivity and Employment Opportunity*

16. Processing of Oil Palm Fruits **269**
16.1. *Milling Technology*
16.2. *Steps Involved in Processing*
16.3. *Oil Storage*
16.4. *Nut Recovery*
16.5. *Kernel Oil Extraction*
16.6. *Refining*
16.7. *Recent Advances in Milling Technology*

17. Food and Non-Food Uses of Palm Oil for Nutrition and Health **279**
17.1. *Food Uses*
17.2. *Non Food Uses*
17.3. *Palm Oil for Nutrition*

18. Waste Utilization and Environment **287**
18.1. *Types of Wastes*
18.2. *Systems Approach for Zero Waste and Zero Discharge in Palm Oil Mill*
18.3. *Management of Environmental Issues*
18.4. *Oil Palm and Carbon Sequestration*

References **305**

Index **311**

Color Plates

Chapter 1

The Oil Palm

Oil Palm (*Elaeis*, Jacq.), popularly known as African or Red oil palm found originally in the wild form in the forests of Guinea Coast was used by the natives all over Africa as vegetable oil before it was developed into an agricultural crop. Today, it is the largest source of vegetable oil in the world used for innumerable purposes. Palm oil, known as Golden Palm Oil, is one of the 17 major oils traded in the global edible oil and fat market. It is consumed as food for the past more than 5000 years. It is a versatile perennial oil yielding tree crop which has a theoretical potential yield of 18 MT of oil/ha/year and a realized potential yield of 12 MT of oil/ha/year in large plantations in Malaysia. It is capable of yielding for an average of 4-6 MT of crude palm oil and 0.4 to 0.6 MT of palm kernel oil/ha/year for a period of 25 to 30 years from fourth year of planting. Crude palm oil from mesocarp and palm kernel oil from nuts are the largest sources of vegetable oils. Palm kernel oil is the largest source of lauric oil. Over a period of four decades, this crop has emerged as the second largest oil yielding crop in the world with the bulk of production from two countries *viz.*, Malaysia and Indonesia. Recently, this crop has reached the top place pushing soya bean oil to second place. Oil palm is one of the world's most efficient oil bearing crops in terms of land utilization, efficiency, productivity, economy, employment, poverty reduction and environment sustainability.

The future global demand for vegetable oils is likely to be met to a major extent by palm oil because of its high oil yield and low cost. Thus, oil palm is the crop of future. It is eco-friendly, environmentally sustainable, nutritive, import substitute and useful for cogeneration of bio-fuel. Above all, it will help to elevate the socio-economic status of the small and marginal farmers.

This crop was introduced in India in 1886 as an ornamental plant in the Botanical gardens of Kolkata and remained only as an ornamental palm. It has the same historical perspective of planting four oil palms in Bogor Botanic Garden of Indonesia in 1846 which became the parental source for oil palm development in South East Asia. Malaysia and Indonesia took up commercial plantings during 1960 and India also started commercial plantations in Kerala during the year 1965 at Thodupuzha. But not much progress was made in India except raising two large scale plantations, one in Kerala and another in Little Andaman in the forest lands. While Malaysia and Indonesia moved fast in oil palm development and attained top two positions in the world, India did not make further headway instead gave negative potential of oil palm development. However, oil palm planted from different sources at Thodupuzha served as source for tenera seed production in the country. Oil palm is one of the world's most efficient oil bearing crops in terms of land utilization, efficiency and productivity.

1.1. Importance of Oil Palm

Oil palm is the highest oil yielding perennial palm compared to any other oil yielding annuals and perennials (Figures 1.1 and 1.2). Palm oil is the major resource for revenue in Indonesia and Malaysia. It is true for all developing and under developed Nations. Oil palm will assist the developing countries to promote poverty eradication and improve income for small holders and uplift the economy of developing Nations. Palm oil production respects and adopts the 3 Ps principles of sustainability-People, Planet and Profit. Dominance of Palm oil is increasing and it continues to be a long term commodity for producers and consumers. Palm oil is a solution provider. Consumers get strategic solutions from palm oil for food security, trans free products and competitive pricing.

1.1.1. Oil Palm for Diversification

The concept of growing irrigated oil palm as small holder's crop in the conventional agricultural crop land is a land mark for crop diversification in Indian Agriculture. Various Expert Committees constituted by the Ministry of Agriculture, Government of India identified a total of about 2.0 million hectares area in 18 states

Figure 1.1: The Oil Palm Stands Tall in Oil Yield Amongst Oil Yielding Crops.

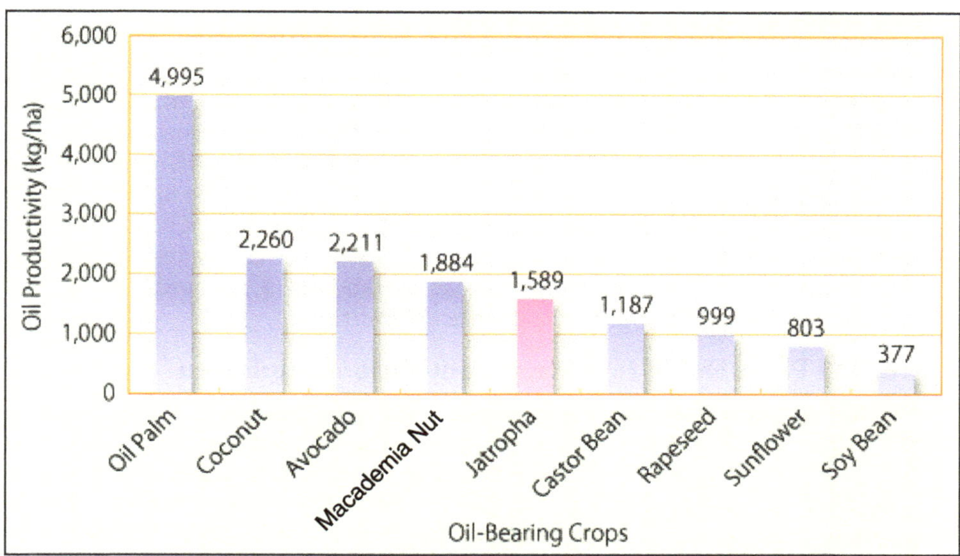

Figure 1.2: Comparitive Productivity of Nine Major Oil Bearing Crops.

of the country suitable for oil palm cultivation under irrigation. About 80 per cent of the area identified, is located in Andhra Pradesh, Karnataka and Tamil Nadu. The remaining areas are in Assam, Arunachal Pradesh, Bihar, Chhattisgarh, Goa, Gujarat, Kerala, Maharashtra, Mizoram, Meghalaya, Nagaland, Odisha, Tripura and West Bengal. Thus, oil palm with adequate irrigation facilities provides scope for diversification from low yielding dry and garden land crop to a high value oil yielding crop. This diversification will provide sustainable income, livelihood security and a favorable ecosystem. In the North Eastern India it will help overcome *jhum* cultivation and provide sustainable income.

1.1.2. Oil Palm for Value Addition

The superior characteristics of palm oil make it an excellent raw material for the production of oleo-chemical products including fatty acids, fatty alcohols, glycerol and other derivatives. These are widely used in the manufacture of cosmetics, pharmaceuticals, house hold articles and industrial products. Oleo-chemicals are used for manufacture of environment friendly detergents, as they are highly degradable.

1.1.3. Oil Palm for Non-conventional Energy

Oil palm wastes like empty bunches; leaves, leaf petioles, male inflorescence *etc.* are a good source for production of non-conventional energy. The fibre and most of the shell produced by the processing mills are used as fuel for the boiler. Even leaf petiole and empty bunches are used for producing bio-energy. Considering its positive characteristics and already established wide ranging benefits, it is hardly surprising that palm oil has progressively gained acceptability and popularity all over the world.

1.1.4. Oil Palm for By-product Utilization

The leaves of oil palm contain tocopherol and this can be profitably exploited industrially. Fibre from fronds and empty bunches are reconstituted to make materials such as medium density fibre boards and chip boards. Fibre from trunk is transformed into beautiful furniture. The empty bunches can be utilized as medium for mushroom cultivation. The meso-carp fibre coming out of the mill is found to have about 10-12 per cent of oil and that can be extracted through solvent extraction technology. Oil palm trunks, fronds and empty fruit bunch can be used to produce pulp, paper and various other products. Malaysian Palm Oil Board (MPOB) had developed lot of technologies for by product utilization.

1.1.5. Oil Palm Helps to Balance out Carbon Emission

While industrial land use generates carbon, Oil palm keeps the land green by contributing up to 21.3 MT of Oxygen per ha into the atmosphere every year. Global carbon storage by oil palm is estimated at 74 MT C y 1 for 12 million ha due to its high annual bio mass productivity. The amount of carbon sequester by 11 year old oil palm hybrids under irrigated condition ranged between 7.98 and 35.44 t/C ha with Papua New Guinea and Ivory Coast Hybrids sequestering the highest and lowest carbon content respectively. With about 3.0 lakh ha of plantings in India, oil palm seems to be a prime candidate for storing carbon and is also eligible for Clean Development Mechanism (CDM). When 2 million ha is going to come under oil palm in the country the eligibility will be much more.

1.1.6. Oil Palm Helps to Save the Environment

Oil Palm is a high yielding crop providing up to ten times more oil per ha of land/year, thus less clearing of forest land is required for cultivation of this crop. In India virtual growth is envisaged in the cultivated lands as well as in the wastelands with irrigation facilities. This will help to create forests, for 30 to 35 years.

If the envisaged area of about two million-hectares is brought under oil palm, over a period of about two to three decades, it will definitely add to the evergreen forests. Oil palm plantations serve as effective sinks for carbon dioxide and other greenhouse gases. Application of chemical fertilizers is minimized due to the reuse of recyclable organic residues obtained from both plantation and the processing units as substitute. Biological method of pest control is employed to keep pesticide use to the barest minimum. Milling and refining techniques are carefully selected to generate minimum waste effluents. The concepts of zero waste and zero burning are widely adopted in oil palm cultivation in Malaysia and elsewhere. The zero-based commitment has resulted in new product development from oil palm by-products including fronds, tree trunks, bunch waste and refinery distillate. The zero burning involves disposal of old palms by shredding and decomposition, by utilizing methods like vermicompost and organic farming which are technically feasible, socially acceptable, economically viable and environmentally sustainable. By doing zero burning and recycling the biomass into the field, the soil is enriched with the nutrients and also helps to improve the physico-chemical and biological properties of **soil.**

1.1.7. Oil Palm: The Most Versatile and Useful Plant

From cooking oil to raw materials for toiletries and cosmetics, from fuel for cars to components for furniture - virtually every part of the oil palm is of use and benefit to mankind, making it as one of the best gifts ever from nature.

1.2. Importance of Palm Oil

Palm oil consumption has recorded history of 5000 years. It feeds more than 3 billion people in more than 150 countries. Unlike other oils, palm oil is highly versatile. It is used in thousands of food and non- food industrial products, bio-fuel and bio lubricants being the latest additions.

1.2.1. Palm Oil for Health and Nutritional Security

The Food and Agriculture Organisation (FAO) and World Health Organisation (WHO) have endorsed palm oil as meeting food standards under the Codex Alimentarius Commission Programme. Palm oil is a balanced vegetable oil and a source of energy. It has excellent health attributes, *e.g.* is cholesterol free, raises the level of good HDL cholesterol, lowers LDL cholesterol, and reduces the tendency for blood clot thereby reducing the risk of heart disease. It is a rich source of beta-carotene, an antioxidant and precursor of vitamin E and tocotrienols. It contains 500-700μg/g of tacoferal, 1000 μgm/gm of tocotrienols and vitamin E. Carotenoids in red palm oil (RPO) can also serve as powerful antioxidants and anticancer agents. It is good for heart. It also has protective role in cellular aging. The recently introduced Palm Vitae is rich in vitamin E which acts as more effective antioxidants in the biological system and may play a protective role in reducing cellular aging, heart disease and cancer.

1.2.2. Palm Oil as Import Substitute

Though the mission oriented approach through Technology Mission on Oil seeds and Pulses (TMO and P) has enhanced the production of oil seeds from 10.83 MT in 1985-86 to 28.83 MT in 2007-08 and presently 27.0 million ha, still there is a shortage of more than about 10 MT of vegetable oil at the per capita consumption of12.89 kg. The per capita consumption further increased to 14.8kg during 2010-2011 and has been rising at the rate of 3 to 4 per cent per annum. But it is still lower than the quantity prescribed by the World Health Organization. Import of vegetable oil is on the increasing side of which the Palm Oil and Soya bean oil occupies the major shares. To bridge the gap, oil palm can play a very vital role with its high oil yield and perennial nature. If concerted effort is made to bring 2 million ha under oil palm, in the identified potential areas, it will be possible to get 6-8 million MT of crude palm oil and 0.6 to 0.8 MT of palm kernel oil within two decades. This will help to increase the domestic availability and reduce the import bill.

Palm Oil supplies 35 per cent of the world's oil needs at the most competitive price. This keeps prices low for consumer and industries. The plantation industry drives economic growth and provides employment opportunities to poorest people and brings in revenue for national development and stability. Oil palm cultivation and palm oil production are highly remunerative activities which help to keep world economy healthy and no doubt that Indian economy also will be healthy.

Chapter 2

Global Vegetable Oil Scenario

Globally 17 oils and fats contribute to human needs. Of these, 13 are vegetable oils, while 4 are of animal origin. Of the vegetable oils, nine namely, palm, palm kernel, soyabean, cotton seed, ground nut, sun flower, rapeseed, coconut, olive make substantial contribution while castor, linseed and sesame oils contribute less than 1 per cent of the total oil production.

2.1. Global Scenario

2.1.1. Global Production

Global production of these nine crops was estimated at 169,225 thousand ton during 2014 (Table 2.1). Major oils making significant contribution are palm oil (35.0 per cent), soya bean oil (26.65 per cent), rape seed (16.0 per cent), sunflower (9.54 per cent) followed by others.

The vegetable oil production during 1960-2014 increased from 16,072 thousand ton to 1,69,225 thousand ton. The rate of growth varied from 1.50 in case of ground nut to 136.65 in Soyabean followed by 46.86 in palm oil.

Out of 4.6 billion ha of total agricultural land available globally, 1.3 billion ha are arable, of which only 258.90 million ha are under oil seed crops. Of the 258.9 million ha under oil seeds. 40.0 per cent is under soyabean, 14.0 per cent under cotton, 13 per cent Rape seed, 10 per cent sunflower, 5 per cent oil palm and 18 per cent under other crops. This 5 per cent of land under oil palm provides 32 per cent of global supply of oils and fats.

Table 2.1: World Production of Nine Major Vegetable Oils ('000' tons)

Vegetable Oils	*1960*	*1970*	*1980*	*1990*	*2000*	*2010*	*2011*	*2012*	*2013*	*2014*	*Per cent Share (2014)*
Coconut	1,949	2,020	2,717	3,387	3,281	3,629	2,990	3,240	3,340	3,018	1.78
Cotton	2,325	2,503	2,992	3,782	3,864	4,601	4,859	5,088	4,969	4,853	2.87
Groundnut	2,587	3,044	2,864	3,897	4,560	4,074	4,030	4,007	3,931	3,874	2.29
Olive	1,339	1,442	1,701	1,855	2,5ffl	3,331	3,440	3,361	2,849	3,302	1.95
Palm	1,264	1,742	4,543	11,014	21,874	45,873	50,518	53,665	56,211	59,228	35.00
Palm Kernel	421	380	547	1,450	2,691	5,229	5,649	5,924	6,241	6,520	3.85
Rape seed	1,099	1,833	3,478	8,160	14,466	23,778	23,657	24,483	25,148	27,187	16.07
Soya bean		6,477	13,382	16,097	25,541	40,181	41,562	41,710	42,731	45,095	26.65
Sunflower	1,788	3,491	5,024	7,869	9,700	12,428	13,098	14,799	14,014	16,148	9.54
Total	**16,072**	**22,932**	**37,243**	**57,511**	**88,517**	**143,124**	**149,803**	**156,277**	**159,434**	**169,225**	**100**

2.1.2. Imports/Exports

Global import and export of 17 oils and fats are given in Tables 2.2 and 2.3 indicate that there is great demand for vegetable oil and palm oil made tremendous impact followed by soya bean and sunflower oils. In the lauric oils palm kernel oil has made more impact than coconut oil.

Table 2.2: World Imports of 17 Oils and Fats: 1998-2014 ('000' MT)

Oils/Fats	*1998*	*2000*	*2005*	*2010*	*2011*	*2012*	*2013*	*2014*
Palm Oil	11,245	15,223	26,602	37,137	38,803	41,363	43,732	44,368
Palm Kernel Oil	1,026	1,242	2,117	3,047	3,060	3,057	3,367	3,157
Soyabean Oil	7,636	6,548	9,611	9,868	9,434	9,161	9,506	9,817
Cottonseed Oil	225	198	147	151	178	212	195	157
Groundnut Oil	254	239	194	226	197	170	192	223
Sunflower Oil	3,001	2,959	3,108	4,770	5,262	7,232	6,499	8,241
Rapeseed Oil	2,215	1,778	1,381	3,330	3,707	4,132	4,153	3,992
Corn Oil	816	790	798	633	838	901	878	726
Coconut Oil	1,978	1,851	2,141	2,333	1,875	1,873	2,022	1,876
Olive Oil	469	510	721	757	793	805	927	892
Castor Oil	249	266	309	451	406	489	533	474
Sesame Oil	22	25	36	37	37	36	37	38
Linseed Oil	113	143	113	102	94	99	85	87
Total Vegetable Oils	**29,249**	**31,772**	**47,278**	**62,842**	**64,684**	**69,530**	**72,126**	**74,048**
Butter	653	627	737	705	720	718	778	859
Tallow	2,320	2,260	2,058	2,075	1,977	1,678	1,591	1,576
Fish Oil	416	819	678	799	794	881	695	748
Lard	163	204	130	119	119	114	112	123
Total Animal Oils/ Fats	**3,552**	**3,910**	**3,603**	**3,698**	**3,610**	**3,391**	**3,176**	**3,306**
GRAND TOTAL	**32,801**	**35,682**	**50,881**	**66,540**	**68,294**	**72,921**	**75,302**	**77,354**

Table 2.3: World Exports of 17 Oils and Fats: 1998-2014 ('000' MT)

Oils/Fats	*1998*	*2000*	*2005*	*2010*	*2011*	*2012*	*2013*	*2014*
Palm Oil	10,898	15,019	26,502	36,539	39,183	40,780	44,042	44,549
Palm Kernel Oil	1,045	1,220	2,095	3,064	3,136	3,067	3,273	3,158
Soyabean Oil	7,934	6,771	9,774	10,150	9,314	9,340	9,649	9,729
Cottonseed Oil	259	196	141	153	192	200	200	171
Groundnut Oil	254	235	169	201	191	191	185	245
Sunflower Oil	2,904	3,054	3,083	4,736	5,338	7,256	6,618	8,162
Rapeseed Oil	2,241	1,783	1,418	3,432	3,773	4,131	4,081	3,968
Corn Oil	823	768	808	634	835	919	865	685
Coconut Oil	1,861	2,046	2,277	2,353	1,881	1,986	2,036	1,865

Contd.

Oils/Fats	*1998*	*2000*	*2005*	*2010*	*2011*	*2012*	*2013*	*2014*
Olive Oil	476	496	748	779	835	926	918	933
Castor Oil	250	276	305	447	401	495	521	489
Sesame Oil	23	25	36	38	35	39	40	38
Linseed Oil	129	122	113	100	94	100	88	91
Total Vegetable Oils	**29,097**	**32,011**	**47,469**	**62,626**	**65,208**	**69,430**	**72,516**	**74,083**
Butter	654	666	744	727	719	790	808	851
Tallow	2,352	2,215	2,069	2,158	1,894	1,636	1,578	1,592
Fish Oil	427	849	639	805	781	914	722	807
Lard	161	193	132	120	126	114	121	122
Total Animal Oils/ Fats	**3,594**	**3,923**	**3,584**	**3,810**	**3,520**	**3,454**	**3,229**	**3,372**
GRAND TOTAL	**32,691**	**35,934**	**51,053**	**66,436**	**68,728**	**72,884**	**75,745**	**77,455**

The Global Exports and Imports of vegetable oils during 2014 are given in Table 2.4.

Table 2.4: Exports and Imports of 17 Oils and Fats ('000' MT) during 2014

Oils/Fats	*Exports*	*Share per cent*	*Imports*	*Share per cent*
Palm Oil	44549	60.13	44368	59.92
Palm Kernel Oil	3158	4.26	3157	4.26
Soybean Oil	9729	13.13	9817	13.26
Cottonseed Oil	171	0.23	157	0.21
Groundnut Oil	245	0.33	223	0.30
Sunflower Oil	8162	11.02	8241	11.13
Rapeseed Oil	3968	5.36	3992	5.39
Corn Oil	685	0.92	726	0.98
Coconut Oil	1865	2.52	1876	2.53
Olive Oil	933	1.26	892	1.20
Castor Oil	489	0.66		0.64
Sesame Oil	38	0.05	892	0.05
Linseed Oil	91	0.12	87	0.12
Total Vegetable Oils	**74083**	**100.00**	**74048**	**100.00**

Palm oil continues to be the most traded oil globally with 60.13 per cent exports and 59.92 per cent imports respectively during 2014. This is followed by Soya bean oil with 13.13 per cent and 13.26 per cent and sunflower oil with 11.02 per cent and 11.13 per cent exports and imports respectively. Other 3 oils globally traded in significant amounts are Rapeseed oil (5.36 and 5.39 per cent), palm kernel oil (4.26 and 4.26 per cent) coconut oil (2.52 and 2.53 per cent) exports and imports respectively.

2.1.3. Future Vegetable Oil Consumption

To estimate future total consumption, one needs estimates of population and of per capita consumption. For consumption, there are two opposing trends: first, as living standards rise in the developing world, per capita consumption is increasing. Consumption calculated from the USDA and UNDP figures showed an increase from 15.8 kg/head in 2003 to 18.4 kg in 2007. Some countries are still well below this Figure 2.1, the consumption of 'others' was 12.4 kg in 2007. Conversely, 'western' consumption of fats is said to be unhealthily high.

Global demand for oils and fats is growing fast since the past 15 years and almost doubled (92.438 metric ton in 2000 to 179.569 metric ton) in 2015 (Figure 2.1). The four major vegetable oils palm (30 per cent), soya (24 per cent), rape seed mustard (13 per cent) and sunflower (7.0 per cent) totally contributed 74 per cent of world production. The growing demand will continue in the coming decades. By 2050 the global demand will rise by 35 per cent over 2013 and may require 250 metric ton. It will be a great challenge to meet the demand keeping the situation of limitation in area expansion, rain fed nature of the bulk of the crop, limited adoption of technologies and non-availability of seeds of high yielding varieties and hybrid development and growing population, increased prosperity in living conditions with high income resulting in increase of per capita consumption, growing need for biofuel and bio-lubricants, new applications for energy, chemicals and technical products with a bio based economy.

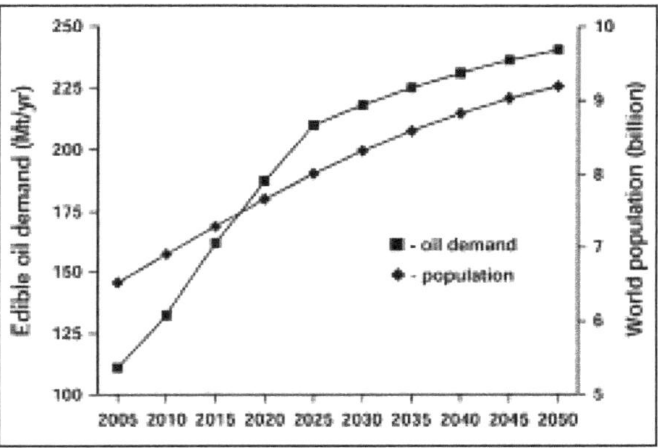

Figure 2.1: Future Global Demand of Vegetable.

2.2. Indian Scenario of Vegetable Oils

India is one of the largest producers of oilseeds in the world which occupy an important position in the Indian agricultural economy. There are nine important oilseeds crops grown in India out of which seven are of edible oils (soybean, groundnut, rapeseed mustard, sunflower, sesame, safflower and niger) and two are of non-edible oils (castor and linseed). In terms of acreage, production and economic value, oilseeds are second only to food grains.

India is a leading player in edible oils, being the world's largest importer (ahead of the and China) and the world's third-largest consumer (after China and the EU). Palm oil (mainly imported) and Soya bean oil account for almost half of total edible oil consumption in India, followed by mustard and groundnut oil.

2.2.1. Area, Production and Productivity

India is fortunate in having a wide range of oilseed crops grown in its different agro-climatic zones. Ground nut, rapeseed/mustard, sesame, safflower, linseed, Niger seed/castor are the major traditionally cultivated oilseeds. The area, production and productivity under vegetable oilseeds between 1950-51 to 2014-15 is given in Table 2.5.

Table 2.5: Area, Production and Productivity of Major Oil Seed Crops

Year	*Area (million ha)*	*Production (million tons)*	*Yield (kg/ha)*	*Coverage under Irrigation (Per cent)*
1950-51	10.73	5.16	481	NA
1960-61	13.77	6.98	507	3.3
1970-71	16.64	9.63	579	7.4
1980-81	17.6	9.37	532	14.5
1985-86	19.02	10.83	570	NA
1990-91	24.15	18.61	771	22.9
1997-98	26.12	21.32	816	24.3
1998-99	26.23	24.75	944	23.2
1999-00	24.28	20.72	853	25.2
2000-01	22.77	18.44	810	23
2001-02	22.64	20.66	913	24.3
2002-03	21.49	14.84	691	22.7
2003-04	23.66	25.19	1064	24.5
2004-05	27.52	24.35	885	26.6
2005-06	27.86	27.98	1004	28
2006-07	26.51	24.29	916	28.3
2007-08	26.69	29.76	1115	27.1
2008-09	27.56	27.72	1006	27.1
2009-10	25.96	24.88	959	25.9
2010-11	27.22	32.48	1193	25.1
2011-12	26.31	29.8	1133	27.6
2012-13	26.48	30.94	1168	N.A.
2013-14	28.53	32.88	1153	N.A.
2014-15	27.58	36.34		N.A.

Note: i) Data for 1950-51 to 1969-70 relate to total of five major oilseeds *viz.* groundnut, Castor seed, sesamum, rapeseed and mustard and linseed. ii) The yield rates given above have been worked out on the basis of production and area figures taken in '000 units. (Source: Agricultural Statistics at a Glance, 2014, Directorate of Economics and Statistics, Ministry of Agriculture, Govt. of India (Website: http://www.dacnet.nic.in/eands).

The area under oilseeds increased from 10.73 million ha in 1950-51 to 28.53 million ha (2.6 times) while production increased 6.1 times to 32.88 million from 481 kg/ha to 1153 kg/ha (2.4 times) during this period. The increase in productivity is mainly due to increase in irrigation potential as well as improved varieties and hybrids.

The pattern of contribution of different vegetable oils in India is similar to that of global. The highest share of contribution is made by palm oil (33.95 per cent) followed by soya bean oil (25.85 per cent), rape seed oil (15.58 per cent) and sunflower oil (9.26 per cent). Other oils form a very low share between 0.36 per cent in linseed oil to 3.74 per cent in palm kernel oil (Table 2.6).

Table 2.6: Share of Major Vegetable Oils in India (2014)

Oils/Fats	*Total Production ('000' tons)*	*Per cent*
Palm Oil	59228.00	33.95
Palm Kernel oil	6520.00	3.74
Soybean Oil	45095.00	25.85
Cottonseed Oil	4853.00	2.78
Groundnut Oil	3874.00	2.22
Sunflower Oil	16148.00	9.26
Rapeseed Oil	27187.00	15.58
Corn Oil	3162.00	1.81
Coconut Oil	3018.00	1.73
Olive Oil	3302.00	1.89
Castor Oil	645.00	0.37
Sesame Oil	783.00	0.45
Linseed Oil	629.00	0.36
Total	**174444.00**	**100.00**

The yield levels of different oil seeds in the world/country with the highest productivity as well as India are given in Table 2.7. While USA recorded highest productivity in groundnut, Germany reported the highest productivity in rapeseed, mustard; Canada in linseed and China in sunflower, Malaysia reported highest yields in oil palm. The differences between global yields in a country with highest productivity are 15-20 metric tons in different crops.

Table 2.7: Oilseeds Productivity (kg/ha) in India vis-a-vis World (2012)

Groundnut	1179	1676	4699 (USA)
Rapeseed-Mustard	1140	1873	3690 (Germany)
Soybean	1208	2374	2783 (Paraguay)
Sunflower	706	1482	2494 (China)
Sesame	426	518	1315 (Egypt)
Safflower	654	961	1489 (Mexico)
Castor		1162	1455 (India)
Linseed	260	752	1358 (Canada)
Oil palm fruit	12380	14323	21901 (Malaysia)

(Source: FAOSTAT. 1012).

The low yield levels in India are due to the fact that bulk of the oilseed crops are grown in marginal and sub-marginal levels under rain fed conditions and often with sub optimal management.

2.2.2. Export/Import/Consumption

Export/import and consumption scenario of vegetable oils in India between 2007-08 to 2012-13 is given in Table 2.8 and summed up figures for 2015 are given in Table 2.9.

Table 2.8: Vegetable Oil Scenario in India (million tons)

Particulars	*2007-08**	*2008-09*	*2009-10*	*2010-11*	*2011-12*	*2012-13*
Total consumption	15.62	17.55	17.93	18.41	19.25	19.66
Production	9.63	8.71	8.80	9.78	9.65	9.32
Imports	5.99	8.84	9.13	8.67	10.2	11.09
Soya oil	0.76	1.06	1.60	0.95	0.97	1.30
Sunflower oil	0.02	0.59	0.62	0.78	1.10	1.00
Palm Oil	5.02	6.88	6.61	6.68	7.12	8.45
Exports	0.42	0.29	0.45	0.41	0.48	0.51
Per capita consumption (kg)**	**12.60**	**13.90**	**14.30**	**14.60**	**14.80**	**15.4**

* October to September. ** Population: 1275 millions.

Table 2.9: Consumption of Various Oils in India in 2011-12 (Qty. in million tons)

Types of Oil	*Domestic*	*Import*	*Total*	*Consumption Per cent*
Palm Oil	0.07	7.50	7.57	46.0
Soybean Oil	1.65	1.01	2.66	16.2
Rapeseed Oil	1.74	0.09	1.83	11.1
Sunflower Oil	0.22	1.08	1.30	7.9
Cottonseed Oil	1.19	-	1.19	7.2
Rice Bran Oil	0.85	-	0.85	5.2
Groundnut Oil	0.39	-	0.39	2.4
Other	0.65	-	0.65	4.0
TOTAL	**6.75**	**9.67**	**16.42**	**100**

India's vegetable oil supply and demand were 4.41and 5.62 million tons in 1983-84 respectively with an import of 1.37 million tons. The supply was always lesser than demand and the import reached 11.09 million tons as on 2012-13 of which palm oil is 8.45 million tons (Table 2.8).

Consumption pattern of vegetable oils indicates that palm oil consumption is the largest(46 per cent) followed by soya bean (16.2), rape seed (11.1 per cent) and the rest of other oils (Table 2.9).

India is a leading player in edible oils,being the world's largest importer and third largest consumer. Indonesia is supplying 73 per cent of India's local market in the form of Crude Palm Oil (CPO) followed by Malaysia mostly in the form of Refined, Bleached and Deodorized Palm Olein. Import of Vegetable oils is rising from year to year to bridge the demand and supply gap.

India is the fourth-largest economy in the world which is growing with an average GDP of 6 per cent. It has a population of 1.06 billion, which is growing at a rate of 1.65 per cent per annum. The country's top 10 per cent of the population consumes 20 kg per capita and the bottom 30 per cent, less than 5 kg per capita of oil. According to the NCAER, there are five classes of consumer households, ranging from the destitute to highly affluent, which differ considerably in their consumption behavior and ownership patterns across various categories of goods. These classes exist in urban as well as rural households while consumption trends may differ significantly between similar income households in urban and rural areas. Consumption of oil is income elastic and it is likely that the per capita consumption will increase on one side and the population expansion will increase the demand further.

2.2.3. Future Projections

Indian vegetable oil industry consists of 15,000 oil mills, 600 solvent extraction units 600 vegetable oil refineries and 250 vanaspati units spread over the country, crushing/processing oil seeds, oil cakes, rice bran and processing vegetable oils. With all that we still depend on other countries to meet our demand for vegetable oils. The demand projection of vegetable oils in India is given in Table 2.10.

Table 2.10: Demand Projection of Vegetable Oil in India

Particulars	*Years*			
	2020	*2030*	*2040*	*2050*
Projected population (billion)	1.32	1.43	1.55	1.68
Per capita consumption considering 50, 60, 70 and 75 per cent above the prescribed consumption levels during 2020, 2030, 2040 and 2050 respectively				
Per capita consumption (kg/annum)	16.43	17.52	18.62	19.16
Vegetable oil requirement for direct consumption (million tons)	21.69	23.13	24.58	25.29
Vegetable oil requirement for non-industrial use (million tons)	3.57	6.34	9.69	10.61
Total vegetable oil requirement (million tons)	25.26	29.47	34.27	35.9
Vegetable oil availability from secondary sources (million tons)	5.05	5.89	6.85	7.18
Total vegetable oil requirement from annual oilseed crops (million tons)	20.21	23.58	27.2	28.72
Total vegetable oilseds requirement from nine annual oilseed crops (million tons)	67.37	71.45	80.65	82.06

Source: Indian Oil seed Scenario: Challenges and opportunities, Strategy paper by Dr. R.S. Paroda (2013).

Taking into consideration a host of factors *viz.*, domestic production, import dependency, trade buoyancy, pattern of per capita consumption, changes in dietary standards, growing trend of out-of-house consumption, the swiftly rising demand for vegetable oils for non industrial uses and production of biofuel; the projections have been made for the Indian vegetable oilseeds. The projections are based on the assumptions that the per capita consumption would be increasing annually at 3 per cent till 2015, followed by an increase at a declining rate of 2.5 per cent from 2015 to 1.75 per cent in 2020, with a further decline in the incremental consumption to negligible levels by the year 2050.

The estimated per capita consumption is accordingly placed at 16.43, 17.52, 18.62 and 19.16 kg/annum for the year 2020, 2030, 2040 and 2050, respectively. A newer dimension of vegetable oil requirement for industrial use is estimated to grow by 15 per cent in 2020, 20 per cent in 2030 and 25 per cent post 2040, thus requiring around 3.57, 6.34, 9.69 and 10.61 million tons in 2020, 2030, 2040 and 2050 respectively. The Indian trade industry, therefore, predicts much greater expansion. The total vegetable oil requirement is thus estimated at 25.26, 29.47, 34.27 and 35.90 million tons during 2020, 2030, 2040 and 2050, respectively, which is a gigantic task for the country to increase its domestic production. The contribution of vegetable oil availability from secondary sources including arboreal tree species (20 per cent) is estimated at 5.05, 5.89, 6.85 and 7.18 million tons during 2020, 2030, 2040 and 2050, respectively. Thus, the total domestic vegetable oilseeds requirement from nine annual oilseed crops is estimated at 67.37, 71.45, 80.65 and 82.06 million tons by 2020, 2030, 2040 and 2050, respectively. As per the population estimates, the Indian middle-class population is expected to touch one billion over the next two decades. The middle class population would be the major consumer of edible oils in the country. Further, the urban population in the country is increasing by more than 3 per cent annually. The report by McKinsey Global Institute predicts that 590 million people or 40 per cent of India's population will live in cities by 2030, up from 340 MToday. This would have tremendous effect on increased consumption of edible oils, looking down to the social status and the size of the strata to the total population in the country. Higher economic growth and concomitant rise in incomes, coupled with change in tastes and preferences in both urban and rural areas are expected to increase the demand for high-value commodities, especially the edible oils.

The table above clearly indicates that the future vegetable oil requirement will keep on increasing which will necessitate the need to either go for import or increase the production of available secondary sources,particularly oil palm which is the highest oil yielding palm.

Chapter 3

Global Palm Oil Scenario

Oil palm a perennial oil yielding tree found in Africa in wild form and the oil from the palm used by natives as cooking oil became a cultivated crop in Asia to start with. Presently, it is cultivated roughly in 18.67 million hectares in more than forty-six countries across the world to produce 56.21MT of palm oil with an average yield of 3.11 metric ton/ha of oil per ha per year (FAO Statistics Division, 2013).

3.1. Harvested Area, Production and Productivity of Palm Oil

The harvested area, production and productivity vary among the oil palm growing countries. The production area in Malaysia has increased from 3.31 million ha in 2001 to 4.689 million ha in 2014 and that of Indonesia from 2.20 to 7.40 million ha for the same period. The area, production and yield of oil palm in major oil palm growing countries are given in Table 3.1.

Of the total area of 18.67 million ha during 2014 under oil palm in the world Indonesia occupies 39.66 per cent followed by Malaysia (25.11 per cent), Nigeria (16.20 per cent) and Thailand (3.55 per cent). Lowest area is reported from Venezuela (0.21 per cent). Together these four countries cover 84.53 per cent of the total global area under oil palm.

The global palm oil production in 1998 was 17.158 metric ton and it increased to 59.299 metric ton in 2014 (Table 3.2). Malaysia was the largest producer with 14.96 metric ton of oil followed by Indonesia with 13.92 metric ton in 2005. But in 2006, Indonesia exceeded Malaysia by producing 16.08 metric ton and that of Malaysia was only 15.82 metric ton. During 2014, Indonesia produced 31.00 metric ton and that of Malaysia 19.61 metric ton. Nigeria, Ivory Coast, Colombia, Thailand and Papua New Guinea also produce palm oil in significant quantities. The global average in the year 2000 was 3.32 tons oil/ha/year. During 2000, the average yield was 3.63 tons oil/ha/year in Malaysia, 3.45 tons oil/ha/year in Indonesia, 4.23 tons oil/ha/year in Papua New Guinea and 4.11 tons oil/ha/year in Costa Rica. This also has considerably increased due to the development and cultivation of high yielding hybrids but the overall average has reduced. As on 2014, the productivity in Malaysia and Colombia was above 4.0 t palm oil/ha and that of Indonesia and Thailand were 3.42 and 3.76 t/ha. respectively.

Table 3.1: Area, Production and Yield of Oil Palm in Major Countries during 2014

Country	*Area (ha)*	*Production FFB (million tons)*	*Yield FFB (kg/ha)*
Brazil	126,559	460,000	11,014
Cameroon	138,000	1,150,000	17,928
China	50,200	650,000	13,471
Colombia	270,000	2,600,000	20,067
Congo, Dem Republic of	277,350	1,085,070	12,460
Costa Rica	77,750	666,084	11,375
Cote d'Ivoire	277,090	1,238,186	15,374
Ecuador	214,570	1,424,000	11,939
Ghana	349,040	1,102,087	7,000
Guinea	313,310	840,010	2,681
Guatemala	70,000	466,600	21,429
Honduras	130,650	2,124,980	16,265
Indonesia	7,407,090	126,591,790	17,091
Malaysia	4,689,321	96,066,760	20,486
Mexico	50,868	678,935	13,347
Nigeria	3,025,950	7,968,440	2,633
Papua New Guinea	157,100	2,158,930	13,742
Peru	49,230	617,634	12,546
Philippines	55,083	437,434	7,941
Thailand	663,707	12,503,447	18,839
Venezuela	40,198	407,111	10,128
Others	557,249	12,891,679	23,135
Total	**18,674,736**	**274,129,177**	**14,679**

3.2. World Major Producers of Palm Oil

The major world producers of palm oil is given in Table 3.2. Indonesia and Malaysia are the largest producers of palm oil and the other World major palm oil producing countries is given in Table 3.2. Indonesia occupied the first place in area and production from 2006 and continued to make fast progress in palm oil production.

Table 3.2: World Major Producers of Palm Oil ('000' tons)

Country	*1998*	*2000*	*2005*	*2011*	*2012*	*2013*	*2014*
Indonesia	5,361	7,050	13,920	24,100	26,900	28,400	31,000
Malaysia	8,315	10,840	14,961	18,912	18,785	192,216	19,617
Thailand	475	525	685	1,530	1,600	1,970	1,820
Nigeria	690	740	800	930	940	960	910
Colombia	424	524	661	941	967	1,041	1,109
Papua New Guinea	210	336	310	520	530	500	520
Ecuador	199	222	319	495	540	495	490
Cameroon	139	140	154	254	245	225	253
Ghana	111	108	117	420	420	410	446
Cote d'Ivoire	269	278	260	400	420	415	370
Costa Rica	105	138	210	250	260	230	210
Honduras	92	97	175	320	395	425	460
Brazil	89	108	160	270	310	340	370
Guatemala	47	65	92	248	310	402	448
Phillippines	49	54	61	87	98	106	115
India	12	28	47	92	96	100	103
Venezuela	44	73	66	60	55	50	47
Other Countries	524	549	593	689	794	924	941
TOTAL	**17,155**	**21,875**	**33,591**	**50,518**	**53,665**	**229,209**	**59,229**

World Palm Oil Production in comparison with world oils and fats from 1960 to 2014 is given in Table 3.3.

The share of palm oil production has increased from 4.5 per cent to 30.45 per cent from 1960 to 2014.The increase has been spectacular from 2000 onwards. Only with 5 per cent global oil palm area this crop could contribute 30.45 per cent of share in world vegetable oil pool.

3.3. Major Countries Exporting and Importing Palm Oil

The major countries which are exporting palm oil are given in Table 3.4. Indonesia and Malaysia are the major palm oil exporting countries followed by other countries in small quantities.

India and China are the major importing countries followed by Netherlands and others (Table 3.5).

3.4. Palm Oil Production in the Asia-Pacific Region

Both Malaysia and Indonesia are the major palm oil producers with total Production of about 28.60 million tons in 2005. The production of palm oil in Asia Pacific countries from 1998 to 2013 is given in Table 3.6. Though oil palm is native of Africa and was introduced to Asia, Malaysia and Indonesia occupy first and second places, respectively in the world today.

Table 3.3: World Palm Oil Production from 1960 to 2014 ('000' tons)

Country	*1960*	*1970*	*1980*	*1990*	*2000*	*2005*	*2010*	*2014*
Malaysia	92	431	2573	6095	10840	14961	16993	19617
Indonesia	141	216	691	2413	7050	13920	22200	31000
Ivory Coast	14	50	182	270	278	260	300	370
Nigeria	638	449	433	580	740	800	830	910
Papua New Guinea	N.A.	N.A.	35	145	336	310	500	520
Columbia	N.A.	27	74	226	524	661	753	1109
Others	421	623	631	1225	2107	6240	4296	6229
Total	**1306**	**1796**	**4619**	**10953**	**21875**	**37152**	**45872**	**59229**
World oils and fats production	**28778**	**39043**	**57049**	**80674**	**114759**	**140983**	**172131**	**200246**
Share of oil palm in world oils and fats production (per cent)	**4.5**	**4.6**	**8.1**	**13.6**	**19.06**	**26.35**	**26.64**	**30.45**

Table 3.4: World Major Countries Exporting Palm Oil: 2014 ('000' tons)

Country	*2010*	*2011*	*2012*	*2013*	*2014*
Indonesia	16,450	17,070	19,094	21,471	22,950
Malaysia	16,664	17,993	17,576	18,147	17,306
Thailand	133	391	304	562	238
Benin	142	143	156	212	222
Colombia	90	159	180	185	246
Papua New Guinea	486	572	525	500	525
Ecuador	146	250	276	213	227
Cole d'Ivoire	201	254	278	201	215
Honduras	156	206	269	295	330
Guatemala	154.2	217.1	268.2	361.8	401.8
India	17.4	31.2	44.3	73.2	5.8
U Arab Emirate	350	290	270	220	293
Other countries	1,499	1,569	1,545	1,516	1,590
TOTAL	**36,487**	**39,146**	**40,785**	**43,957**	**44,549**

Source: Oil World Annual, 2015.

Table 3.5: Major Countries Importing Palm Oil (M'000' tons)

Country	*2010*	*2011*	*2012*	*2013*	*2014*
India	6,649	6,745	7,817	8,471	7,931
China	5,823	6,221	6,591	6,190	5,634
Pakistan	2,100	2,014	2,036	2,399	2,432
Bangladesh	1,065	946	1,013	1,250	1,246
USA	948	1,088	991	1,373	1,188
Nigeria	799	1,128	903	1,195	1,500
Spain	618	538	516	709	1,229
Netherlands	1,929	1,880	2,478	2,870	2,355
Italy	981	872	961	1,261	1,689
Other Countries	16,289	17,354	17,997	18,431	19,165
TOTAL	**37,201**	**38,786**	**41,303**	**44,149**	**44,368**

Source: Oil World Annual, 2015.

Table 3.6: Palm Oil Production in the Asia Pacific Region ('000' tons)

Country	1998	2000	2005	2010	2011	2012	2013
Indonesia	5,361	7,050	13,920	22,200	24,100	26,900	28,400
Malaysia	8,315	10,840	14,961	16,993	18,912	18,785	19,222
Thailand	475	525	685	1,340	1,530	1600	1970
Papua New Guinea	210	336	310	500	520	580	500
Philippines	49	54	61	92	87	98	106
India	12	28	47	83	92	96	100

3.4.1. Oil Palm in Indonesia

Four Oil palm seedlings were brought from Africa to the botanical garden, Bogor, Indonesia and planted as ornamental plants in 1848. Those four palms have become the source of parental palms in all the South East Asian countries. The first commercial plantation was started in Sungei Liput (Aceh) and Tanah Itam Ulu (North Sumatra). By 1917, the total area under oil palm plantation was about 1605 ha and in nineteen forties it was 110,000ha. By 1969, Indonesia produced 180,000 metric ton of palm oil and about 40,000 metric ton of palm kernel oil. The massive development of oil palm was started in the eighties when about 100,000-200,000 ha of new oil palm plantations were developed each year.

Small private companies and farmers came into this business in 1975. The development of oil palm slowed down in the late nineties when economic crisis occurred. The government of Indonesia established a scheme called National Estate Scheme (NES), where state owned plantation companies helped farmers to grow oil palm. The plantation companies provide seedlings, technical assistance and finance to small holders. Their produce would be purchased by the mills, thus providing the farmers direct access to the mills. As a result, more and more farmers and small companies were attracted to oil palm cultivation. Thus, after 1975 the development of area under oil palm was carried out by three agencies *viz.*, large companies (both foreign and domestic), state owned companies and small holders. Presently about 32 per cent of the total planted area belongs to small holders, about 50 per cent to large companies and about 18 per cent to state owned companies. Out of the total planted area of 5.15 million ha as on 2005, about 4.15 million ha comprises of matured plantation and about 1.0 million ha is under immature plantations. The increase in area was very fast which is reflected in the production. Present area is 7.40 million ha. Indonesia excelled Malaysia in production of palm oil by producing 31.00 million tons of crude palm oil as on 2014. Indonesia emerged as the world's top most producer of palm oil since 2006. Still there is lot of scope for new areas coming under Oil palm. The projected plan for 2025 is 26.0 million ha.

3.4.2. Oil Palm in Malaysia

Oil Palm was introduced in Malaysia at the start of the 20th century. It was identified as an economically important crop in 1903 by the Department of Agriculture which started its commercial exploitation in 1917. The oil palm seeds for Malaysian industry first arrived as ornamental plants on the Malaysian shores in 1887. These were derived from the famous four palms of Bogor garden. From a few insignificant acres of oil palm cultivated in the early 1900s, the Malaysian oil palm industry today has grown by leaps and bounds and the country is now the world's largest producer and exporter of palm oil. Oil palm cultivation covers 11 per cent of the total land area of Malaysia. This represents 62 per cent of the total agricultural land under cultivation.

The Malaysian oil palm industry has witnessed unprecedented growth in the past four decades. From modest 50,000 ha by 1960, the area under oil palm cultivation increased to about 1.0 million ha by 1980, two million ha by 1990 and 4.55 million ha by 2013. The oil palm industry has evolved over years into integrated enterprise, transforming the country's rural landscape. It continues to play a pivotal role in the socio economic development of the country. Even though the development was slow in fifties, the decision taken to diversify rubber and providing land to landless accelerated the expansion of oil palm. A significant development took place when the Federal land Development Agency (FELDA) took up settling small holders by clearing the jungle. The FELDA is the largest oil palm cultivating public sector agency. This agency has developed basic infrastructure like roads, public amenities such as schools, hospitals, clinics, places for worship and recreational and commercial facilities. Malaysian oil palm industry is a mixture of big plantations owned by companies as well as small holdings. The research contribution through Malaysian Palm Oil Board (MPOB), previously known as Palm Oil Research Institute of Malaysia (PORIM), advanced quality planting materials with high FFB, high Oil Extraction Ratio and processing technologies helped in substantial development of oil palm in the country.

The two most significant contributions from R&D were the replanting of dura palms (thick shelled fruit) with tenera palms (thin shelled fruit) in 1960s which led to the highest oil extraction ratio and the introduction of pollinating weevil, *Elaeidobious kamarunicus* in the early 1980s which increased fruit yield significantly. In 1984 it was estimated that the palm oil and palm kernel oil output increased over the previous year by 24 per cent and 17 per cent respectively, and this was mainly accredited to the role played by the pollinating insects.

The distribution of the oil palm planted area in Malaysia is in the ratio of approximately 60 per cent, 30 per cent and 10 per cent between Peninsular Malaysia and the East Malaysia's states of Sabah and Sarawak, respectively. The ownership distribution is also in a similar percentage ratio between the private sector plantation companies, the government schemes and small holders *i.e.*, land ownership of below 40 ha. The industry provides employment both directly and indirectly to about 1.5 million people. The area has increased from 2.20 to 4.689 million ha from 2001 to 2014 and accordingly the production also increased from 11.840 million tons to 19.47 million tons for the above period.

The technological breakthroughs and continuous improvements in the upstream sector including production of superior oil palm planting materials, systematic agronomic practices and mechanization improved the oil palm productivity. The two main industry Productivity indicators *e.g.* fresh fruit bunches (FFB) and oil extraction rate (OER) have gradually improved. Efforts are on way to achieve the Vision 35:25 which is to achieve 35 tons FFB and 25 per cent OER by 2020 from the present level of 20:20.

3.4.3. Oil Palm in Thailand

The oil palm industry in Thailand is small compared to its southern neighbours, Malaysia and Indonesia, but nevertheless important for the country's agricultural diversification and the economy of the provinces in which it is grown. From an initial 3,200 ha in 1969 the industry expanded to about 2, 22,074 ha in 1997, growing at about 9.4 per cent p.a. in recent years, as a result of expansion into new land and conversion of rubber holdings. Low rubber prices have accelerated the growth of oil palm especially in the south while marginal areas in the East are being re-examined for oil palm extension because of climatic considerations.

At present main growing areas are located in the southern regions, with the largest belt stretching between Krabi and Surat Thani districts, where annual rainfall ranges from about 1,800 - 2,100 mm. The dry season, with less than an average of 100 mm rain per month, lasts from December to March and is the principal constraint to yield here and to productive expansion to other parts of Thailand. The Thai oil palm industry is mainly a smallholder industry with only a fraction grown by large concerns in the plantation scale. The smallholder crop is sold to the plantation mills of 15 - 30 tons/h capacity, as well as to independent mills of 5-20 tons/hr capacity. There are small milling facilities which buy loose fruits also. All the palm oil produced is consumed locally but is insufficient and Thailand imports about 20,000 - 30,000 tons of palm oil annually, principally from Malaysia.

Besides the climatic constraint to expansion, lack of suitable land and, until very recently, poor planting materials were limiting the growth of the industry. Statisticians may find a good correlation between the smuggling of cooking oil and the smuggling of oil palm seeds at the Malaysia - Thai border. The smuggled seeds may be of genuine good quality, explaining the large volume consumed by the urban island of Penang, but sadly for the industry do also include DXP of the type Dibawah Pokok " and Dibelakang pondok. The former is likely when there is a Malaysian interest or advisory in the Thai planting, the latter when seeds are bought outright from dealers and suppliers. Presently (2014) the area has come up to 6.63 lakh hectares and the production is 2.5 MT.

3.4.4. Oil Palm in Myanmar

The available information suggests that the Oil palm was introduced into Myanmar in 1921-22, brought in all probability from Indonesian Deli dura stock and test planted by the Government of the time. The first commercial plantation of some 120 ha was established at Yebyu in the Tanintharyi Region in 1926. Oil palm extension was implemented by the Ministry of Agriculture and Irrigation (MoAI) as a government project at various locations from 1970s to 1980s. Privatisation of the expanding industry commenced in 1993 when government estates were leased to private companies. In 1999 the then Government initiated a military sponsored industrial oil palm development as a part of its national self-sufficiency plan. By 2000, some 50,000 ac (±20,200 ha) of oil palms had been established. The plantations are principally found within a narrow belt of coastal lowlands in Myeik and Kawthoung Districts, and to a lesser extent in the drier Dawei District. Data from DICD for 2014 provided indicates that a total of almost 1,000,000 ac (almost 405,000 ha) have been allocated by the Government of Myanmar to 44 Oil Palm Plantation companies to develop plantations in the Kawthoung, Myeik and Dawei Districts of the Tanintharyi Region. Of that land area, almost 350,000 ac (±142,000 ha) have been planted, some 283,000 ac (±115,000 ha) in Kawthoung, 46,000 ac (±19,000 ha) in Myeik, and 17,000 ac (±7,000 ha) in Dawei Districts. Of the 44 companies concerned, apparently 43 are Myanmar owned (three foreign companies have Joint Venture Agreements (JVA) with local companies), and one is the result of FDI.

Myanmar imports 400,000 tons of palm oil annually to fulfil domestic demand for edible oils. Domestic consumption of edible oils amounts to 900,000 tons annually. Since 2011, the Myanmar authorities have granted private companies permission to import palm oil from Malaysia and Indonesia in a bid to meet the demand. Malaysian entrepreneurs have introduced their palm oil to the Myanmar market over the past few years.

Realising the necessity for increasing the vegetable oil production, the Myanmar government started to allot land for oil palm cultivation to private companies in 70's. The responsibility of research and development was entrusted with MPCE (Myanmar Perennial Crops Enterprises) in 1994. -Tanintharyi is the only area with the right soil and climate conditions to grow oil palm in Myanmar. Fuelled by a need to meet Myanmar's demand for cooking oil and reduce the high cost of palm oil imports (which cost the country $376 million in 2012 from Indonesia and Malaysia), to date over 140,000 hectares of oil palm have been planted and 400,000 hectares allocated to over 40 local and three international companies. By 2001, the area reached 21,000 hand subsequently to 1,09,000 ha in 2005-06 and the projected area for 2010-11 was 2,31,000 ha (Table 3.7). The oil palm development was supported by the FAO especially in providing the technical inputs.

Table 3.7: Area Covered from 2000 to 2011

Year	*Area in ha.*
2000-2001	21,000
2005-2006	1,09,000
2010-2011	*2,31,000

The Government's strategy is to increase palm oil production in Myanmar to:

- ★ Fulfil the needs of growing local edible oil consumption
- ★ Substitute the cost of importation and capital outflow
- ★ Improve rural industrial and social infrastructure
- ★ Create inflow of FDI into the oil palm industry.

A thirty year long-term development plan, was enacted in 2000, to initially plant 500,000 ac (202,343 ha) of oil palms by 2030.

Even as a small country like Myanmar could realize the need to grow oil palm to meet the vegetable oil need of the country, it is needless to emphasise that India should move fast in oil palm development to produce palm oil at least to reduce the import of vegetable oil.

3.4.5. Oil Palm in Africa

The production and the processing of oil palm occupy an important position in the society and economy of many African countries. Oil palm was grown already during pre-colonial period and palm oil had been traded with Europe since the 18th century. The oil palm plays a vital role in the socio-economic life of African countries. It provides farmers with regular incomes and is a part of the traditional staple foods. Oil palm grows naturally in certain ecological areas of the African continent. In the sixties the continent provided almost 83 per cent of the world oil palm production which has fallen to 13 per cent, whereas, Asia has captured the greater part of this production and produces 82 per cent of the world oil palm today. In Africa, oil palm is grown in countries like Benin, Burundi, Cameroon, Central African Republic, Republic of Congo, Cote d'Ivoire, The Republic of Gabon, The Republic of Ghana, The Republic of Guinea, Republic of Madagascar, The Federal Republic of Nigeria, The Republic of Sao Tome and Principe, The Republic of Tanzania and The Republic of Togo. Though the continent holds all the necessary trumps to produce more, the state of the African palm grove neither meets the expectations nor the demands of the continent.

3.4.6. Oil Palm Industry in Latin America

The first commercial oil palm plantation in the region goes back to 1943 when the United Fruit Company established a commercial plantation in San Alejo, Honduras, and another a year later in Quepos, Costa Rica. Since then, other countries started planting oil palm extensively in Central and South America. The country with the largest industry in the region in terms of area planted is Colombia followed by Ecuador and Honduras. In terms of land availability, there is a huge potential to expand the industry in Latin America, but the lack of clear economic targets and strong government incentive policies still keep the region at a much slower development pace in comparison with Asia. Nevertheless, the growing (CPO) prices are expected to boost the expansion of the industry in the region. The total area of 8,92,053 ha of oil palm developed in Latin America through 2006 represents only around 8.6 per cent of the world industry, which is estimated at more than 10.4 million ha.

Most of the oil palm estates in Mexico, Central America and the Caribbean, are relatively young. Approximately 33.8 per cent of them are seven years old or younger. Honduras and Gautamala have had recent vigorous development in the region, with 84,463 ha and 45,576 ha planted respectively. These two nations planted 10,904 and 8,655 hectares respectively in the last two years, followed by Costa Rica and Mexico. Although Nicaragua has large areas suitable for oil palm cultivation and could expand its industry substantially, fewer hectares were planted there in the last two years; a similar area was developed in Mexico, while Panama and the Dominican Republic had the lowest development of the region.

The expansion of oil palm estates in South America was more uniform than in Mexico, Central America and the Caribbean and the age of distribution of the plantings is quite similar up to the age of 22 years. In other words, at the beginning, small areas were developed, and then plantation growth took place at a relatively steady pace. Most of the crop expansion during the last two years occurred in Colombia (84,275 ha) and Ecuador (19,629 ha), while the expansion pace in Brazil (5800 ha) and Venezuela (4395 ha) was much slower. The country with the least development was Peru, with only 1,418 ha planted during the same period. The percentage of old groves to be renovated in South America is similar to the percentage for Mexico, Central America and the Caribbean, at 9.2 per cent and 8.9 per cent respectively.

Without any doubt, the oil palm industries in Colombia and Ecuador are the largest in Latin America and continue to expand at a vigorous pace in comparison with other countries in the region. However, the overall potential for expansion in terms of land availability is higher in Colombia than in Ecuador being 1.7 million ha versus 340,000 ha. In addition, the Colombian industry is best prepared for expansion due to the experience of its planters, who are organized into a strong growers federation that covers all aspects of the industry including research and development. Furthermore, Colombia is the only Latin American country with reliable manufacturers of crude palm oil extraction and refining equipment, which makes the country even more competitive despite its social and political problems.

Peru is a net importer of vegetable oil despite its potential of more than 2 million ha suitable for oil palm cultivation; however, Peruvian Government is pursuing the immediate development of 17000 ha by small holders and also encouraging the private sector to plant large oil palm plantations under special land allotments. As a country with abundant and relatively cheap labour the land of Peru seems to be a good alternative for investors for future expansion programs.

Honduras, Mexico and Nicaragua also have sizeable areas for expansion: 282,000, 200,000 and 150,000 ha respectively. However, land ownership in Mexico is fragmented into smallholdings making the establishment of large private plantations difficult; nonetheless, operations with extraction mills purchasing fresh fruit bunches from smallholders is an ongoing successful business. Perhaps Nicaragua, with substantial land available and low wages, is the country offering the best perspectives in Central America. With the exception of Nicaragua, Central America is less attractive than South America for investors seeking large areas due to higher land prices and wages.

The only South American country with suitable land for oil palms but without a single hectare planted is Bolivia. In this country 120,000 ha or more can be planted with a high yield expectation and low costs due to its abundant cheap labour.

Brazil constitutes the last frontier for oil palm expansion worldwide, with a huge area of over 30 million ha of land suitable for oil palm cultivation. However, its development faces several challenges, starting with the Amazon protection law that limits land use to 20 per cent of new forest land and up to 50 per cent of land cleared before 1998. Hence, current development is taking place mainly on grassland areas in the northeastern part of the State of Para. Furthermore, establishing large plantations in Brazil in remote areas will require the development of infrastructure, which may significantly increase investment levels. The potential and planted areas in Latin American countries are given in Table 3.8.

Table 3.8: Latin America Potential Oil Palm Areas (hectares)

Country	*Planted*	*Potential*
Bolivia	0	120000
Brazil	77864	30000000
Colombia	303743	1750000
Costa Rica	48406	90000
Dominican Rep.	8329	15000
Ecuador	212821	340000
Guatemala	45576	100000
Guvana	1100	50000
Honduras	84463	282000
Mexico	24104	200000
Nicaragua	10933	150000
Panama	6786	20000
Peru	16707	2250000
Venezuela	49530	125000
Surinam	1691	10000
Total	**892053**	**35502000**

Most of the crude palm oil (CPO) volume in America is produced by Colombia, with an output of 37 per cent of the entire Latin America oil palm industry, estimated at 1.93 million tons in 2006 followed by Ecuador, Costa Rica, Honduras and Brazil with outputs of 18, 11 and 9 per cent respectively. The remaining countries have shares lower than 9 per cent with Nicaragua having the lowest comparative production (<1 per cent) (Table 3.9).

Table 3.9: Historical Crude Palm Oil (CPO) Production in Latin America

Country	*(000's of tons)*						
	2001	*2002*	*2003*	*2004*	*2005*	*2006*	*Per cent*
Colombia	547.6	528.4	526.6	630.4	672.6	710.4	37
Ecuador	205.4	238.1	261.9	279.2	319.3	346.0	18
Costa Rica	149.9	128.4	155.0	180.0	210.0	220.0	11
Honduras	130.0	126.5	158.0	170.0	175.0	190.8	10
Brazil	110.0	118.0	129.0	142.0	160.0	170.8	9
Guatemala	70.1	86.0	85.0	87.0	92.0	95.7	5
Venezuela	52.0	55.2	41.1	60.6	65.8	70.0	4
Mexico	34.0	36.0	39.5	41.0	42.6	43.5	2

Contd.

Peru	34.0	30.0	27.0	28.0	29.0	30.6	2
Dominican Republic	26.0	25.2	27.0	27.6	29.0	29.6	2
Panama	11.8	11.5	11.8	13.0	13.6	13.9	1
Nicaragua	8.0	8.0	8.0	8.4	8.6	8.8	0.5
Total	**1378.8**	**1391.3**	**1469.9**	**1667.2**	**1817.5**	**1930.1**	**100**
Increase		**1%**	**6%**	**13%**	**9%**	**6%**	

The case of Costa Rica is particularly noteworthy; approximately one fifth (48,406 ha) of the planted area of Ecuador (212,821 ha), Costa Rica is producing the equivalent of 60 per cent of Ecuador's CPO production, which signifies an approximate comparative productivity of 5.3 versus 1.8 tons of CPO/ha (Table 3.10). These figures demonstrate the importance of technology and management for enhancing the competitiveness of producing countries.

Table 3.10: Competitiveness of Leading Countries in Latin America

Country	*Harvested Area (ha)*		*CPO Production (000 tons)*		*Per ha Yield (tons)*	
	2006	*2013*	*2006*	*2013*	*2006*	*2013*
Colombia	219468	250000	710.400	1041.000	3.2	4.16
Ecuador	193100	218833	346.000	494.000	1.8	2.31
Costa Rica	41582	74500	220.000	230.000	5.3	3.12
Honduras	73559	125000	190.800	425.000	2.6	3.40
Brazil	72060	108635	170.800	340.000	2.4	3.13

Source: Oil World Annual 2006, Fedepalma Colombia www.fedepalma.org, Ancupa, Ecuador and ASD, Costa Rica.

Current total CPO exports in Colombia in Latin America are estimated at 607,900 metric ton and the main exporters are Colombia, Costa Rica, Ecuador, Honduras and Guatemala; the remaining countries in the region produce palm oil for their own consumption only (Table 3.11). Basically, most of the exporters go to Mexico and Europe, and to a lesser degree to other countries within the Latin American region. Variations in CPO exports between years in the different exporting countries are mainly due to changes in domestic consumption and use of the commodity.

Table 3.11: Crude Palm Oil Exports by Year in Latin America ('000' tons)

Country	*2001*	*2002*	*2003*	*2004*	*2005*	*2006*
Colombia	90.1	85.3	115.2	204.0	200.4	185.0
Costa Rica	73.2	80.4	106.0	122.7	146.6	110.0
Ecuador	32.1	31.7	44.0	57.2	102.1	124.9
Honduras	55.9	57.2	109.8	108.2	106.1	115.0
Guatemala	48.9	57.8	58.2	66.9	64.2	73.0
Total	**300.2**	**312.4**	**433.2**	**559.0**	**619.4**	**607.9**
Increase/Decrease (Per cent)		**4**	**39**	**29**	**11**	**-2**

3.5. Future Contribution from Palm Oil

Of the two main oils *e.g.*, palm and soya bean; demand for the latter is driven primarily by demand for meal, rather than for vegetable oil. In the past, the rate of growth in demand for Soya bean meal has been approximately the same as the rate of growth in demand for vegetable oil, so that Soya bean oil has retained a more or less constant share of the vegetable oil market. Thus base assumption is that Soya bean oil will continue to make upto 30 per cent of total production. Other oils remain constant at today's production levels, with palm oil making up the difference. Palm oil is the oil with the lowest production cost and hence considered that palm oil is now the 'marginal' oil. The move away from trans-fatty acids also favours palm oil, as it can be used without hydrogenation as the solid fat component in many formulations. An alternative scenario therefore has palm oil meeting the entire additional requirement, with the other oils all remaining constant at today's levels, including Soya. The demand for palm oil will be at least 93 million tons and more likely between 120 and 156 million tons (medium estimate).

It is proposed to bring an area of 20.70 million ha by 2015 in different oil palm growing countries. Almost more than 200 million seeds per year were required to bring the proposed area by 2015.While Malaysia will go only for replanting and rehabilitation, Indonesia is likely to double the area under oil palm. This country will emerge as the single largest producer of oil palm in the world. oil palm cultivation is expanding in every other country including India. Oil Palm cultivation is gaining importance all over the world since palm oil is considered as versatile oil, can be produced at competitive cost and is being used as food, nonfood and industrial purposes. The biofuel using palm oil is getting more popular.

3.6. Global Oil Palm Research

3.6.1. Global Research on Oil Palm

Oil Palm Research in the world is being conducted both by public and private sector having large scale oil palm plantations. Their research results on crop improvement, production, protection, processing, value addition, product diversification, by product utilization *etc.*, are being used for their plantations and also published in the journals and presented in the conferences for the benefit of others.

a) Research Institutes Involved in Oil Palm Research

The major research institutes conducting research on oil palm are given below:

i) Africa

- ★ AFOPDA-African oil palm Development Association
- ★ IRAD-Institute de Recherche Agricole pour le Development, Cameroon
- ★ NIFOR-Nigerian Institute for Oil Palm Research
- ★ OPRI-Oil Palm Research Institute, Ghana

- ★ INRAB-Institut National de la Recherche Agronomique du Benin
- ★ CNRA-Centre National de Recherche Agronomique, Cote d'Ivorie

ii) South America

- ★ ASD-Agricultural Services and Development, Costa Rica
- ★ CENIPALMA-Centro de Investigacion en Palma de Aceite, Colombia
- ★ Embrapa-Empersa Brasileira de Presquisa Agropecuaria, Brazil
- ★ FEDEPALMA-Federacion Nacional de Cultivadores de Palma de Aceite, Columbia

iii) Papua New Guinea

- ★ OPRS-Dami Oil Palm Research Station/Kimbe, West New Britain Province
- ★ PNGOPRS-Papua New Guinea Oil Palm Research Association/Kimbe, West New Britain Province
- ★ OPIC-Oil Palm Industry Corporation/Port Moresby

iv) France

- ★ CIRAD-Centre de Cooperation Internationale en Recherche Agronomique Pour le Developpement, France
- ★ IRD-Institut de Recherche pour le developpement, France
- ★ CPR-South China Tropical Crops Research Institute/Zhan County, Hainan Province, China

v) Myanmar

- ★ ARCPC-Applied Reserach Centre for Perennial Crops

vi) Phillipines

- ★ PICRI-Phillipines Industrial Crops Research Institute/Kabcan, Cotabato, Mindanao

vii) Thailand

- ★ STHHRC-Surat Thani Horticultural Research Centre/Surat Thani Province
- ★ PSU-Prince of Songkla University/Hai Yai, Songkla Province
- ★ OPGA-Oil Palm Growers' Association

viii) Indonesia

- ★ IOPRI-Indonesian Oil Palm Research Institute, Indonesia
- ★ IPARD-Indonesian Planters' Association for Research and Development
- ★ IBRIEC-Indonesia Biotechnology Research Institute Estate Crops, Bogor, West Java CAB-Centre for Agricultural Biotechnology/Cibinong, Bogor, West Java

★ PPKS-Pusat Penelitian Kelapa Sawit, Indonesia

ix) Malaysia

★ PORIM-Palm oil Research Innstitute of Malaysia/Kuala Lumpur

★ FELDA-Federal Land Development Authority/Kuala Lumpur

★ MARDI-Malaysian Agricultural Research and Development Institute

★ MPOB-Malaysian Palm Oil Board

x) India

★ IIOPR-National Research Centre for Oil Palm, India -up graded as Project Directorate on Oil Palm Research, and further upgraded as ICAR Indian Institute of Oil Palm Research, Pedavegi, A.P, India.

3.6.2. Recent Advances in Global Research on Oil Palm

a. Crop Improvement

The first breakthrough in the oil palm research was the identification of single gene responsible for shell thickness of tenera, pisifera and dura which led to the production of thin shelled tenera hybrids with thick mesocarp by crossing the selected duras and sterile pisiferas. Now superior planting materials are being produced through hybridization using elite duras and pisiferas.

Presently, the features which are being looked for hybridization are broad and proven genetic base, better adaptability over a range of environments, high precocity, high yield potential and high oil yielding extraction. Malaysia's slogan is 35:25 *i.e.*, 35 metric ton FFB/ha/yr and 25 per cent oil extraction ratio (OER). Many of the research organizations in the field of seed production had gone to the third cycles of improvement which are giving more than 30 metric ton FFB/ha/year with an OER of more than 25 per cent. United plantations, Malaysia had developed third round evaluation of DXP, which gives the mean yield of 31.75 metric ton FFB/ha in the seven years of research. The FELDA Yangambi materials yielded 30 metric ton FFB/ha, while the hybrid from Deli - Nigeria X Yangambi material has yielded 8.19 t oil/ha/yr and the DXP (Yangambi) standard cross yielded 6.15 metric ton oil/ha. FFB yields of over 35 t/ha/yr have been recorded on commercial fields in coastal and inland estates indicating the good adaptability of Golden Hope planting materials to these areas. Sawit Kinabalu oil palm D X P planting materials in Trial 3 with 9 Progenies yielded 6.46 metric ton oil/ha while Trial 46 with 38 Progenies yielded 7.09 metric ton oil/ha/year. Sampoema DXP planting materials in Indonesia had an oil yield level ranging from 7.0 to 8.5 MT/ha/yr. during 5-7 years and 9.5 to 11.5 metric ton/ha/yr from 7-9 years of planting (Asmono, 2007).The Genetic Resources program involves the collection of germplasm in the wild; evaluation, utilization and conservation *ex-situ*. The yield improvement over a period of time in Indonesia is given in Table 3.12.

Table 3.12: Yield Potential of Oil Palm Planting Materials from 1960-2006 (Indonesia)

Year	Planting Materials	FFB metric tons/ha/ year	OER Per cent	Oil Yield t/ha/year
1960	D XD, DXT, TXD	23.1	18.8	4.3
1970	DXT, TXD, DXP	23.9	22.6	5.4
1980	DXP	27.2	23.5	6.4
1990	DXP	29.8	23.8	7.0
2000	DXP (S1, Average year 6-8, N.Sumtra)	28.4	26.5	7.5
	DXP-Clones	38.4	30.7	9.0
Now	DXP (S3, Year 3, Alluvial, relatively drought, South Sumatra)	32.9	27.4	9.0
	DXP (S2, Year 3, Deep Peat Soil- Riau)	26.6	23.5	7.0
	DXP (S2, Year 3, Alluvial, -Riau)	28.0	26.3	9.3
	DXP (S1, Year 6-9, North Sumatra	34.9	32.3	11.0

Some of the achievements are the production of elite dura x pisifera (DxP) and *Elaeis oleifera* planting materials are deli X AVROS, PSI -dwarf stature, PS 1.1 - dwarf stature, PS2 - high iodine value, PS3 - high kernel, PS4 -high carotene (*E. oleifera*). Exotic genes are being utilized to improve the genetic variability of current breeding populations and provide wider perspective in breeding and selection.

The best experimental plot was reported to produce 8.6 metric ton/ha/yr. Selected progenies within the MPOB-Nigerian population have the potential of producing 12.2 metric ton/ha/yr of oil. In other selection programs elsewhere, the best palm was reportedly yielding 13.6 metric ton/ha/yr. Extensive efforts are still required however to reach the maximum theoretical yield of 18.2 metric ton oil/ha/yr. The role of germplasm is thus crucial for future developments of the oil palm industry.

b. Production of Planting Materials

During 2007, the oil palm scenario has changed dramatically for planting materials, since the development of oil palm was taking place at a rapid pace in Indonesia. In 1995, there were three major seed producers, producing 61 million seeds of which 50 million was from Oil Palm Research Institute (IOPRI).Now there are seven seed producers producing 115 million seeds. The seed production in Malaysia increased marginally from 50 to 75 million from 1995 to 2007 and the number of seed producers remained constant. In Asia Pacific, Papua New Guinea produces 30 million, Thailand 11 million and India 2 million seeds. Central and South American countries produces 40 million seeds by six producers of which ASD Costa Rica alone produces 30 million seeds. In African countries, Ivory Coast is the major seed producer. Other seed producing countries are Benin, Nigeria, Cameroon, Ghana and Zaire, which account for 17 million seeds. About 3 MTissue culture plantlets are produced in the world of which Malaysia produces 2 million, Costa Rica 0.5 million and Indonesia 0.5 million (Raja Naidu, 2007). United Plantations, Malaysia is the first company in the world to produce 1 million biclonal seeds. About 2.5

million interspecific hybrid seeds are produced in the world. The main producers are Equador (CIRAD partners) 1 million, Brazil, 1 million, Columbia (Lacabana) 0.3 million and Columbia (Indupalme) 0.2 million (Raja Naidu, 2007). It has been estimated that annually 202 million seeds annually are required of which Indonesia alone produces 152 million seeds since larger areas are likely to come under oil palm. Malaysia will concentrate mostly on replanting.

c. Breakthroughs in Biotechnology

The potential of genetic engineering in the production of novel crop plants should be fully exploited by the oil palm industry in order to remain competitive. The oil palm is the most productive of all oil crops and this inherent high productivity can be channeled towards the development of high value products such as novel oils, nutraceuticals and pharmaceuticals. As early as 1983, the Palm Oil Research Institute of Malaysia (PORIM), now known as the Malaysian Palm Oil Board (MPOB), identified biotechnology as a promising technology and made initiatives towards developing and exploiting this cutting edge technology for the progress of the oil palm industry. The basic tools and techniques for genetic manipulation for producing high oleate and high stearate transgenic palms are now in place at MPOB. The necessary target genes, the mesocarp-specific promoter and the techniques for oil palm transformation and regeneration into transgenic plants are available. The production of the first transgenic oil palm containing the Basta herbicide resistance gene in 1999 indicated that the key milestone for future utilization of this technology had been achieved. An Arabidopsis laboratory is being established at MPOB to support the ongoing genetic engineering programme as it will provide a rapid way to test the various oil palm promoters and to ensure that the gene constructs can produce the desired characteristics prior to transforming oil palm. The availability of molecular diagnostic tools such as genetic markers can help alleviate some of these difficulties and aid in the production of elite planting materials. Methods employed in these efforts include genetic mapping, partial cDNA sequencing, cytogenetics and DNA microarray analysis. In the oil palm, the two single gene inherited traits of importance to plant breeders are shell thickness and fruit colour. Research is to be done to find these markers, which are more tightly linked to these characters, which would provide a handle to breeders to distinguish virescence from nigrescence fruits as well as dura, pisifera and tenera types at the nursery stage.

The oil palm genome research by MPOB and plantation companies in Malaysia have shown encouraging achievements in sequencing the functional genes of oil palm for desired traits not only for higher productivity but also in increasing the marketable products. MPOB research has successfully generated over 400 sequences representing 90 per cent of the functional genes in oil palm. This has led to a better understanding of the pros and cons of each sequencing platform and their complementarity in obtaining the full picture of the transcriptome in oil palm.

d. Tissue Culture

The *in vitro* propagation of oil palm involves the exploitation of rapid multiplication and regeneration potential of plant cells in the laboratory in order to generate large clonal planting material. A data tracking system has been developed by MPOB, which involves a database system that is interfaced with a barcoding system. This system is a convenient and comprehensive system for monitoring the production of tissue culture clones and allows for a better analysis of production parameters. The development of the suspension culture technique for the micro propagation of oil palm has proven to be more efficient than the current solid culture system in terms of production capability and conservation of labour and space. The liquid suspension culture system has been further improved using bioreactor technology. Using this system, physical and nutrient conditions of the culture can easily be manipulated and optimized. There has been a thirteen-fold weight increment in these cultures over that of liquid suspension cultures. These improvements when utilized properly could be the driving force to make clonal palms, the future planting materials.

The sustainability of oil palm production can be addressed through new genomics-led research. More efficient use of genetic resources is one of the most important approaches to enhance productivity. The 1.8 Gb *E. guineensis* genome sequence recently released by the Malaysian Palm Oil Board provides an accurate record of genes, predicted proteins and other genomic elements and is a valuable reference for crop improvement. The sequence will provide a framework for faster and more effective breeding methods. MPOB houses more than 100,000 wild germplasm accessions and these are an indispensable resource for genomics-guided breeding.

One of the most important economic traits of the oil palm is how the thickness of the shell correlates to fruit size and oil yield. The genome sequence facilitated the identification the Shell gene, responsible for the three fruit forms: Dura (thick-shelled), Pisifera (shell-less) and Tenera (thin-shelled). Tenera palms provide the optimum productivity with about 30 per cent more oil than Dura palms. Currently, it can take four to six years to identify whether an oil palm plantlet is of the desired fruit form. The discovery of the Shell will allow for the three fruit forms to be distinguished in the nursery long before they are field planted, thus enabling significantly enhanced breeding operations. The Shell marker will also be an important quality control tool in commercial seed production as up to 10 per cent of the plants may be the low yielding Dura due to uncontrollable wind and insect pollination. The unraveling of the oil palm genome sequence will pave the way for the discovery of other important agronomic traits like fruit colour, disease resistance and low height increment (Choo Yuen May, 2013).

3.6.3. Recent Advances in Crop Production

a. Agronomy

The nursery management studies conducted at United Plantations recommends three rounds of culling at the main nursery. The first round of culling is at 6 months, but the real culling is done at 9th and 12th month, where the culling of abnormal seedlings like erect habit, poor vigour, juvenile, flat top, squat habit, wide internodes, acute angle of insertion of leaflets, severe leaf spot disease, collante, narrow leaflet and closed internodes. It was estimated that the total percentage of culling will be 8.9 per cent. Nursery manager should pay sufficient attention in the nursery and not to delegate it to the labour. Cutting the cost and trying to reduce the input at the critical stages of the nursery is a recipe for disaster. An innovative technique has been developed where young palms are planted directly on to the residue piles in order to improve accessibility and efficiency of nutrient utilization. This technique offers greater synchrony between nutrient release and plant uptake in terms of space and time compared to the standard replanting practice. A hectare contains nutrient equivalents of 642 kg N, 58 kg P, 1384 kg K and 156 kg Mg. Oil Palm Nutrient System (OPENS) can be used successfully to predict the maximum site yield potential, detection of the most limiting nutrient and to estimate the amount of fertilizers to be used to achieve maximum yields.

A protocol to produce Fertilizer Management Map to cater to site specific fertilizer application has been developed for adoption by the Industry. The irrigation study at Univanich has indicated that the four selected progenies with and without irrigation have got differential oil yield level. While progenies without irrigation yielded 6.7 metric ton oil/ha and with irrigated condition the oil is increased to 9.5 metric ton/ha/year.

Revolutionizing the crop and livestock integration in oil palm area is an advanced step to optimize productivity of oil palm plantations with crop and livestock integration. The crop integration activity can only be extended to a mature oil palm age if, the oil palm is planted following the double avenue planting system. Several forage crop integrations have been elaborated that include Napier grass, forage sorghum and kenaf. For every crop details are given on planting, maintaining, harvesting, processing and the nutrient values. Intensive livestock integration is the other important component to compliment this project. Planting of improved pasture has made possible to increase the stocking rate of livestock integration. Intensive integration of cattle and goat are used to demonstrate the potential of suggested production system. In conclusion, systematic approach of forage crop and livestock integration has great potential to oil palm growers and the country as a whole. It is able to increase productivity of oil palm plantation and enhance national food production. Future works to highlight the benefits of nutrient cycling in integrated crop-livestock and oil palm integration system should be conducted to optimise gains from this system (Kami/Azmi Tohiran *et al.*, 2013).

To ensure economic and environmental sustainability, yield must be maximised over the entire life of an oil palm planting. Much work has been done on breeding for higher yield, and on optimizing fertiliser inputs. Planting at 160 palms/ha with later thinning by 25 per cent gave 18 per cent greater yield over 18 years than planting at 143/ha without thinning, provided that thinning was done in year 8. If thinning was delayed until year 12, the yield increase was negligible.

In a comparison of irrigation methods, drip irrigation proved to be the best. With a mean annual water deficit of around 290 mm, drip irrigation at 450 litres/palm per day gave an average yield increase of 10 metric ton FFB/ha yr from mature palms. There were large differences between DxP progenies in response to drought, with yield reductions ranging from zero to 50 per cent. Planting drought tolerant material will helped to increase yields where a regular dry season occurs, and irrigation is not possible. Yield loss at replanting can be reduced by underplanting. In commercial practice this gave 36 per cent more FFB compared to clear felling over the first 5 years after replanting, with little difference thereafter. Thinning and replanting 50 per cent of the stand at 10 years, while retaining the other 50 per cent in a two-tier canopy, gave 9 per cent greater yield over 18 years than a standard planting. Either method allows the possibility of continuous production, and recycling of biomass nutrients after felling of the old stand, reducing the need for conventional fertilizers (Cortley and Palat, 2013).

Oil palm biomass produced as by-products in the plantations is pruned oil palm fronds (OPF) that are available throughout the year during harvesting of FFB and pruning of palm tree. The OPF biomass available annually is about 14.75 metric ton ha^{-1} yr^{-1} annually from natural, plantations, which contributed to about 136kg N, 10.4kg P, 183kg K and 16.5kg Mg ha^{-1} yr^{-1} annually from natural plantations. When recycled in the plantations it plays an important role in conserving soil fertility in oil palm ecosystem. At replanting, the above-ground biomass from the old palms, mainly trunks and fronds is about 85 t ha^{-1} dry matter which contributed to about 577kg N, 50kg P, 1255kg K and 141kg Mg ha^{-1}. The palm biomass during replanting provide significant amount of nutrients to the succeeding palms that can reduce the chemical fertilizer input.

Quantification of the amount of Potash (K) budget that is imported and exported in the oil palm ecosystem can be estimated based on the theoretical calculation from the total mature area of oil palm, total requirement of K for matured area, the processed FFB, amount of K in the fresh fruit bunches and the amount of K in palm oil mill wastes. The amount of K removed from the ecosystem which is considered as waste can be estimated from the calculation.

In 2012, the total matured area of oil palm in Malaysia was registered at 4,352,872 ha and the FFB processed was 92,328,847 metric ton. Based on these figures, theoretically the total requirement of K fertilizer for the matured area of oil palm amounted to about 886,440 metric ton K_2O, or equivalent to about 1,477,400 metric ton of Muriate of potash - MOP (60 per cent K_2O). The total amount of K that was removed from the processed FFB was about 361,929 metric ton K (or 436,124 ton K_2O) or equivalent to 726,845 metric ton of MOP. The calculated total amount of K available from the EFB of the palm oil mill waste was estimated to be about 267,753 metric ton K_2O or equivalent to 446,255 metric ton of MOP fertilizer and that available in the POME was about 131,859 metric ton of K or equivalent to 264,817 metric ton of MOP fertilizer or 30 per cent and 18 per cent respectively, of K requirement for mature oil palm in Malaysia. The industry needs to find ways on how to manage and recycle significant amounts of plant nutrients especially K that are available from the mill wastes which is about 48 per cent from the industry's requirement of K and in monetary value is worth more than RM 1.03 billion annually.

Potential palm oil mill waste management practices such as composting of empty fruit bunches and liquid waste effluent generated from palm oil mill are quite useful. Over 25 years, oil palm plantations potentially accumulate up to 1569.32 t C ha^{-1} with good agricultural practices such as frond stacking and EFB usage, but whether all of this carbon is retained is uncertain. Numerical validations of the AAR carbon model are similar to estimates by other authors. (Cheahl, H.H. Gani and K.J.Goh 2013).

b. Crop Physiology

Understanding of basic physiological processes of the oil palm and relating it to the production and management of the crop continues to be a challenging and active area of investigation. Light use efficiency and partitioning of assimilates are two areas which need to be emphasized during crop development. Humidity and radiation are the two climatic factors affecting the yield. Low humidity restricts stomata opening and hence affecting the process of photosynthesis. These findings have major implications for oil palm plantations grown under non-traditional environments. Early yield benefits can be obtained when optimum leaf area is reached as soon as possible after field planting and under high radiation conditions of S.E. Asia, it is acceptable to assume an optimum leaf area index value between 5.5 to 6 (Breure, 2007). Climate based growth models are very important in such conditions. Studies should also focus on the characterization of the physiological responses to drought and define criteria for irrigation scheduling. Mini rhizotron tubes with camera attachment are currently being used to monitor root turn over, which would be helpful in knowing the amount of carbon needed to maintain the root system.

c. Recent Advances in Crop Protection

i) Pest Management

Insect parasitoids, *Bacillus thuringiensis* and *Beauveria basiana* play an important role in the control of bagworm. A strain of *Metarhizium anisopliae* has been proven effective against the rhinoceros beetle. Three strains of viruses against Oryctes rhinoceros have been isolated, characterized and identified in Malaysia. Of the three, one strain has been effective to control rhinoceros beetles causing 70 per cent mortality to larvae and 87 per cent mortality to adult beetles. Outbreak of hairy caterpillar, *Dasychira sp.* was found in Sabah, Malaysia. It has been suggested that a pest monitoring and surveillance system should be implemented. It was also important to avoid excessive destruction of ground cover and to plant flowering plants like *Euphorbia heterophylla, Antigonon leptopus* and *Turnera* sp. to maintain a congenial habitat for the natural enemies of the leaf eating caterpillar pest of oil palm. Oil Palm being a perennial tropical tree crop has got greater biodiversity than the cereals, vegetables and other short term cropping system of the world. A typical oil palm estate teems with 268 sp. of flora and fauna which include microbes, insects, anthropods, reptiles, fish, birds and small mammals including relatively rare leopard cat. In addition rearing of cattle, sheep and deer add to the diversity of life in the plantations.

ii) Diseases Management

Among the major diseases of oil palm, vascular wilt disease is the most widely studied; and a wealth of knowledge is available about this disease. Reliable screening techniques have been developed at the nursery stage permitting selection for resistance to this disease. As for Ganoderma basal stem rot disease is concerned, successful inoculation of oil palm seedlings with the pathogen is a recent development. This development should form the basis for a screening technique to look for the resistance. Likewise, effective screening, for resistance to red ring disease, sudden wither disease spear rot-bud rot is virtually absent. Active research by various groups is going to gain a better understanding on the host pathogen interaction and also the role of the environment on disease expression. This information would be useful to eventually control these diseases. Currently Ganoderma Basal Stem Rot has been reported to be present in all the oil palm growing areas of Asia, Africa, south and Central America. The Vascular wilt disease, although initially restricted to Africa, is already established in south and Central America, but not in Asia. Spear rot, bud rot, Sudden wither disease, red ring disease are confined to south and Central America only. With this scenario, every care must be taken to avoid possible introduction of the pathogens to areas currently free from them. Strict quarantine measures must be taken, especially on the movement of planting materials across continents.

d. Recent Advances in Harvest and Post-Harvest Technology

Oil Palm has different products which can be utilized in food, nutritional, industrial and eco-friendly products. Value addition to palm oil, oleo chemicals, esters, bio-lubricants and bio-diesel *etc.* technologies have been developed for utilization Of POME, Empty bunches, leaf fronds and stems of oil palm after one cycle. Such technologies are available with MPOB, Malaysia.

Harvesting of young palms is done using Chisel and older palms with sickles attached with long telescopic light weight poles. Because of high labour cost and constraints in the supply of labour, mechanization is essential for the future sustainability of the oil palm industry. The most adopted machine in Malaysia is the prime mover with the grabber and hi lift. A commercial prototype of oil palm harvesting machine has been developed in Malaysia (Figure 3.1) which takes 2.5 to 3.5 minutes for a complete cycle of harvest which includes positioning the cutter, cutting the frond and setting the grapple to engulf the fruit, cutting the bunch and bringing it down into a container. It is also suited to harvest tall palms of more than 7 meters and is manned by one person only. Two stroke petrol engines of 1.3 HP fitted harvesters by Cantos are now available for harvesting the oil palm bunches.

Malaysia has also developed a mini -tractor -trailer system of evacuation with a cuter and carrier gang of three people. By this the worker productivity is increased by 33 per cent. Mechanical fertilizer applicator and weed control using 60 to 80 hp has also been developed.

Extensive research works have been done on Bio-gas utilization in palm oil mills. Innovations have also been made on Palm Oil Milling Technology like, Continuous Sterilization System, Plant-wide Automation System, Environment friendly Clarification Process, Trash Removal System, Rolek TM Palm Nut Cracker, and Dry Separation of Kernels and shells via a Four stage Winnowing Column; Palm Oil Mill Effluent Treatment and Palm oil Milling Technology Centre (POMTEC).

Figure 3.1: Power Operated Harvestor.

3.7. Oil Palm Research in India

Oil Palm Research in India was started at a small scale at the erstwhile Central Plantation Crops Research Institute-Research Centre, Palode during 1986. Subsequently four Research Centres were started under the All India Coordinated Research Project on Palms at Vijayarai (Andhra Pradesh), Gangavathi (Karnataka), Mulde (Maharashtra) and at Aduthurai (Tamil Nadu) during the VIII Five Year Plan. The National Research Centre for Oil Palm (NRCOP) was started in February, 1995. This NRCOP was upgraded as Project Directorate of Oil Palm Research (DOPR) in the year 2008.It is now upgraded as ICAR-Indian Institute of Oil Palm Research.

The major achievements of research on oil palm are given below.

a) Oil Palm-Germplasm and Crop Improvement

The oil palm germplasm assemblage being maintained at NRCOP, Pedavegi and Palode, consists of 128 indigenous and exotic accessions collected. Screening of 240 African dura palms for drought tolerance based on physiological and biochemical characters was done. Evaluation of inter-specific hybrids at Palode resulted in identification of three promising dwarf palms that can be used for further improvement.

Genetic diversity study by RAPD analysis for five accessions (Palode: GD3, 240D x 281D and 80D X 281D and two Costa Rican: 98C -254D and 98C208D) with the help of 33 primers indicated that no two palms within the accessions were found genetically similar. DNA fingerprinting of different germplasms also revealed that no two palms were genetically similar even within the same accessions. Wide genetic diversity was found among the different accessions by Randomly Amplified Polymorphic DNA (RAPD) analysis. Fatty acids composition of oil from Oleifera palms indicated that palms no.Eo- 02, Eo-04, Eo-05, Eo-11, Eo-17, Eo-18, Eo-19, Eo-20, Eo-22 and Eo-23 could be recommended for inter-specific hybridization. Oil from Oleifera palms was analyzed for quality. Iodine value as well as Gas Chromatography analysis revealed wide variation in the fatty acid compositions of oil among the different Oleifera palms. However, a few Oleifera palms were found superior in terms of unsaturated fatty acid content in the oil.

Attempts were made to germinate 111 seeds by wet heat treatment for 60 days. The first germination was observed after 17 days of incubation and a total of 33 per cent germination was recorded after 30 days.

b) Hybrid Seed Gardens

Commercial hybrid seed production was started at Thodupuzha, Kerala State, utilizing the indigenous Duras and Pisiferas from 1982 onwards. Advanced generation parent materials were evolved through reciprocal recurrent selection and planted at Palode which had now started seed production. Selection of duras was done based on yield and bunch analysis data. Parental duras and pisiferas (tenera x tenera) were supplied to establish 3 seed gardens by Government Department one each at Rajahmundry (Andhra Pradesh), Taraka (Karnataka) and Thodupuzha (Kerala) and at M/s. Nava Bharat Enterprises, Lakshmipuram (Andhra Pradesh). Seed processing and germination facilities were developed and techniques have been perfected to achieve 95 per cent germination. Using the elite parent materials evolved at Palode, commercial hybrid seed production at the rate of 5 lakh seeds annually has been accomplished. Proven Dura parent materials were introduced from ASD Costa Rica and were planted at Pedavegi and Palode and evaluated for identifying best duras for DxT seed production. Presently it is possible to produce 2.46 million seeds from these indigenous seed gardens annually as against the full potential of 4.9 million seeds. Recently 4 more seed gardens have been proposed to be established.

c) Nursery Management

Nursery management techniques for raising healthy oil palm seedlings were developed. Soil-sand-Farmyard Manure or oil palm waste compost in 1:1:1 proportion was found to be an ideal potting mixture for growing seedlings. A fertilizer dose of 10-5-5 g N, P_2O_5 and K_2O per seedling was found optimum. Application of fertilizers in equal splits at three monthly intervals was suitable for optimum performance.

d) Nutrient and Water Management

A fertilizer dose of 1200-600-1200g N, P_2O_5 and K_2O per palm per year resulted in better growth characters, maximum flowering and early yield in the third year of field planting under rain fed conditions. 1200:600:2700 g NPK/Palm/year for Coastal Tamil Nadu, Andhra Pradesh and Konkan Coast, and 1200:600:1200 g NPK/Palm/year for Tungabhadra Command area were recommended based on Coordinated

Research programme. Boron deficiency is very frequently noticed for which 50 to 100 g borax application thrice in a year is recommended.

Palms irrigated through drip system recorded maximum yield followed by jet and basin. Maximum yield was recorded in irrigation level IW/CPE= 1 followed by IW/CPE=0.8 and IW/CPE=0.6. Through fertigation with 1200:600:2700 NPK/palm/year could produce 24.15 MT FFB/ha/year. Annual carbon sequestration by oil palm was 11.73 and 5.51 t/ha under irrigated and rain fed conditions respectively.

About 2/3rd of the inorganic nutrient requirement of palm could be substituted through organic compost made from oil palm waste. Technique of vermicomposting using oil palm wastes has been standardized.

e) Oil Palm Based Cropping System

Intercrops *viz.*, Maize, Tobacco, Chilly, Black Gram, Horse Gram, Red Gram, Groundnut, Sunflower, Sesame, Brinjal, Elephant Foot Yam, Cauliflower, Cabbage, Drumstick, Tomato, Ridge guard, Water melon, Banana, Pineapple, Turmeric, Cocoa, *etc.* were found to be promising as inter crops in West Godavari District of Andhra Pradesh. Maize, Banana and Tobacco were found to be the most profitable and compatible inter crops in oil palm based cropping system during juvenile phase.

Mixed cropping in Oil Palm involving multi-species crop combinations with Cocoa, Black Pepper, Cinnamon, Betel vine, Piper nigrum, Guinea grass, Heliconia, Red ginger and Anthurium have been established to intensify productivity and improve soil and water conservation in the matured plantation.

f) Crop Physiology

The root distribution pattern studies in an adult plantation irrigated with basin irrigation system indicated that the root biomass was maximum at 10-20 cm depth followed by 20-30 and 30-40 cm depth. The density of roots was maximum at 1 m distance from palm base and decreased with further distance from palm base.

Photosynthetic activity measured in different leaves of Oil Palm canopy revealed that the maximum rate was observed in the 9th leaf followed by 17th, 25th and 33rd leaf. The leaf temperature was maximum in the 1st leaf and decreased with leaf age. Sap flux studies in oil palm in relation to evapo transpiration and vapour pressure deficit could give vital leads in developing an approach for monitoring environmental response in oil palm. Palode and Ivory Coast crosses recorded maximum photosynthetic rates and its associated parameters followed by ASD Deli X Avros. The Bunch Index was highest in Ivory Coast followed by Palode.

Disaster management technique of replanting the uprooted palms due to cyclone was standardized with proper care and management and the palms were brought to normal yield within 3 years.

g) Pest Management

Infestation of Rhinoceros beetle was brought down from 8.25 per cent to 1.8 per cent by release of *Baculovirus* infected beetles. Application of Quinalphos and Lambda *cyhalothrin* was found effective in controlling the slug caterpillar. *Beauveria bassiana* (10-1) was also effective but time consuming compared to insecticides. The slug caterpillar was found migrating from coconut gardens causing damage to oil palm. Heavy incidence of Slug caterpillar was observed. A leaf webber *Ambadra sp* was found to be causing severe defoliation in oil palm. Bagworm, *Melisa plana* was found to be endemic in older oil palm gardens. Scale and mealy bugs were recorded as minor pests feeding on the FFB of the oil palm.

House crow and Jungle crow in Krishna district, Parakeets in West Godavari district, Mynah in Vizianagaram district were the predominant avian pests causing moderate to heavy damage. Nylon nets of mesh size of 10 cm^2 green and violet colours were found effective in controlling house and jungle crows. Use of bamboo noose traps was found to reduce the burrowing rats. Single application of warfarin could reduce only 50 per cent of rat incidence within one week period compared to Zinc phosphide. Two application of zinc phosphide with no gap between two applications was found effective. A wild boar scaring device has also been developed.

h) Disease Management

Nursery diseases can be well managed by clipping of severely affected leaf portion followed by application of mancozeb @ 0.2 per cent or carbendazim @ 0.1 per cent at monthly interval. Techniques for the control of inflorescences diseases *viz.*, bunch failure, bunch rot and bunch end rot have been standardized. Stem surgery technique for management of stem wet rot and upper stem rot was found successful. A technique for the control of bud rot, observed in all oil palm growing states, has been evolved.

i) Post-Harvest Technology

A processing mill of 1 MT FFB/hour was developed at Palode with ICAR -CSIR collaboration and was scaled up to 5 t/hr capacity. Among the different storage conditions of "pure form" caroteniods, storage in the deep fridge showed minimum degradation followed by 'in dark at room temperature". Reuse potentiality of better performing adsorbents were studied and no significant difference in recovery was observed between the fresh and reused Fullers earth, though average recovery from the fresh adsorbent was higher.

Palmolein was found to be the balanced oil with respect to saturated and unsaturated fatty content (46.27: 53.73). Oil from *oleifera* palms was found better among the different categories of palm oil in terms of oleic (Cl8:l) and linoleic (C18:2) acids content. Palm kernel oil was more similar to coconut oil than that of other oils. The effect of high temperature on FFA content, Iodine Value and Peroxide Value showed that the lipase activity was consistent and steady up to 50 C even after 7 days of incubation of the fruits. When the harvested fruits were stored at high temperature, unsaturation of oil increased and there was no significant change in Peroxide value.

A cost and time saving indirect method of oil estimation has been developed. Graphical Comparison between direct and indirect Oil extraction units for direct and methods of Oil Extraction indirect methods of oil extraction. The pulping characteristics of Oil Palm fronds and EFB were studied for the preparation of paper boards and paper board files. Technology for extracting fibre from oil palm empty bunches was standardized.

The studies on gasification of oil palm waste indicated that the fronds cut into a size of 5 cm are suitable for gasification, while shredded EFB requires densification for proper gasification.

Cooling pads and geo-textiles were prepared from EFB fibers. Performance evaluation of EFB fibre mats as cooling pads in concrete terraces indicated considerable reduction of room temperature during summer months.

A modified mini palm oil mill is being evaluated for efficiency and capacity. A mobile Oil Palm waste shredding unit has been designed. A suitable alternative has been found out to make window curtains from the rachis of Oil Palm which is cheap and eco-friendly Oil Palm Frond Curtain.

Chapter 4

Introduction of Oil Palm in India

Oil Palm in India is as old as that in Malaysia and Indonesia, while, four ornamental oil palm seedlings planted in Bogor Botanic Garden of Indonesia in 1856 became the parental source material for South East Asia, the ornamental oil palms planted in 1886 in the Botanic Garden, Kolkata remained as ornamental palms only. The commercial plantations were taken up in Malaysia and Indonesia during 1950 to 1960 and there was progressive increase in area and production. In India too, first commercial planting was taken up at Thodupuzha in Kerala state in 1960 with planting materials received from various countries but not much head way was made after that. However, these palms have been a good source of parental planting materials for establishing seed gardens for producing tenera hybrid seeds.

4.1. Why Oil Palm in India?

Nine annual oilseed crops grown in 27.74 million ha in the country produce about 27.73 MT of oilseeds which give about 7 million ton of vegetable oil but could not meet the demand resulting in the import of about 8.6 million ton of oils and fats. Domestic demand is increasing every year, since oil consumption is income elastic and population continues to increase. Fast shrinking of agricultural land for food production warrants replacement of some low value crops with high value crops. Scope for horizontal expansion of land is limited due to faster urbanization, industrial growth and population explosion. Hence, there is an urgent need for diversification by introducing high value crops.

Oil palm gives the highest oil yield of 4-6 MT/ha/year/with a global average of 3.20 MT/ha/year which no other known oilseed crop is known to produce. The highest theoretical oil yield of this crop is projected to be 18 MT/ha/year with the advanced planting materials. Large scale plantations in Malaysia have yielded about 8-12 million ton oil/ha in some good plantations.

Oil palm is an eco-friendly crop and adds lot of organic matter to the soil for its enrichment. Thereby soil health is improved and there is possibility of reducing inorganic fertilizers if the bio-mass is properly recycled into the plantation. Palm oil is a source of nutrition and health due to its high calorific value and richness in vitamin A and E contents because of which it can contribute substantially to the nutritional security of the world. Palm kernel oil is emerging as a major source of lauric oil. Palm oil is emerging as one of the major sources of Bio-fuel/bio-diesel/bio-lubricant. Oil palm cultivation elevates the socio-economic status of the farmer with its high returns and sustained income generation for a long period. It provides opportunity for higher employment generation in the plantations as well as in allied agro-based industries. It provides the possibility of co-generation as an alternate source of energy.

4.2. Early Introduction of Oil Palm

Oil palm was first brought to India as a botanical collection at the National Botanical Gardens, Kolkata in 1886. Subsequently, Maharashtra Association for Cultivation of Sciences (MACS) introduced African dura palms in Pune during 1947-54. These palms were planted on canal bunds, home gardens and to some extent on forest lands. Even today we can see some palms in Pune city as well as on canal bunds.

The possibility of growing oil palm in India was investigated by Mr. D.H.Uraquart, former Director of Agriculture, Gold Coast in 1950, Mr.L.Davidson, in 1965, Coconut Development Directorate, Cochin, in 1976, and Mr.H.Q.R Reddy and his team in 1972. All these earlier missions looked for forest land for oil palm development and suggested clearing reserves for the purpose.

The Government of Kerala established an Oil palm research station in 1960 at Thodupuzha in about 40 ha and imported seeds of duras, teneras and tenera x tenera hybrids were planted. The varietal collection maintained at this station formed the basic source for later research and development including indigenous tenera seed production.

4.3. Commercial Plantations of Oil Palm

The cultivation of oil palm on a plantation scale was attempted in India in the forest lands on the basis of feasibility studies conducted by oil palm experts from abroad. Accordingly, two large scale plantations were set up in the public sector on a commercial scale, one in Kerala and the other in Little Andaman.

4.3.1. Kerala (1971-82)

A large scale planting of oil palm was launched from 1971 to 1982 in Kerala covering an area of 3645 ha with imported seeds from Malaysia, Nigeria, Papua New Guinea and Ivory Coast. The plantations were established on deforested land where the soil was laterite and shallow. The area was also characterized by having a prolonged dry spell of not less than four months in a year. Though the initial performance of the plantation was far below the expectations, the yield level was brought up to 2.7 MT of oil per ha in certain areas of plantations. However, on the whole, the present average yield was below one ton of oil per hectare per year. It should have be possible to increase the yield if practices like proper manuring, soil moisture conservation and summer irrigation were adopted in all the plantations. There are also two processing units for the extraction of crude palm oil of which one is of 10 MT/hour and the other is 2 MT/hour capacity. Since these processing units are outdated and inadequate to process the available fresh fruit bunches, a most modem mill of 20 tons FFB/hour capacity was commissioned in July 1999.

4.3.2. Little Andaman (1976-85)

The Andaman Forest and Plantation Development Corporation raised oil palm in 1563 ha in Little Andaman between 1976 and 1985 with the planting materials received from Nigeria, Malaysia, Ivory Coast, Papua New Guinea and Zaire. These plantations were initially not managed well probably due to oil palm being a new crop. Consequently, the average productivity obtained was very low. In recent years, one processing unit of 4.5 tons/hour capacity with screw press was put up for ensuring timely processing. There was good scope for getting at least 15 tons FFB/ha/year from these plantations if practices like proper manuring at appropriate time, adoption of soil-water conservation methods, organic residue management, growing cover crops *etc.*, were adopted.

The CPO recovery has increased from 433 metric ton in 1990-91 to 5987 metric ton in 2006-07. This can be increased further with timely harvest, processing and adopting proper harvest standards. A study team headed by Sri. K.M.Tiwari, Ex-President, Forest Research Institute, Dehradun, conducted feasibility studies for oil palm plantations in Andaman and Nicobar Islands in June 1986. The team identified 60,000 ha in various Islands, which could be put under oil palm or other plantations without affecting the natural ecosystem of the Islands. However, this report did not get much importance since forest land conversion was not favored by the Government of India.

4.3.3. Oil Palm in Karnataka

A small-scale plantation of 4.5 ha of oil palm was planted by the Forest Department of Karnataka at Sambaje Reserve Forest near Sulliya hills (Dakshin Kannada) during 1969 with dura population. The palms nearer to water source produced good-sized bunches. However, no attempt was made to manage the plantation which was left as forest crop. Since the performance of oil palm was not satisfactory as a forest crop without adequate management and the Government did not encourage the conversion of forest, the oil palm development came to a halt till the concept of irrigated oil palm was introduced.

4.4. Reports on Oil Palm Cultivation in India

Indian Oil Palm has a long history starting from 1959 till date. Many missions/ committees were constituted to identify potential areas for cultivation of oil palm in the country. Basically all these committees identified areas in the forest lands based on their experience on oil palm grown all over the world. The concept of irrigated oil palm was mooted as a small holder's crop during 1986 and subsequently various committees identified potential areas for irrigated oil palm.

The first mission headed by D.H.Urquhart, former Director of Agriculture, Gold Coast investigated prospects of growing oil palm in India in 1959 and indicated that there are regions of South India where climate and soil are admirably suited to grow oil palm. He also suggested trying oil palm in Tripura.

Later, L. Davidson made a study on the prospects of oil palm development in Kerala in 1965. He identified Kuthali Estate Reserve as best and also indicated that Konny Reserve Forest along Achancoil River, Vazhachal Forest Reserve, Renny area and Nilambur as suitable areas for oil palm cultivation. He envisaged a total of 8000 ha of forest area for this purpose. Coconut Development Directorate, Ministry of Agriculture and irrigation, Cochin, prepared a project in 1976 for raising 2400 ha oil palm in India in which 1000 ha were identified in Kanyakumari.

A mission led by H. G. R. Reddy, Regional Industrial Advisor on Agro-industries and Light industries (UNIDO/ECAFE) made a survey for development of oil palm industries in the world. The mission also visited India in 1972 and reported a strong possibility for an extensive development of oil palm industry in India. He had suggested that a detailed technical feasibility study should be carried out before the establishment of oil palm plantation in Reserve Forest areas of Kuthali, Vazhachal and Konny/Ranny in Kerala. He also suggested a detailed technological and economic feasibility study for establishing oil palm industry in Tamil Nadu. All these earlier Committees/Missions had identified the potential areas in the forest lands. Based on these reports a few plantations were taken up in the forest lands.

4.4.1. CPCRI Committee (Dr. K.V.A. Bavappa)

A Committee appointed by Central Plantation Crops Research Institute, Kasaragod in 1985, comprising of Dr. K.V. Ahmed Bavappa, its Director, Dr. P. Rethinam, Project Coordinator (Palms) and Dr. K. U. K. Nampoothiri, Senior Scientist identified potential areas in the southern states of Tamil Nadu, Andhra Pradesh, Karnataka and Maharashtra. They indicated the possibility of growing oil palm in about 2.4 lakhs ha under irrigated condition nearer to the Irrigation Project areas.

4.4.2. Identification of Potential Areas in Karnataka State (Dr. P. Rethinam)

This Committee appointed by the Government-of Karnataka under the chairmanship of Dr. P. Rethinam, the then Project Coordinator (Palms) with Shri M. Gurusamy, the then Joint Director of Horticulture, Government of Karnataka, and Dr. K.U.K. Nampoothiri, Senior Scientist and Head, CPCRI RS,Palode, Sri K.M. Bhojappa, Professor and Head, Horticulture Department, U.A.S., Bangalore and Sri R.G. Bhat, Director, Project Formulation Division, Planning Department, Bangalore identified an extent of 3,00,000 ha in the five major irrigation project areas *viz.*, Thungabhadra, Upper Krishna, Malaprabha, Ghataprabha and Cauvery as potential areas for oil palm in Karnataka by 2000 AD. Out of 3 lakh ha proposed, an area of 1.37 lakhs ha was to be introduced as a new crop in the command areas of upper Krishna project, and of Varuna canal in Cauvery basin as well as the existing government lands in various command areas. The remaining 2.63 lakhs ha was to be covered by replacement of existing crops in the already developed command areas.

4.4.3. Potential of Oil Palm Cultivation in India (Committee headed by Dr. K.L. Chadha)

Realizing the potential of oil palm in India, the Government of India for the first time constituted a high level working group in 1986 at the instance of Dr. K. L. Chadha, the then Horticulture Commissioner, Government of India, as Chairman including Agriculture Production Commissioners of the states and Managing Directors of Oil Palm India Limited, Andaman and Nicobar Forest and Plantation Corporation; Director, Central Plantation Crops Research Institute, Kasaragod as members. This Committee was assisted by Dr. P. Rethinam the then Project Coordinator in identifying the potential areas in different states in consultation with state Agriculture/Horticulture officers and in the preparation of the report. This committee also made several significant suggestions, which helped in establishing a sound oil palm industry in India.

Table 4.1: Potential Areas for Cultivation of Oil Palm in India

States	*Area (Lakh ha)*
Andhra Pradesh	2.50
Assam	0.10
Karnataka	2.50
Kerala	0.05
Maharashtra	0.10
Odisha	0.10
Tamil Nadu	0.25
Tripura	0.05
West Bengal	0.10
Total	**5.75**

Source: Chadha Committee Report, 1988.

i). The Committee identified 5.75 lakh ha as potential areas in nine states along the Western and Eastern Coasts of India of these, Andhra Pradesh and Karnataka hold key position (Table 4.1).

ii). The Committee also suggested strategies for covering 2.6 lakh ha of area under oil palm plantations by 2000 AD in the identified areas.

iii). It recommended sizable pilot planting coverage at least 1000 ha each in Andhra Pradesh, Karnataka and Tamil Nadu and taking up development work from 1993 to cover 2.0 lakhs ha to produce 1 million MT of oil palm and 0.1 million MT of Kernel oil per year by 2009.

iv). The committee also suggested utilizing the existing source for seed production (CPCRI, Regional Station, Palode) and import of quality planting materials, setting up of three new seed gardens and standardizing tissue culture protocol for mass multiplication of planting materials.

v). As far as processing is concerned, two strategies were suggested, *viz.*, a small scale processing unit to cover 200 ha and a large scale unit of 5 to 30 ton FFB/ha.

vi). Regarding organizational setup, three types *viz.*, Private Companies/ Corporate companies for large compact areas and organization similar to that of FELDA of Malaysia involving small holders and setting up of factories similar to that of sugar factories with assured procurement and paying promptly were suggested.

vii). The committee also suggested setting up of National Research Centre on Oil Palm to upgrade technology required for oil palm in the country.

4.4.4. Other State Committees (Dr. P. Rethinam and Dr. K.L. Chadha)

Subsequently Dr. P. Rethinam, the then Assistant Director General, Plantation Crops, Indian Council for Agricultural Research (ICAR) along with the experts from Andhra Pradesh identified additional potential areas 1.5 lakh ha in the districts of Nellore, Ongole, Guntur, Vijayanagaram, Vishakapatnam and Srikakulam. Later he reported the potential areas in Chittoor District of AP also. The potential area in AP was thus identified to be more than 4.0 lakhs ha.

He along with officers of Horticulture Department of Gujarat State identified 61,350 ha potential areas in the state and some areas were identified in North Eastern India. In all, an area of 8.01 lakh hectares (Table 4.2) were identified as potential areas in 11 states for taking irrigated oil palm over a period of two decades.

Table 4.2: Potential Areas for Oil Palm Cultivation in India

States	*Area Identified (ha)*		*Total (ha)*
	*1986-87**	*1988-91***	
Andhra Pradesh	2,50,000	1,50,000	4,00,000
Assam	10,000	-	10,000
Gujarat	-	61,350	61,350
Goa	-	10,000	10,000
Karnataka	2,50,000	-	2,50,000
Kerala	5,000	-	5,000
Maharashtra	10,000	-	10,000
Odisha	10,000	-	10,000
Tamil Nadu	25,000	5,000	30,000
Tripura	5,000	-	5,000
West Bengal	10,000	-	10,000
Total	**5,75,000**	**2,26,350**	**8,01,350**

* Chadha Committee 1988. ** Other Committees.

Dr. K.L. Chadha, Deputy Director General Horticulture along with Dr. P. Rethinam, Assistant Director General Plantation Crops, ICAR also identified potential areas in Tirunelveli District and cautioned the need for assured irrigation for the entire crop period of 30 years for successful oil palm cultivation. Tenera Oil Palm company which took up oil palm in Karumkulam and Karassery ended up in failure since irrigation was not adequate. Many farmers took up oil palm cultivation depending on shallow wells also ended up in failure.

All these committees identified potential areas taking into account the climate, soil and irrigation projects proposed and underground water potential which can be tapped over a period of time. The Committees also stressed the need for irrigation throughout the year for 30 years of the crop period.

4.4.5. Potential and Prospects of Oil Palm Cultivation in India (Working group headed by Dr K.L. Chadha)

The first two years of Tenth Five year Plan were slow as far as Oil Palm development was concerned but later it started picking up since the price of oil went up and large number of farmers came forward for taking up oil palm cultivation. At this point of time, the Government of India again constituted a working group under the chairmanship of Dr. K. L. Chadha in the year 2005, to have a critical assessment of problems and prospects of oil palm cultivation in India, and to identify new areas, to assess the requirement of import of planting materials, processing facilities and to prepare an action plan for future development of oil palm from Eleventh Five Year Plan onwards.

Table 4.3: Reassessed Potential Areas as per Chadha Committee 2006

State	*Potential Area Identified Earlier (ha)*	*Reassessed Potential Area 2006 (ha)*	*Area Covered (ha)*
Andaman and Nicobar	-	0	1,593
Andhra Pradesh	4,00,000	4,00,000	53,370
Assam	10,000	0	0
Chhattisgarh	0	40,000	4
Goa	10,000	2,000	894
Gujarat	61,350	90,000	725
Karnataka	2,50,000	2,50,000	12,398
Kerala	5,000	6,500	5,501
Maharashtra	10,000	0	1,000
Mizoram	0	61,000	1,000
Odisha	10,000	25,000	2,014
Tamil Nadu	30,000	1,62,000	14,894
Tripura	5,000	0	120
West Bengal	10,000	0	0
TOTAL	**8,01,350**	**10,36,500**	**93,513**

This working group made a critical assessment of problems and prospects and identified newer areas for oil palm cultivation. The potential areas identified earlier as well as the reassessed potential areas are given in the Figure 4.1. This was one of the most comprehensive reports which assessed the potential areas for oil palm as 1.036 million ha) in 14 states (Table 4.3).

While the area projections further increased in the states of Tamil Nadu and Gujarat, states like Mizoram and Chhattisgarh were also added. The states like Assam, Tripura, Maharashtra and West Bengal which did not show interest in oil

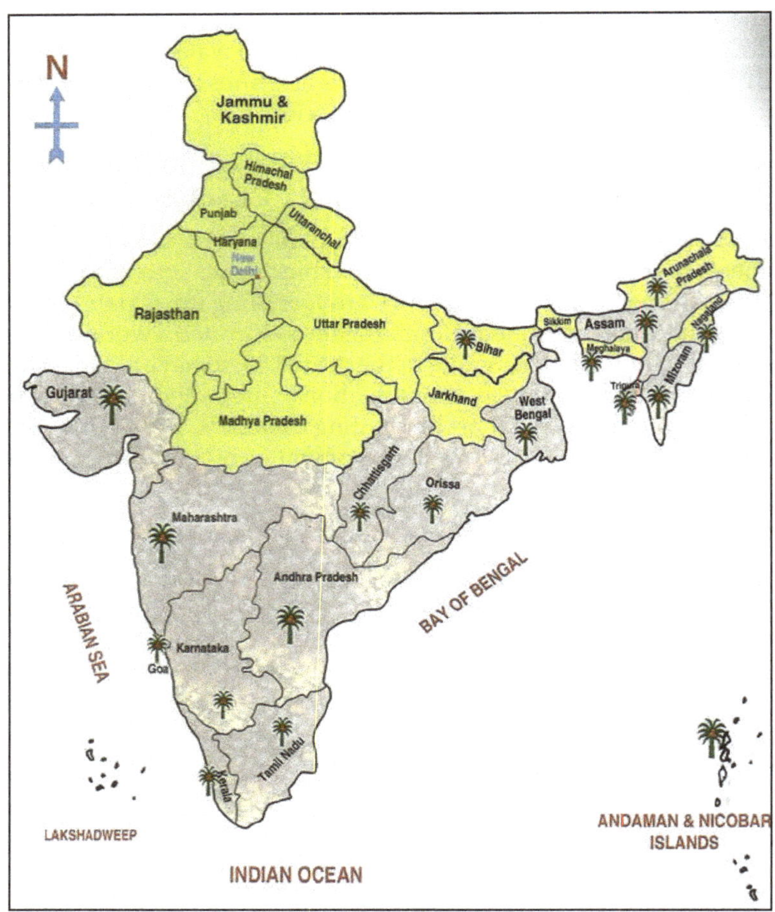

Figure 4.1: Potential Areas Identified for Irrigated Oil Palm in India. [*Source*: Rethinam Committee Report (2011)].

palm cultivation were not included for taking up oil palm cultivation on commercial scale.

The working group also indicated that an area of 23,975 ha be covered in 2006-07. It is also proposed to cover 2, 24,100 ha during XI Five Year Plan and 3,01,396 ha during XII Five Year Plan. Largest areas were projected in Andhra Pradesh followed by Karnataka, Tamil Nadu and Gujarat.

4.4.6. Committee Constituted by Government of India in 2011 (Headed by Dr. P. Rethinam)

This committee identified four new potential states, *viz.* Bihar, Arunachal Pradesh, Nagaland and Meghalaya, added the four already identified states *viz.* Assam, Maharashtra, Tripura and West Bengal and identified additional potential areas in the existing states where OPDP/RKVY was going on. Based on the performance of oil palm in different Districts/Taluks/Mandals, the additional potential areas were identified. For example almost the entire state Andhra Pradesh, Karnataka and Tamil Nadu were observed to be suitable for oil palm cultivation except the hilly areas. However, specific fields suitable for oil palm in these Districts/Taluks/Mandals should be based on the micro level survey and GIS mapping essentially to identify the ground water potential. It was suggested that the Processors/Department may promptly take up GIS/Remote Sensing based micro level survey before identifying/allotting new areas for oil palm cultivation.

Table 4.4 gives the consolidated statement of 10,36,500 ha potential areas of oil palm cultivation in eighteen states in the country identified for irrigated oil palm by various committees.

Table 4.4: Potential Areas for Oil Palm Cultivation in India

States	*Area Identified (ha)*				*Total Reassessed Potential - Area (ha)*
	Asper Chadha Committee 1988	*Asper other Committees 1988-91*	*Reassessed as per Chadha Committee 2006*	*Additional Areas Rethinam Committee 2012*	
Andaman and Nicobar Islands	NIL	NIL	NIL	3000*	3000*
Andhra Pradesh	2,50,000	1,50,000	4,00,000	67,800	4, 67,800
Arunachal Pradesh	NIL	NIL	NIL	25,000	25,000
Assam	10,000	NIL	NIL	25,000	25,000
Bihar	NIL	NIL	NIL	2,00,000	2,00,000
Chattisgarh	NIL	NIL	40,000	20,000	60,000
Goa	NIL	10,000	2,000	NIL	2,000
Gujarat	NIL	61,350	90,000	83,400	1,73,400
Karnataka	2,50,000	NIL	2,50,000	10,000	2,60,000
Kerala	5,000	NIL	6,500	3,500	10,000**
Maharashtra	10,000	NIL	NIL	1,70,000	1,70,000
Meghalaya	NIL	NIL	NIL	50,000	50,000
Mizoram	NIL	NIL	61,000	72,000	1,33,000
Nagaland	NIL	NIL	NIL	50,000	50,000
Odisha	10,000	NIL	25,000	38,500	63,500
Tamil Nadu	25,000	5,000	1,62,000	41,015	2,03,015
Tripura	5,000	NIL	NIL	7,000	7,000
West Bengal	10,000	NIL	NIL	50,000	50,000
Total	**5,75,000**	**2,26,350**	**10,36,500**	**9,16,215**	**19,52,715**

* In Andaman the plantations are more than 25 years and needs no planting. ** PCK and Kuttanad of Kerala has potential for growing oil palm as a Corporate and farmers crop respectively.

4.5. Efforts Made for Development of Irrigated Oil Palm

Several initiatives were taken to develop irrigated oil palm under various programmes detailed below:

4.5.1. DRDA and Nava Bharat Plantations in West Godavari District, Andhra Pradesh

The very first irrigated oil palm as small holder's crop was planted in 1988 at Pedavegi village,West Godavari District by a farmer in about 1 acre under DRDA programme. Subsequently a group of small farmers also took up oil palm cultivation in Pedavegi Mandal and at the same time a large scale planting of oil palm was taken up by Mr.A.S.Chowdhry, Nava Bharat Agro Products at Lakshmipuram by converting Tobacco land to oil palm, during 1987-88. The Nava Bharat oil palm plantation raised with indigenous planting materials from ICAR-IIOPR Regional Station Palode, as a model plantation for all the people who were taking up oil palm in the country. Mr. A.S. Chowdhary distributed oil palm seedlings free of cost to the farmers around his plantation. A Progeny evaluation trial was also laid out with progenies developed at Palode Regional Station. Subsequently a seed garden was also established in his farm with promising duras and tenera X tenera crosses for pisifera source. Many foreign experts and the processors who are taking up Oil Palm got convinced of the performance of oil palm at this plantation.

4.5.2. Oil Palm Demonstration Project (OPD)

As suggested in the Chadha Committee Report 1988, three large scale demonstrations of 1000 ha each were taken up in three states *viz.*, Krishna, West and East Godavari Districts of Andhra Pradesh, Shimoga District of Karnataka as small holders crop jointly developed by DBT and respective Departments of Horticulture and in Konkan Region of Maharashtra it was taken up as corporate crop by Konkan Development Corporation during 1989-90. The two oil palm Demonstration areas of 1000 ha each in AP and Karnataka were successful and developed confidence that the yield level up to 20 to 30 MT FFB/ha/year is possible if plantations are properly taken care. However, the corporate plantation in Konkan region in Maharashtra was not maintained well right from planting and did not succeed.

4.5.3. Oil Palm Development Project (OPDP)

The Department of Agriculture and Cooperation, Ministry of Agriculture and Cooperation, Government of India under the Technology Mission on Oilseeds (TMO) formulated Oil Palm Development Project (OPDP) during the Eighth Five Year Plan.

Though Oil Palm Development Project (OPDP) as small holders' crop was sanctioned in 1990-91, the real implementation started in 1991-92.During the Eighth Five Year Plan it was envisaged to take up Oil Palm Development in 80,000 ha but the achievement was just one third (33360 ha) due to Government of India formalities in communicating the sanction. Other starting problems like the State Governments' acceptance, identifications of Entrepreneurs and allotting areas to them, recruiting staff for development by the Processors, acquiring know how on oil palm cultivation, convincing farmers to take up this new crop *etc.* took considerable time.

During Ninth Five Year Plan also 80,000 ha, was targeted again, but only 17,286 ha only could be achieved due to low FFB price which was the result of decline in vegetable oil price drastically. Farmers were asking for Minimum Support Price for FFB and Govt. of India took some time in announcing the price. This had resulted in removal of oil palm trees and new farmers too had not come forward to take up Oil Palm cultivation.

The Oil Palm Development Project (OPDP) taken up during 1990-91 as a small holder's irrigated crop in the cultivated lands with subsidies for planting materials, plantation maintenance, drip irrigation *etc.* made a humble beginning with the nurseries raised by the respective state department of Agriculture/Horticulture. Only later the respective state governments implementing the OPDP selected potential processors based on their merits, areas of operations were allocated, entered into Memorandum of Understanding, allocated the activities to be implemented by them including raising seedlings, identifying farmers jointly with officials of government departments, providing technic5al support right from planting to harvest, collection of FFB at collection centres and making payment at the price fixed by the price fixing committee and formulated Oil Palm Act to regulate the implementation smoothly.

The target could not be achieved in all the Plan periods due to many reasons which are listed in the constraints.

The table above shows the OPDP started during Eighth Five Year Plan, initially had low progress due to the fact that oil palm as small holders crop was totally a new introduction and farmers as well as officers were having a lot of doubts about technologies, yield performance, marketing, price, *etc.* Subsequently when the oil palm development started picking up there was a fall in price of vegetable oils which affected the price of FFB and so in the subsequent Five Year Plan period it was very slow. Again when the support price was given by the government along with additional increase in subsidy many farmers started taking up oil palm cultivation. Thus, the area almost doubled in the Ninth Plan period. Now with this momentum gained one should proceed fast by encouraging the farmers with additional increased subsidies and providing infrastructure facilities for irrigation and harvesting.

The performance of Andhra Pradesh is far better than any other state in the oil palm development in the country and an area of 1.66 lakh ha has been covered by the state. State-wise and year-wise production of FFB and CPO are given in Tables 4.7–4.9.

The area planted till 2015-15 is presented in Table 4.5

Table 4.5: Present Area under Oil Palm Cultivation in India

Sl.No.	*States*	*Potential Area Identified by Dr. Rethinam and Dr. Chadha (Ha.)*	*Area Planted Till 2015-16 (Ha.)*	*Balance Potential Area for Oil Palm Development (Ha.)*
1	Andaman and Nicobar islands	3000	1593	1407
2	Andhra Pradesh	419500	160948	258552
3	Telangana	50000	14954	35046
4	Arunachal Pradesh	25000	0	25000
5	Assam	25000	10	24990
6	Bihar	200000	0	200000
7	Chhattisgarh	48000	855	47145
8	Goa	2000	891	1109
9	Gujarat	260250	5380	254870
10	Karnataka	260000	47462	212538
11	Kerala	6500	6024	476
12	Maharastra	180000	914	179086
13	Meghalaya	50000	0	50000
14	Mizoram	61000	38888	22112
15	Nagaland	50000	0	50000
16	Odisha	56000	23642	32358
17	Tamil Nadu	205000	12722	192278
18	Tripura	7000	530	6470
19	West Bengal	25000	0	25000
	Total	**1,933,250**	**314,814**	**1,618,436**

The Plan wise targets and achievements are summarized in Table 4.6

Table 4.6: Plan-wise Target and Achievements of Area Average

Plan Period	*Target (ha)*	*Achievements (ha)*
VIII	80,000	33,366
IX	80,000	17,280
X	52,000	46598
XI	177736	105942
XII	200032	108439
Total	**389236**	**203186**

The above table indicates that the overall area expansion has steadily increased from 2002-03 to 2008-09: though the targets could not be achieved until 2004-05 there was a spectacular increase of area expansion in the next three to four years. Over a period of seven years, 93525 ha was planted as against a target of 112800 ha from 2002-03 to 2008-09. Andhra Pradesh exceeded the target during 2004-05 to

Table 4.7: Production of FFB and CPO in India from 1992-93 to 2015-16 (unit 1000 tons)

Year	*Production MT*	
	FFB	*CPO*
1992-1993	21.23	-
1993-1994	24.62	1.13
1994-1995	NA	3.00
1995-1996	9.14	
1996-1997	14.23	7.92
1997-1998	47.41	10.06
1998-1999	67.37	10.94
1999-2000	144.43	18.16
2000-2001	142.82	25.29
2001-2002	128.87	22.61
2002-2003	157.74	28.17
2003-2004	168.42	28.82
2004-2005	176.14	30.84
2005-2006	2 44.69	51.33
2006-2007	258.90	43.96
2007-2008	259.01	45.40
2008-2009	309.99	52.93
2009-2010	396.70	66.49
2010-2011 (App.)	612.85	107.86
2011-2012 (App.)	746.75	128.81
2012-2013 (App.)	839.45	144.05
2013-2014 (App.)	963.05	168.53
2014-15	1068.37	179.23
2015-16	1115.32	187.46
Total	**7917.50**	**1368.16**

2007-08. Out of the target of 57000 ha for the seven years AP has completed 69214 ha. This showed that the farmers of Andhra Pradesh have been keen to develop oil palm. Karnataka the second potential state planted 14457 ha out of a target of 18900 ha. Tamil Nadu planted 8644 ha out of the target of 15300 ha. At this rate it would have taken another 10 years to complete 1 million hectare. This period can even be reduced if the farmers get fair price for the FFB and policy level decisions are taken appropriately.

In 2009-10, the price of the palm oil fell drastically and thereby the FFB price fell from Rs. 6200 to Rs.3400. This drastically affected the area expansion and a few farmers started removing the oil palm. In addition the Government of AP also restricted the area expansion drastically and thereby more than 22000 seedlings raised could not be planted. Though Chadha Committee had given a planting schedule of 549471 ha to be planted by 2016- 17, it could not to be achieved because of slow implementation.

Table 4.8: State-wise Production of Oil Palm Fresh Fruit Bunches Under Oil Palm Development Program (OPDP) (Metric tonnes)

State	*200-01*	*2001-02*	*2002-03*	*2003-04*	*2004-05*	*2005-06*	*2006-07*	*2007-08*	*2008-09*	*2009-10*	*2010-11 P*
Andhra Pradesh	125000.00	108757.00	109688.61	129841.00	138929.00	203000.00	215000.00	220200.00	259495.00	347892.00	585009.00
Karnataka	3998,13	3562.50	3676.65	3483.00	4127.34	4528.00	5415.00	5763.68	6685.00	6385.00	7498.95
Tamil Nadu	2250.00	229.58	434.00	415.00	680.72	1202.00	1584.02	1659.29	1968.00	2079.57	2781.38
Gujarat	0.00	24.90	29.42	28.00	6.10	20.00	10.25	12.36	2.45	5.73	23.00
Odisha	265.00	0.00	0.00	0.00	0.00	0,00	0.00	0.00	1100.00	3464.00	0.00
Goa	1300.00	1603.00	1879.05	1851.00	1993.93	2143.00	1951.00	2040.04	2335.90	1704.12	1934.75
Tripura	0.00	75.00	80.00	503.00	465.00	NR	447.12	NR	0.00	NR	
Kerala	0.00	4820.43	32991.84	32295.00	29939.63	33795.00	34496.00	29329.81	38377.89	35161.60	41058.00
Andaman & Nicobar	10010.82	9800.92	8957.00	0.00	0.00	0.00	NR	NR	NR	NR	NR
Mizoram	0.00	0.00	0.00	0.00	0.00	0.00	0.00	3.10	7.78	10.20	40.00
Total	**142823.95**	**138873.33**	**157736.37**	**168416.00**	**176141.72**	**244688.00**	**253903.39**	**259008.28**	**309992.02**	**396702.22**	**638145.10**

Table 4.9: State-wise Quantity of Crude Palm Oil Production (CPO) Produced Under Oil Palm Development Program (OPDP)

State	*2000-01*	*2001-02*	*2002-03*	*2003-04*	*2004-05*	*2005-06*	*2006-07*	*2007-08*	*2008-09*	*2009-10*	*2010-11 P*
Andhra Pradesh	15729.00	18974.00	18960.00	21457.00	23905.00	43600.00	35509.00	38000.00	43593.00	57402.00	96468.00
Karnataka	731.97	573.58	600.64	646.00	681.01	793.00	974.00	1037.46	1203.00	1165.00	1349.81
Tamil Nadu	86.06	0.00	0.00	NR	110.49	178.26	248.66	272.72	365.51	364.62	466.04
Gujarat	3.56	2.94	2.94	3.00	NR	NR	NR	Nill	0.00	0.00	0.00
Odisha	42.35	0.00	0.00	0.00	0.00	0.00	0.00	NR	1.65	841.00	0.00
Goa	207.75	243.00	330.85	324.00	348.93	379.00	345.00	360.83	392.76	311.00	348.25
Tripura	20.64	0.00	0.00	NR	NR	NR	NR	NR	NR	NR	NR
Kerala	6667.26	980.00	6572.00	6387.00	5792.94	6478.00	6888.00	5732.40	7370.50	6604.50	6681.00
Andaman & Nicobar	1800.76	1840.32	1696.00	0.00	0.00	0.00	0.00	NR	NR	NR	NR
Total	**25289.35**	**22613.84**	**26168.43**	**28817.00**	**30838.37**	**51328.26**	**43964.66**	**45403.41**	**52926.42**	**66488.12**	**105513.10**

P: Provisional

Source: State Govt. and Directorate of Oilseed Development, Hyderabad, compiled by Department of Agriculture and Cooperation (TMOP).

Even though the progress made so far in terms of area expansion, production of FFB and extraction of oil are not impressive, the project has proved the following:

- ★ All the OPDP states are suitable for oil palm cultivation with the existing climatic conditions of very high summer temperature and very low winter temperature.
- ★ The oil palm can be successfully grown in a wide range of pH 5.0 to 8.5, maximum temperature of 40 to 45 °C, minimum temperature of below 16 C (even up to 8 to 10 °C for some days, as in North Eastern part).
- ★ Many of the progressive farmers have got 30 to 40 metric ton of FFB/ha/year which means 5 to 8 million ton of CPO/ha/year. Highest yield of 54 MT FFB/ha/year was recorded in Mysore district of Karnataka State recently. Those farmers who are not taking care of oil palm in terms of optimum management are getting low yields. The farmers who are getting higher yield expanded the area further.
- ★ The economic development due to oil palm cultivation is very well seen in the Coastal Andhra Pradesh
- ★ The farmers could also raise as many intercrops as possible during juvenile phase of oil palm to make good of income loss. The setbacks of price fall for FFB during 1999-2000 and 2008-09 due to global vegetable oil price fall reflected in low area expansion.
- ★ Processing facilities in the country have made encouraging progress as shown below (Table 4.10). Many processing industries started with 5 t/h capacity have been scaled up to process more fruits. So for 509.3 t/h capacity of processing facilities are available and more processing unit are likely to come in some states.

Table 4.10: The Details of the Extraction Capacity of the Processing Mills and Location Furnished Below

Sl.No.	*Name of the Unit*	*Location*	*Capacity (Tonnes/Hr)*
	Oil Palm FFB Processing units in India		
	ANDMAN and NICOBAR ISLANDS		
1	Andman and Nicobar Islands Forest and Plantation Development Corporation	Little Andamans	5
	Sub Total		**5**
	ANDHRA PRADESH		
2	A.P Cooperative Oil Seeds Growers Federation Ltd.	Pedavegi, West Godavari	24
3	Laxmibalaji Oils	Takarakandi village Kurupum, Vizianagarm	10
4	3F Oil Palm Agro Tech.	Yernagudem, West Godavari	30
5	3F Oil Palm Agro Tech.	Kottapet village Kaligiri Mandal, Nellore	10
6	Godrej Agrovet	Pothepalli, West Godavari	40
7	Godrej Agrovet Ltd.	Chinthampalli	60

Contd.

Sl.No.	Name of the Unit	Location	Capacity (Tonnes/Hr)
8	Navabharath Agro Products Ltd.	Jangareddygudem, West Godavari	90
9	Ruchi Soya Industries Ltd.	Peddapuram, East Godavari	75
10	Ruchi Soya Industries Ltd.	Ampapuram, Krishna	50
11	Radhika Vegetable Oils (Pvt.) Ltd.	Garividi, Vizianagaram	15
12	Sri Srinivasa Palm Oil Mills	Rajam, Srikakulam	5
13	Agro Cooperative Ltd.	Asilmetta, Visakhapatnam	5
14	Subrahmanyeswara Agro Products Pvt Ltd.	Ambajipeta, East Godavari	10
	Sub Total		**424**
		GOA	
15	Godrej Agrovet Ltd.	Valpoi, Sattari	5
	Sub Total		**5**
		GUJARAT	
16	Kalyan Agrl. Crop Sales and Processing Co-op. Society Ltd.	Maroli Bazaar, Navsari	2.5
	Sub Total		**2.5**
		KARNATAKA	
17	Bhadravathy Balaji Oil Palms Ltd.	Bhadravathy, Shimoga	10
18	Directorate of Horticulture, Government of Karnataka	Kabini, Mysore	2.5
19	Simhapuri Agro Tech.	Devangere	5
20	3F Oil Palm Agro Tech.	Koppal	5
	Sub Total		**22.5**
		KERALA	
21	Oil Palm India Ltd.	Anchal, Kollam	20
22	United Oil Palm Planters and Extractors	Kuravilangad, Kottayam	0.3
	Sub Total		**20.3**
		MIZORAM	
23	Godrej Agrovet	Bukvannei village Kolasib District	5
	Sub Total		**5**
		ODISHA	
24	Lakshmi Balaji Oils Pvt. Ltd.	Attada Village, Kerada Panchayat	5
	Sub Total		**5**
		TAMIL NADU	
25	Cauvery Oil Palm Ltd. ' Godrej	Varanavasi, Perambulur	5
	Sub Total		**5**
		TELANGANA	
26	T S Oil Fed	Ashwarapet, Khammam	15
	Sub Total		**15**
		Grand Total	**509.3**

If only the Government of India determined to take up 2.0 million ha in the next 10 to 15 years with strong political will and commitments, the country can produce almost 6 to 8 million ton of crude palm oil and 0.6 to 0.8 million ton of palm kernel oil within another 15 to 20 years, besides 1.1 to 1.6 million ton of palm kernel cake, 9.3 million ton of palm biomass waste, 42 million ton of POME, 1176 million ton of bio gas, lot of employment generation (1.0 million), produce considerable quantity of electricity this providing livelihood and social security to millions of farmers.

4.6. Reservations in Oil Palm Cultivation in India

Even though several committees have made definite recommendations for oil palm cultivation in India, the expansion of area has not taken place as recommended due to there are several reservations among farmers, technocrats, bureaucrats and even planners whether investments in oil palm cultivation in the above areas are justifiable and profitable. Unfortunately in taking decisions with regard to oil palm cultivation, neither the ground realities are encouraging nor even opinion of experts available in the field. In several crops where we made similar decisions based on the judgment, we succeeded. For instance cultivation of grape in tropical areas of India, even though not in range of optimum zone,has given the highest yields in the world and optimum quality. Extending coconut cultivation which is a plant of the coastal region in non-conventional areas like Bihar, M.P., NE states has met with good success, there, therefore does not seem to be much doubt about the success of oil palm cultivation in India. However, it will be quite pertinent to discuss various reservations expressed in this regard.

4.6.1. Is India Climatically Suitable?

One of the first reservations is the suitability of newly selected areas for oil palm cultivation. In this connection it is pertinent to point out that areas have been selected keeping in view optimum conditions of oil palm cultivation. Oil palm grows well in localities where the annual temperature varies between 20-30 °C with well distributed rainfall of 2500-4000 mm per annum. However, it can withstand higher rainfall as well as 3-4 months of drought without any drastic reduction in yield. Where rainfall is not enough, availability of irrigation facilities has been kept in view.

Oil palm is sun loving requiring 1800 hours of sunshine and yields are not so attractive under heavy shade. Altitudes below 450 M are considered to be the best for oil palm cultivation. Moist deep loamy soil rich in humus with good water permeability suits as the palm best. India with diversified agro-climatic regions definitely is suitable for growing oil palm as evidenced by the existing performance in the areas identified for this crop. Dr. Richardson, consultant, Costa Rica has stated that the oil palm is adapted to purely tropical conditions. It has been grown successfully on a very large variety of soils. The oil palm responds well to good management practices, but is basically a rustic crop. Once it has reached the bearing stage it will produce reasonable volumes of fruits even under poor management. With the performance of existing oil palm plantations as irrigated crop in India yielding 20 to 30 metric ton FFB/ha/year there should not be any further doubt about the suitability of oil palm crop. There are many farmers getting above 40 million ton FFB/ha/Year and the highest yield of 50 metric ton FFB was recorded in Karnataka.

4.6.2. Can Oil Palm be grown Under Irrigated Conditions?

Another important question being raised by several people is that when all over the world it is grown in areas with year round precipitation why India should grow oil palm under irrigation? Such doubts are logical considering that the major oil palm growing countries like Malaysia, Indonesia, Thailand *etc.* have fair rainfall distribution to grow successful crop so that there is no need for irrigation but in India the rainfall distribution is seasonal with less number of rainy days which is adequate for oil palm. It is also true that there is drop in yield from 30 to 6 metric ton FFB/ha/year if deficit raises from 0 to 600 mm as observed by IRHO.

Though available results on irrigation experiments are only few in the world, these have definitely shown that the yield increase is from 60 to 90 per cent due to irrigation. The mini sprinkler system used in Unilever plantations in Ivory Coast in about 200 ha where 4-5 months of dry spell is experienced, has given encouraging results. The irrigation experiments conducted at Grand Drewen (Ivory Coast) indicated that providing 150 mm water/month yielded 23.5 MT/ha/year while no irrigation gave only 10.5 metric ton FFB/ha/year. At Lame (Ivory Coast) where water deficit is around 254 mm spray irrigation over three years increased yields to 32.7 MT/ha as against 18.5 metric ton/ha from non-irrigated areas. Hence, there is no doubt that the crop will respond to irrigation in areas where there inadequate and uneven rainfall conditions prevail.

Another question raised is whether we will be depleting the underground water heavily since this crop requires fairly large quantities of water. We have the experience of growing banana and sugarcane using water throughout the year. Moreover systems like drip irrigation with adequate mulch at base if adopted will definitely help to reduce the quantity of water and also conserve the moisture.

Further, irrigating 150 palms over an area of one hectare in a basin system of irrigation in the command areas at weekly interval with 400-500 l of water per palm may require about 3.12 lakh l of water per annum which is comparable with that of many other crops grown in this country. Since the projects like Upper Krishna, Malaprabha, Ghataprabha are meant for light irrigated crops, oil palm will definitely fit in these areas. Water logging,could also be avoided in the command areas by growing this crop and avoiding flood irrigation as in the case of paddy.

The high initial cost of drip system is always talked about. Oil palm with only 150 palms/ha may require about Rs. 15,000 per ha for installing a drip system which is not a very costly proposition as compared to that of other countries. However what is required is to ensure that the requirement of water by the palm can be met with by this system coupled with mulching.

Dr. Richardson in his report also indicated that the Indian farmer with their considerable experience in irrigation would take care of the irrigation needs of oil palm. It has been proved by the farmers by getting higher yields with drip irrigation.

4.6.3. Will Oil Palm Replace other Economic Crops?

Another major reservation is about crop replacement in lieu of oil palm. The Working Group on oil palm carefully analyzed crop replacement. Most of the areas identified in A.P. are upland areas where new irrigation facilities are being developed by DRDA programs. There is, therefore, need to switch over to irrigated crops from dry land crops. This is true in case of Upper Krishna Project in Karnataka and barrage areas. In Tungabhadra project area of Karnataka sugarcane is grown and major sugarcane factories have been closed and farmers are switching over to some other crops. In Bhadra project area of Karnataka paddy lands are being converted into arecanut which is not a desirable proposition since we have self-sufficiency in arecanut. Under such conditions there is a need to change over the cropping system. Oil Palm is a high value oil yielding perennial crop with low input and high output as has been proven beyond doubt.

4.6.4. Can Oil Palm Yields be Comparable to those Obtained in Traditional Oil Palm Growing Countries?

This is another question of paramount importance in the minds of growers and planners. Yield levels are important parameters for performance of oil palm. Oil palm begins producing harvestable fruits 3 years after planting. Now there are new tenera hybrids and compact seeds which can give yields within 30 months. The number and size of fruit bunches and their oil content improve as the palm matures into full yielding crop by about 8 years after which yields more or less stabilize for the remaining economic life of 25-30 years. Fruit yields are affected by climatic and agronomic factors. A period of low rainfall may introduce water stress which affects fruit yields. It is not surprising therefore, that a high proportion of research work in the oil palm industry is devoted to yield improvement. Yield was significantly higher when the industry discovered tenera hybrid palms in comparison to the dura variety which was the standard planting material in the 1950's.

The performance of oil palm in A and N Islands and Kerala can be rated as satisfactory despite poor growth and low yields. The major limitation in OPIL estates has been a severe water deficit and heavy gap filling in the initial years. Water deficit is caused by lack of rainfall from December to March, shallow gravelly soils with low water holding capacity, steep slopes promoting water run-off and very high levels of evapo-transpiration during dry season. Another major cause of low yields has been non availability of insect pollinators. However, it is heartening to note that Oil Palm India Ltd. introduced the pollinating weevil *Elaeidobius kamerunicus* a few years back. This has increased the fruit set considerably. Accordingly the yield level in Kerala had reached 14.96 MT/ha of FFB in 1971 plantation and 12.47 and 12.73 MT in 1972 and 1973 plantations respectively. These are expected to rise further provided adequate management practices like application of fertilizer, pruning etc are taken up regularly. Similarly in Andaman and Nicobar, in some old plantations, the oil yield level has now reached 3 MT/ha.

The yield levels can be rated quite satisfactory when we consider the limitations of lack of elite planting material, limited management practices and inadequate expertise in oil palm cultivation. Even these so called low yields are more economical than several other crops grown in these states. It has been seen that many farmers of East and West Godavari and Krishna District have obtained more than 30 MT FFB/ha. The farmers social status and economy had considerably increased.

Yield levels can be rated satisfactory when these are compared with the yield levels in other countries. For instance the area under oil palm in Malaysia during 1985 was reported to be 1.5 million ha with a production of crude oil to the extent of 4.13 metric ton. This shows that the average yield of palm oil in Malaysia for the year is 2.86 metric ton/ha. It is likely that while reporting area some non-bearing areas might also have been taken into consideration. However, in Malaysia advanced management practices and planting techniques have resulted in average yield per hectare of up to 25 metric ton FFB. The authors visited several Estates in Malaysia and most of these reported yields/ha ranging from 20-30 MT FFB/ha/year. It was observed that in most of the oil palm growing countries the average oil yield is 3.0 MT/ha/year which is attainable in India. Many of the Indian farmers are also getting more than 4.0 metric ton oil/ha/year. So the reservations of getting poor yields in India are not called for.

4.6.5. Is Oil Palm Cultivation Economically Viable?

A vital question which arises in the minds of the prospective growers is whether the economics of oil palm cultivation is comparable with other crops they grow. The working group of oil palm indicated that for raising one hectare of oil palm up to maturity (5 years) under high rainfall conditions without irrigation during 1986, an investment of about Rs. 30,000/- is required. The recurring costs from sixth year onwards in maintenance, harvesting and processing are estimated at Rs. 15000/ ha up to 20 years. At modest production level of 2.5 -3 ton/ha oil from ordinary plantations in India the farmers can get a gross annual income of Rs. 45,000 to 60,000 and a net income of Rs. 30,000 to 45000/ha. This is more than the annual income from irrigated rice and wheat. The Technology Mission on Oilseeds worked out a cost benefit ratio of 1.0;1.78.Further under better management conditions the yield level will be more as experienced by farmers and hence there is strong justification for bold investment in oil palm cultivation.

Moreover, during the first year, nearly 80 per cent of the area would be available for cultivation of inter crops. In the second and third year, the inter space will be progressively reduced as the canopy spread of oil pal will be more and only shade loving crops can be raised. Though costs and return of these intercrops will depend on the crops chosen, there is scope for growing a variety of inter crop like oil seeds and pulses among field crops and banana, mulberry, vegetables and flowers amongst horticultural crops. Hence, growing of oil pal under irrigation should definitely be profitable provided right type of planting material is used with adequate management practices. Also the profit margin from oil palm will be definitely higher than most of the annual field crops.

4.6.6. Why Oil Palm should be Grown when Palm Oil is Available for Import from Other Countries?

This question remains in the minds of many people including policy makers and public servants. Yes, it is always easy to import all food commodities including sugar and oil spending huge foreign exchange. There is no need to grow many crops in this country. But why the Govt. of India has been preparing Five Year Plans and gives importance to Agriculture. Pulses, Sugars, Vegetable Oils *etc.* can be imported to our requirements and meet the demand. Why oil palm cultivation alone should not be promoted when we know that it is only oil palm which can give highest vegetable oil per hectare per year as obtained by farmers with 15 to 25 MT of FFB/ha/year which is equivalent to 3 to 5 MT of Crude Palm Oil. The oil palm growers are also happy with the remuneration that they are getting but for the sudden price falls.

It is also a fact that oil palm, a perennial oil yielding crop with its life span of 25-30 years, highest oil yield of 4-6 MT CPO/ha/year can produce about 3-4 million ton of CPO and 0.3-0.4 million ton of palm kernel oil/year if the potential area of 1.0 million ha is brought under oil palm as envisaged within a period of 15 years. But then it is also a fact that 27 million ha of annual oil seeds produce only 7.0 MT of oil. If only the policy makers decide to bring at least 2 million ha under oil palm it will definitely be possible to produce 7-8 million ton of palm oil/year.

By importing palm oil and vegetable oils from other countries, India, spends valuable foreign exchange of Rs. 60000/Crores annually. Many intermediaries get benefit in this other than the exporting countries.

4.6.7. Are Incentives Required for Oil Palm Development?

Oil palm being a perennial crop committed to land for about 30 years with a juvenile phase of 3-5 years, the farmers may have to forego the full utilization of land for generating the income in the early years. The full benefit from oil palm will be available only from sixth year after planting. Hence farmers need encouragement by way of incentives similar to that of rubber plantation scheme of Ministry of Commerce which gives incentive to farmers *viz.* cash subsidy, quality planting material and 5 per cent interest subsidy on NABARD loan. These may form suitable incentives for oil palm also. The Chadha Committee also recommended certain incentives like 50 per cent of investment cost of Rs.30,000 ha. For initial plantation raising, as subsidy, free crop insurance to cover crop failure for a period of 5 years after the palms begin yielding, free flow of finance through commercial banks on soft loan basis, a nominal levy on vegetable oil, simplified loan and incentive procedures. Proposals for development of oil palm are being processed by Technology Mission on Oil seed, Ministry of Agriculture, Govt. of India, keeping the above points of view to the extent possible. The farmers also need to be given constant technical guidance on production aspects including the supply of quality planting material.

4.6.8. Is Palm Oil Consumption Safe?

Another question which is often discussed by present day oil consumers is whether palm oil consumption is safe. Research carried out elsewhere had established that palm oil is safe for human consumption. It contains 46 per cent saturated fatty acids, 43 per cent unsaturated fatty acids and 11 per cent poly unsaturated linolic acid which is essential for human and animal nutrition. Palm oil is also rich source of Vitamin E, the Tocopherols and Tocotrienols anti-cancer agents. It has anti thrombolitic effect as well. Unrefined palm oil is the richest source of Beta carotene, having the capacity to prevent cancer of epithelial cells. 90 per cent of the palm oil is used for edible purpose and only 10 per cent for non edible purpose. Palm oil is used for making margarine, soup mixes, shortening, vanaspati, cooking and frying oil, non-dairy creamers and for making vitamin E. The oil is also used for making soaps, fatty acids and methyl esters. Red Palm oil or red palm olein is a healthy oil and can be used for cooking and salad oil. The very fact that we import more than 10 million of palm oil shows the acceptability of this oil in the Indian households.

4.7. Problems that have Affected the Area Development in Oil Palm

- The major goal for introducing Oil Palm in the Technology Mission on Oilseeds in 1986 was to increase the vegetable oil production through this crop. But while formulating the project it was only restricted to promote small holders and SC/ST.
- Slow pace of area expansion with low budgetary provision: Only 2.60 lakhs ha could be covered against assessed potential of 19.60 lakhs hectares over a period of nearly 3 decades.
- There was no uniformity in implementing program in different states like seedling price, price fixing for FFB, supply of inputs *etc.*
- Uncertainty regarding area expansion program/targets leading to improper planning by the processors involved in promoting Oil Palm. Sudden reduction in area expansion in the middle of the year resulting in huge number of old seedling to be maintained by the Processors.
- Lack of confidence due to fall in prices; delayed support price, not having a price policy for a long time.
- Monitoring and evaluation by outside agencies on low contract tender system did not bring out the real situation.
- Low yield realization in general due to various reasons, mainly irrigation and nutrition which are the two important requirements. Resource constraint farmers could not adopt the same.
- Crop nutrition is inadequate and imbalanced due to financial difficulties experienced by the farmers.
- An area of 30,000 ha of planted Oil palm had been removed prior to 2006 by the farmers due to various reasons including price fall twice. Some more areas would have been removed after 2006 also thus discouraging other farmers.
- In states like Odisha and Gujarat Oil Palm was planted by farmers directly promoted by the Govt. Dept. of Horticulture and no company was allotted for processing. When the palms started yielding there was no processing unit or buyer of FFB and this resulted in the removal of large number of palms with the farmers also losing interest in Oil Palm cultivation. Even though this problem was settle by allocating areas to three to five companies the fear of farmers could not be wiped out and hence oil palm development took place in slow phase.

- ★ Similarly, large area of Oil Palm was uprooted in Tamil Nadu due to various reasons like withdrawal of one company, planting oil palm in all areas without adequate water to achieve the target, low pricing for FFB compared to Andhra Pradesh *etc.*
- ★ Oil Palm Development Programme in many states was implemented as a routine programme rather than a special one. There was considerable delay in dispersal of subsidy amount to farmers and processors.
- ★ Import Policy: Import of edible oil was brought under OGL in 1995. Present import duty on CPO is 2.5 per cent and on Olein is 10 per cent. Domestic prices of oil palm are significantly affected by cheaper imports from Malaysian and Indonesia hence fluctuate considerably.
- ★ Inadequate financial support by the Government. The total expenditure for the last 6 years on oil palm development has been Rs. 170 crores at an average of Rs. 28 crores p.a. reaching Rs. 40 crores p.a. in the last 2 years only. A minimum amount of Rs. 645 crores is required as share of GoI for assistance for new plantations to achieve the Chadha Committee's target of 2.20 lakh hectares by the end of XIth Plan (*i.e.* 1.29 lakh Ha. in 2010-11 and 2011-12).
- ★ Resources and security related issues *viz.* credit from Commercial Banks and NABARD. Even today commercial banks are not coming forward to give loan for oil palm.
- ★ Delay in implementation of crop insurance schemes.
- ★ Enactment of legislation Oil Palm Act in many states *etc.*
- ★ The Market Intervention Scheme (MIS) has been applied to Oil Palm but it has not been facilitative or encouraging.
- ★ Lack of Farmer and Processor friendly Price Fixing norms.

4.8. Steps Suggested to Sustain Oil Palm Development

- ★ A strategy should be evolved for the development of road map for promoting Oil Palm production in the country to reach 1.0 million ha. in 18 potential states by 2023-24 and 2.0 million ha within 20 years *i.e.* 2033-34.
- ★ The Road map thus developed should be implemented fully with strong political will. An organisation like NDDB with autonomous powers to be set up to take up time targeted oil palm development on all aspects right from seed to oil as well as marketing and value addition.
- ★ Higher target of area expansion with high financial allocation for Oil Palm should be made.
- ★ Small and marginal farmers are resource poor farmers and not able to create irrigation facilities by their individual resources and buy inputs. Cluster approach/Group farming concept has to be created and community irrigation to be created. This is very important for states like Odisha, Eastern and North Eastern States.

- ★ Deepening of the existing irrigation tanks available in these areas in Odisha and North East will help to provide adequate irrigation to Oil Palm.
- ★ There is lot of land allotted to SC/ST in contiguous areas in all states which are lying vacant for want of irrigation facilities and due to lack of power supply. Their livelihood security can be assured through Oil Palm cultivation. The companies allotted in the respective areas can coordinate taking up oil palm cultivation for which the respective State Government with SC/ST Corporation should collaborate and coordinate.
- ★ Relaxation of land ceiling norms like coffee, tea, rubber *etc.* and treating oil palm as a plantation crop will help to promote oil palm in large areas.
- ★ Simplify procedures of area allocation to the companies involved in area expansion. Each company should be allotted adequate areas/districts contiguously so that they can go for larger area coverage of setting bigger processing units (more than 40 metric ton FFB/hour to have co-generation). It is better not to thinly spread more companies with less potential areas rather than a few companies with more compact areas.
- ★ Advanced planting material subsidy/OFI/Revision of seedling price to match with growing cost escalations.
- ★ Enhance cultivation subsidy of Rs. 15,500 per hectare to Rs.40, 000 per hectare by expanding eligibility limits up to 20 hectares to individual farmers.
- ★ Increase in subsidy on micro irrigation system from 2 hectare per farmer up to 20 hectare per farmer specifically for Oil Palm. The possibility of providing micro irrigation by Government and recovering the cost borne by farmers only when the fruits are supplied to the Processors. The processors can recover and remit to the Govt. account regularly in installments.
- ★ Establish Oil Palm Board to consider planting, expansion and development of oil palm production, processing *etc.* Malaysian Palm Oil Board (MPOB), Malaysian Palm Oil Promotion Council (MPOPC), RSPO, and FELDA systems to be studied and suitable systems adopted.
- ★ As land resource is static, high density planting with compact planting material should be promoted so that more FFB/Oil/hectare can be obtained. Many Compact high yielding clones and advanced Planting Materials are already available from the Oil Palm growing regions of the world. These materials should be imported to get higher oil yield rather than waiting for a long period for developing our own materials.
- ★ Cultivable wastelands available in plenty in the identified potential states for growing oil palm with underground water potential can be allotted to Oil Palm Companies of the respective areas on long term lease on PPP mode to bring more areas under Oil Palm cultivation.
- ★ The indirect contribution of Oil Palm derivatives in the form of Palm kernel cake as cattle feed and sludge as fish feed may be born in mind while reckoning the value of the crop. Similarly bio-mass energy production

and bio gas production through Palm Oil Mill Effluent (POME) may also be looked into as it helps as alternative renewable energy source.

- ★ Opportunity may please be given to private entrepreneurs who are involved in Oil Palm development for establishment of R&D Centers, establishing seed gardens as joint venture with leading oil palm seed producing countries. Uniform price for oil palm seedlings based on nursery costs of each location to be decided.
- ★ Improve Oil Palm yield with an aim to reach productivity of 5 MT oil per hectare or 20 MT/hectare of FFB/year. Integrated approach to crop production *viz.* planting material, fertilizers, inter crops, drip irrigation *etc.* may be developed. Malaysia, Indonesia, Thailand, Coasta Rica *etc.* are having advanced planting materials which can give even 8 MT of palm oil per hectare/year. Supply of inputs like fertilizers should be done by the Processors to ensure application of fertilizers in time and in adequate quantities.
- ★ Announcement of import policy on long term basis taking into consideration its effect on local oil production, marketing and pricing.
- ★ Implement MIS as an effective intervention instrument and in time to create confidence in the farmers mind.
- ★ Review of FFB price fixing formula based on location specific expenses in respective States.
- ★ Implementation of crop insurance scheme. Sensitize commercial banks and NABARD for promotion of oil palm cultivation.
- ★ Difficulty in harvesting of FFB from adult palms being expressed by farmers can be solved by establishing Harvest Banks with modern tools and experienced/trained man power and helping farmers in the harvest process on charge basis.
- ★ Enactment of Oil Palm Act as mandatory in all States. Only 4 States *i.e.* Andhra Pradesh, Tamil Nadu, Goa and Mizoram have introduced legislation. Oil Palm Act should have provisions to recover the Government assistance if oil palm is uprooted/diverted without proper justification.
- ★ Intensive training at all levels and demonstrations in farmers' field are necessary.
- ★ Efficient monitoring and evaluation mechanism are necessary.
- ★ Bridge anticipated gap in seed production. Indigenous production is estimated as 5.94 million: the requirement of sprouts to achieve Chadha Committee target up to 2011-12 is 18.64 million. Higher area expansion needs large quantity of seeds and the domestic production may not be adequate and there import of quality seeds of different tenera combinations is necessary.
- ★ Expanding of germplasm base, particularly in respect of dwarf and stress tolerant varieties/hybrids. Procure good quality material through import:

Joint venture with reputed foreign companies and diversification of import sources should be encouraged.

- ★ Special package may be announced to boost area expansion program for Oil Palm Development on Fast Track Mode to save Foreign Exchange on Import of Edible Oil and to become self-reliant.
- ★ The crop may be promoted on large scale as the crop is Eco friendly contributing large quantity of Biomass and highly advantageous to farmers in terms of cultivation, maintenance and remuneration.
- ★ Subsidy on Palm Oil Processing Mills may be enhanced as a promotional activity and to reduce the financial burden for a long period of getting fruits for processing.
- ★ If another 4.0 million ha of cultivable wastelands and wastelands with underground water potential is identified, this will go a long way in increasing vegetable oil production and which can be used as bio fuel.
- ★ Khozy River valley (Bihar) farmers who lost their cultivable land due to flooding and breaching can be provided livelihood security through group farming of oil palm cultivation.
- ★ Aliapet island in Baruch District of Gujarat having about 6000 ha can be brought under oil palm as Corporate crop.
- ★ North Eastern States are having potential of more than 3.0 to 4.0 lakhs ha have to be dealt separately by giving 100 per cent subsidy for inputs supply and the entire labour cost to be borne by the farmers.
- ★ Rectifying the defects in Oil palm fresh fruit bunches (FFBs) pricing formula.
- ★ Timely release of Government dues/subsidies to farmers and developers.
- ★ Relief from additional VAT burden on sale of Crude Palm Oil from Andhra to Telangana State.

4.9. Way Forward

Indian Oil Palm has to go a long way to achieve 1.03 million ha identified in 13 states from the present area of 292001 ha in 1.03 million ha. The Expected Outcome from Two Million ha of Oil Palm

By growing Oil Palm in two million ha we may get the following:

- ★ Crude Palm Oil (CPO) @3 to 4 metric ton/ha/yr : 6.0 to 8.0 million ton
- ★ Palm Kernal Oil(PKO) : 0.6 to 0.8 million ton
- ★ Palm Kernal Cake (PKC) : 1.10 to 1,60 million ton
- ★ Palm waste : 9.32 MT
- ★ POME : 42 MT
- ★ Bio Gas :1176 M3
- ★ Electricity : 1.38 million MW hr
- ★ Employment Generation : 1.0 million

Value added products from empty bunches, fibre, wood *etc.*

Provide nutritional, lively hood, social security

Increase in farmers income

Substantial addition to vegetable oil pool of the country and thereby reduction in imports.

If the Oil Palm development is taken on war footing in the identified potential states of the following projections are possible (Table 4.10).

Table 4.11: Projections Likely to be Achievable by 2023-24 and 2033-34 over a Span of 10-20 years through 1.0 Million ha and 1392 million of Oil Palm respectively

Particulars	*Achievable Projections 2023-24*	*Achievable Projections 2033-34*
Oil palm area development (million ha)	1.16	19.92
FFB production (million MT)	76.07	326.82
CPO production (million MT)	13.69	58.83
PKO production (million MT)	1.52	6.64
Biomass production (million MT)	1.9	8.17
Processing mill capacity in MT/hr	3614	8737
Revenue to Government throught VAT (Rs in million)	40553	174234
Employment generation (nos. in million)	0.70	1.16
Carbon sequestration (nos. in million)	106	17.45

4.10. To Sum Up

1. 27.0 million ha of nine/oil seed crops produce about 9.0 MT of oil per year but 2.0 million ha of oil palm could produce 8 million ton of Crude Palm Oil, 0.8 million ton of palm kernel oil, palm kernel cake, bio-mass for bio-energy, eco-friendly bio diesel *etc.*
2. Concerted, time targeted Mission oriented Project to be formed with full Political commitment, involving people who can dedicate themselves fully in implementing the programming with adequate powers and funds to run the project.
3. When we are spending about Rs. 65,000 Crores annually for importing vegetable oil, there is no harm to spend at least Rs.1000 crores every year for oil palm development and processing to augment vegetable oil production, uplifting Socio-economic situation of farmers, generating good employment potential and production eco-friendly fuel and energy.
4. It should be borne in mind that Oil Palm cultivation was taken up to boost vegetable oil production in the country and not only as small holders' crop but also as big corporate plantations. Hence it is necessary to mobilize all possible ways and means for faster and larger oil palm development.

Chapter 5

Climate and Soil Requirements

Climate and soil are the two important parameters for successful oil palm cultivation. These are discussed in this chapter.

5.1. Climate

Oil Palm is characterized as a humid tropical palm. The hot humid equatorial climate with no dry period is best suited for oil palm. The optimum climatic requirements under which oil palm is grown are described below.

5.1.1. Rainfall and Distribution

Oil palm is quite sensitive to soil moisture extremes namely drought and moisture saturation and relative humidity. A well-distributed rainfall of about 2,000-3,000 mm per annum is considered to be optimum, though the plant can tolerate up to 10,000 mm of rainfall per annum. Oil palm requires a minimum rainfall of 125 mm/month preferably every month. In areas where 2 to 4 months drought exists, there will be lot of yield fluctuations from year to year.

5.1.2. Temperature

Temperature is an important climatic factor affecting Oil palm yields. For optimum growth of oil palm, specific temperature requirements are very vital. The mean monthly minimum temperature should not be lower than 18°C and should be in the range of 22°C-24°C. Similarly, the mean maximum temperature should be in the range of 29°C-33°C. Areas with very hot day temperature and cool nights give reduced yields due to floral abortion. However, now it is grown even under 42°C to 45°C successfully in Andhra Pradesh. Some new varieties such as Bamenda x Ekona, Tanzania x Ekona and Deli x Ekona perform quite well in the tropic Uplands at an altitude up to 1000 feet and in latitude over 10° where the temperature drops below 19°C during several hours in some months of a year.

5.1.3. Humidity

As oil palm is a humid tropical crop, high humidity of more than 80 per cent is beneficial.

5.1.4. Wind

Wind causes shredding of the palm leaflets and reduces photosynthesis. Strong winds can cause frond breakage especially to genotypes having larger and longer leaves. Strong recurrent winds do not blow off the palm, but can cause extensive leaf damage and tilt the palms from the vertical position.

5.1.5. Sunshine and Solar Radiation

Oil palm is a strong light demander. Sunlight is necessary for the pants to perform photosynthesis and produce Carbohydrate that are used to produce bunches and oil. It puts on vigorous growth and gives high yield only under conditions of copious sunshine. In areas with high light intensity, the fruits produced are smaller but with higher oil content than fruits produced in areas of low insulation. Solar radiation below 350 Cal cm^{-2} day^{-1} affects the growth and yield. The number of sunshine hours necessary to achieve full yield potential have often been estimated to be 5 hours/day or 1,800 h/annum, while less than 1,500 h/annum are considered as a limiting factor. Shading reduces growth and net assimilation rate in palms of all ages; particularly in the adult palms, production of female inflorescences is reduced.

While higher yields are obtained under climatic conditions stated above, oil palm has also been profitably cultivated in regions, where rainfall is either poorly distributed or is very high and where temperatures are low in certain months or sunshine hours well below the optimum. This is because oil palm is well adapted to a wide range of situations. Even the dry period for three months does not markedly reduce the health of palms. Further oil palm is a high producer of oil even under the poorest plantation conditions. The yield reduction due to drought was marked, ranging from 25 to 30 metric ton/ha of FFB in Malaysia and from 8 to 11 metric ton/ha of FFB in Nigeria. Even under such adverse conditions, 2.0 metric ton/ha of palm oil and 0.5 metric ton of kernel oil production are possible.

The climatic conditions in the oil palm growing countries as shown in the Table 5.1. However, in India though ideal climatic conditions are not there, oil palm is grown with certain modifications. Since the rainfall is inadequate it is grown under irrigated conditions.

5.1.6. Climatic Zones and Oil Palm Growing States in India

For resource development, the country has been broadly divided into fifteen agricultural regions based on agro climatic features, particularly soil type, climate including temperature and rainfall and its variation and water resources availability. Of these oil palm is grown in almost 11 agro climatic regions (Figure 5.1) leaving only five regions *viz.* Western Himalayan Region (1), Upper Gangetic Plains Region(5),

Trans Gangetic Plains Region(6), Central Plateau and Hills Region(8) and Western Dry Region (14).

Table 5.1: Climatic Conditions in Oil Palm Growing Countries in the World

Country	*Mean Max. Temp. (°C)*	*Mean Min. Temp. (°C)*	*Mean Sunshine Hours (h/day)*	*Mean Rainfall (mm/yr)*
Brazil	25.8-32.3	17.7-21.6	6.1	1899-3306
Cameroon	26.9-32.1	22.0-22.9	3.6	3508
Columbia	33.0-34.6	21.4-23.2	3.5-7.7	1661-3354
Congo	28.5-30.8	19.4-20.3	5.6	1835
Costa Rica	27.8-32.0	18.0-22.3	2.7	3307
Ecuador	27.2-30.2	19.3-21.3	2.2	3188
Honduras	27.8-32.2	18.1-21.1	6.9	2666
India	30.2-34.3	21.4-24.6	6.0-7.5	950-1200
Indonesia	29.6-32.8	22.1-22.8	6.9	1912-3627
Ivory Coast	27.2-32.2	20.9-22.7	5.0	1990
Malaysia	32.0-33.6	22.6-23.8	5.3-6.1	1678-3651
Nigeria	27.6-32.7	21.3-22.4	4.6	1162
Sierra Leone	28.2-33.3	19.8-21.8	5.4	2822

Source: Corley and Tinker (2003), The Oil Palm (Fourth Edition), Blackwell Science Ltd., U.K.

Oil palm being a humid tropical crop, the potential areas for growing Oil palm in different states of the country were identified keeping the broad guidelines barring the rainfall and its distribution, which is not adequate in most parts of the country. The irrigated oil palm concept was put forth because of shortage of rainfall as well as its irregular distribution. The climatic conditions prevailing in different oil palm growing states in India are given hereunder.

a) Andaman and Nicobar Islands

Andaman and Nicobar Islands have a warm tropical climate, with the presence of irregular rainfall during the south-west monsoon. Sea breezes are also common in the tropical climate of Andaman and Nicobar.

The climate of the Islands is characterized with a minimum of 23°C temperature; the maximum temperature in the Andaman and Nicobar climate being 31°C. The climate of Andaman and Nicobar Islands is moderate, *i.e.* the temperature is neither too hot nor too cold.

The relative humidity in Andaman and Nicobar Islands is 70 to 90 per cent. There is no extreme in the climate, except the rains and storms. The Islands experience monsoon season in two phases: *i.e.* May to mid-September and November to mid-December. The island witnesses north-easterly gale from November to December, while during May to October, it experiences south-westerly gale. Only between Januarys to April the Islands experience calm weather.

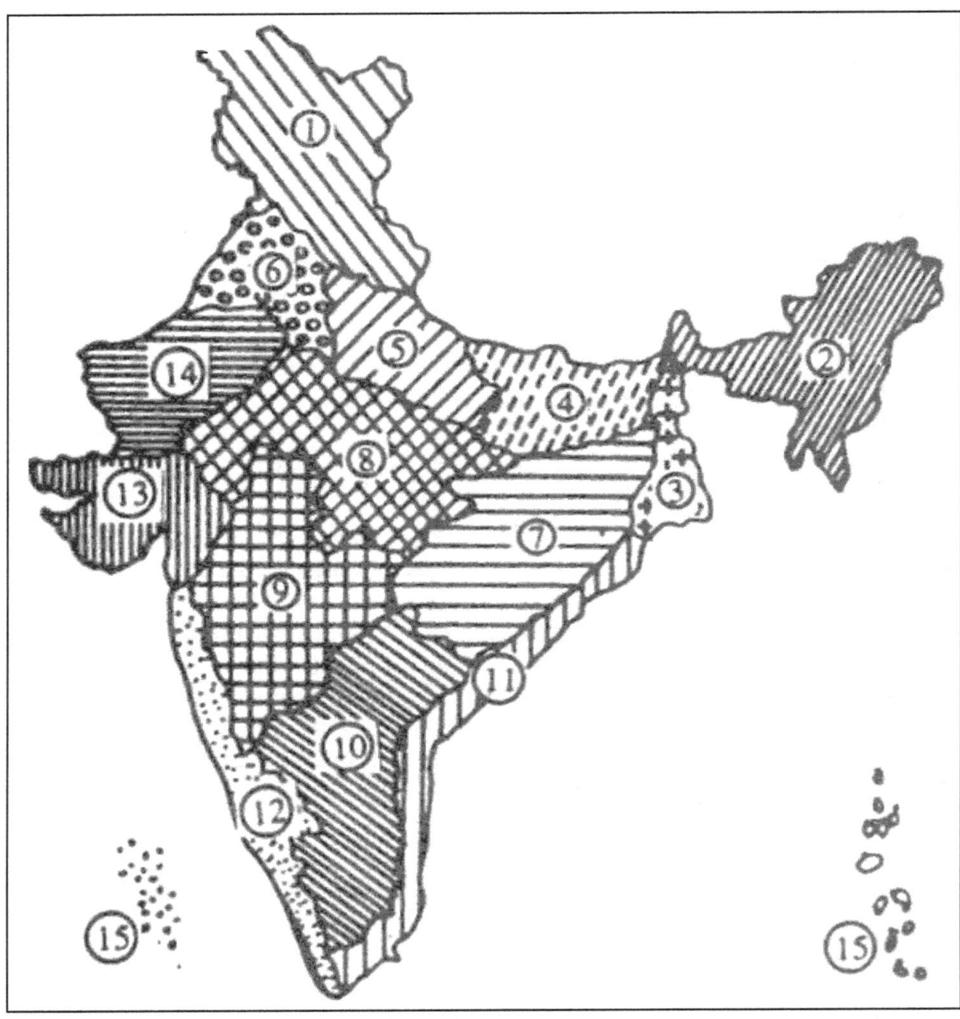

Figure 5.1: Agro-climatic Zones of India.

1. Western Himalayan Region, 2. Eastern Himalayan Region, 3. Lower Gangetic Plains Region 4. Middle Gangetic Plains Region, 5. Upper Gangetic Plains Region, 6. Trans Gangetic Plains Region, 7. Eastern Plateau and Hills Region, 8. Central Plateau and Hills Region, 9. Western Plateau and Hills Region, 10. Southern Plateau and Hills Region, 11. East Coast Plains and Hills Region, 12. West Coast Plains and Hills Region, 13. Gujarat Plains and Hills Region, 14. Western Dry Region, 15. The Island Region.

The average annual rainfall of Andaman and Nicobar Islands is 318 cm. Rough weather prevails during the monsoon season with temperature, ranging between 22.5° C to 29.9° C.

b) Andhra Pradesh

The coastal districts starting from Chittoor to Srikakulam along East Coast and Khammam and Nalgonda districts have been identified for growing oil palm, as these have fairly high humidity coupled with very high summer temperatures.

Humid sub-tropical conditions prevail in the areas where oil palm is grown in Andhra Pradesh. The annual maximum temperature ranges from 28.6-41.1°C in the coastal districts, *i.e.*, Krishna, West and East Godavari of Andhra Pradesh, which are on par with that of temperature prevailing in Khammam, Guntur, Prakasam, Nellore and Chittoor. In the recent years the maximum temperature had gone up to 48° C for a few days. In Srikakulam, Vizianagaram and Visakhapatnam the temperatures are lesser compared to that of above mentioned ones (26.8-34° C). The minimum temperatures range from 16.7-28° C. The trend of the minimum temperature follows that of maximum temperature in these districts. Amongst the districts, Chittoor is an exception in recording lowest minimum temperature (14.6° C). April and May are the hottest months in the state, irrespective of the district.

In general, all the coastal districts along with Khammam, Vizianagaram and Visakhapatnam receive rainfall in the range of 1044.8 to 1158.0 mm with the maximum rainfall is in Vizianagaram, while the least in Khammam. Guntur, Chittoor and Prakasam receive rainfall of 889.1, 547.7 and 750.9 mm. respectively. The rainfall is well distributed from June to November in all the districts. It is evident that the rainfall is not adequate for growing oil palm as rain fed crop and hence, without adequate irrigation facilities, one should not venture for growing oil palm.

Relative humidity was observed to be lowest in Khammam (56.8 per cent) and highest in Nellore (86.8 per cent). The RH ranged from 70-84 per cent in the districts of Srikakulam, Vizianagaram and Visakhapatnam.

Oil palm is a strong light demander. It puts forth vigorous growth and yields good only under conditions of copious sunshine. In areas with high light intensity the fruits produced are smaller but with higher oil content than fruits produced in areas of low sunlight. The number of sunshine hours necessary to achieve full yield potential have often been estimated at 1800 h/annum while less than 1500 h/annum are considered a limiting factor. There is no dearth of sunshine hours in all the potential areas identified for growing oil palm in the state.

The total annual evaporation ranges from 1650 to 2250 mm. High evaporation rates coupled with deficit rainfall during summer months (particularly April and May) warrant assured irrigation facilities.

c) Arunachal Pradesh

The weather of Arunachal Pradesh differs with altitude, *i.e.* areas situated at higher altitudes observe tundra climate while those at lesser elevation enjoy temperate weather, and those are sea-level experience sub-tropical climate. In the

Upper Himalayas, where elevation is really high, alpine or tundra type of weather is experienced while temperate climate is enjoyed in the Middle Himalayas. At sea-level, the climate is humid sub-tropical, specifically, hot summers and pleasant winters. It receives heavy showers about 2,000 to 4,000 mm between the months of May and September yearly. The summer season extends from March to May, making the plain areas quite warm. Winter officially arrives in November and prevails till the month of February. During summers, the maximum temperature recorded in the foothills is 40 °C; however, average temperature during winters varies from 15 °C to 21 °C. The areas at high altitude become immensely cold during winters.

d) Assam

The Brahmaputra river basins extending on both the sides of the river have been identified for growing oil palm. Parts of Assam are situated in the sub-tropical zone characterized by hot and wet summers along with dry and cool winters. The average maximum temperature ranges from 24 C in February to 35 C in May/June. The minimum temperature falls below 10°C during January/February.

All the districts identified receive more than 1700 mm rainfall per year. Goalpara receives the maximum rainfall (3089 mm). The major amount of rainfall is received during April to October months.

Relative humidity exceeds 80 per cent in all the districts identified. The RH does not go below 75 per cent even during dry months.

Average sunshine hours during December to May and October to November are 6-7 while these are 4-6 during June to September.

e) Bihar

The climate of Bihar is diverse with an annual rainfall of 1326 mm in Plateu region and 1186 mm in Plains. Bihar's climatic conditions can be divided into four segments: summer, winter, monsoon and post monsoon. Winter season is marked with low temperatures which may be as low as 5-10 °C degrees whereas summer is marked with high temperatures and a characteristic hot wind 'loo'. Bihar gets a reasonably good rainfall. The summer starts from April and lasts up to the middle of June. Temperature rises from being pleasant March (mean temperature about 24 °C). Days in April are considerably hot with an average temperature of 30 °C. Temperature touches 35 °C and it may reach as high as 42 °C or more in June, making it the hottest month in the whole year. Summer in Bihar, just like most of the North Indian States, is marked with dust-storms and thunder-storms. Dust storms usually occur in May and most of them have a velocity of about 48-64 km/h. Hot winds called 'Loo" blows through the Bihar plains during April and May with an average velocity of 8-16 km/h.

Rainfall during summer is scanty, with an average of 45 mm. Monsoon starts by mid June and lasts till end of September. Temperature drops with the advent of Monsoon but, the humidity increases. Monsoon may come as early as last week of May or as late as first or second week of July. July and August generally receive maximum rainfall. The oppressive heat of the rainy season may persist till August. Average precipitation during June is about 185.5mm, which rises to an average of about 340 mm in July. It drops down again in August to an average of about 259 mm. There are three distinct areas in Bihar which receive high precipitation, more than 1800 mm/annum. Two of these regions lie on northern and north western wings of Bihar and the third region lies on the Netarhat in the Chota Nagpur Plateau. Monsoon is accountable for more than 90 per cent of the total rainfall in Bihar.

The average temperature is about 31 °C. The south-west monsoon that causes rainfall in Bihar generally withdraws in the first week of October. Bihar is influenced by the typhoons which originate in the South China Sea. The months of September and October are marked with the frequency of the tropical cyclones in Bihar. The average temperature during these months is about 27 °C. Bihar continues receiving precipitation during this period as well. It may receive an average precipitation of 240 mm in September, while in October it drops down to about 39 mm.

Winters start in early November and extend up to mid-March. Days are cold yet pleasant with warm sun and bright light but, temperature falls as soon as the sun sets. Nights are generally colder than days. The mean temperature in November is about 21 C, while in December it may drop to an average of 17 °C. Temperature falls still down in January, which marks the coldest month in the whole year. Average temperature in the month of January is about 16 °C, though it may drop down to 4 C-10 C. Days in March are generally pleasant. Bihar receives little rain during winter, with an average of about 3-4mm.

f) Chhattisgarh

This newly formed State was part of undivided Madhya Pradesh. The two agroclimatic regions identified for Oil Palm are Chhattisgarh plains and Bastar Plateau. The maximum temperature in the identified area range from 30.1 °C to 32.55 C and the minimum from 17.4 to 21.1 C. The rainfall is fairly good but irrigation is a must. The details are given in Table 5.2.

Table 5.2: Agro-climatic Zones and Districts Proposed for Oil Palm cultivation in Chhattisgarh

Agro-climatic Zones	*Districts*	*Temperature (°C)*		*Rainfall (mm)/Month*	
		Max	*Min*	*Max*	*Min*
Bastar Plateau	Dante Wada, Jagdalpur and remaining parts of Kanker District	31.5	18.08	127.09	7.05
Chhattisgarh plains	Bilaspur, Durg, Korbam Mahasamund, Raipur, Narharpur and Kanker block of Kanker district, parts of Raigarh District	32.55	21.15	115.72	5.39
Northern Hills	Jashpurnagar, Korea, Sarguja and Dharamjaigarh tehsil of Raigarh district	30.16	17.40	117.73	3.05

g) Goa

Being in the tropical zone and near the Arabian Sea, the climate of Goa is warm and humid for most of the year. The month of May is the hottest, with day-time temperatures touching 35 C. The heat is coupled with high humidity.

The temperature in winter *i.e.,* late November to Mid-February (Min. 3 °C-Max. 11 °C) while in summer *i.e.,* mid-March to end of June (Min. 25 °C - Max. 45 °C). The monsoon arrives around early June and provides a much needed respite from heat. Goa receives the full blast of the Indian monsoon with sudden downpours and tropical thunderstorms.

There are no extremes in temperature during monsoon. The monsoons are the main feature of the climate of Goa. The average rainfall is approximately 325 cms, the average daily hours of sunshine are nine to ten hours in summer and three to five hours during the monsoon. During the two months preceding the onset of the monsoon the humidity increases dramatically. During the monsoon, 250 cm to 300 cm of rain is normal, although in the Western Ghats the downpour is considerably high than on the coast. Once the monsoon has run its course the skies clear and the weather becomes pleasant. By mid-March the humidity starts to rise as the monsoon begins to approach again.

Monsoon: July to End September (65 cm). The monsoons lasts till late normal September. Goa has a short cool season too which lasts between mid-December and February. These months are marked by cool nights with temperatures of about 20°C and warm days of about 29°C. Humidity remains in moderate range. The nights are a few degrees cooler further inland, due to altitudinal gradation.

h) Gujarat

Sub-humid, sub-tropical climate prevails in oil palm growing regions of Gujarat. Among the different regions, middle Gujarat is the hottest one (41.7 °C) followed by South Gujarat (40.6 °C). Not much difference is observed in the minimum temperature amongst the region (10.5-11.2 °C). In Aliabet Island, the maximum and minimum temperatures are 39.9 and 10.0 °C, respectively.

The high rainfall areas of Gujarat receive 1000 mm to more than 2000 mm per year. In South Gujarat, it ranges from 472 mm in Acchalia to 1510 mm in Dabhol and Tharre of middle Gujarat receive annual rainfall of 776 and 1046 mm., respectively.

Relative humidity ranges from 51 to 95.4 per cent. It is maximum in high rainfall areas of South Gujarat and middle Gujarat and least in South Gujarat. The RH of Aliabet Island ranges from 59-90 per cent. Surat, Valsad., Baroda and Ahmedabad districts have been identified for growing oil palm.

i) Karnataka

In Karnataka five major irrigation project areas of the Cauvery Basin Project comprising of Mysore and Mandya; the Bhadra Project of Shimoga, the Tungabhadra Project areas of Bellary and Raichur; the Upper Krishna Project areas of Gulburga and Belgaum and the Malaprabha and Ghattaprabha Project as well as barrage areas of Dharwad were identified for growing oil palm.

The climate of Karnataka can be classified as humid to sub-humid, sub-tropical. The maximum temperatures range from 31.5 to 45.6 °C. In Belgaum and Tungabhadra Dam, temperatures of 36.9 and 39.4 °C are recorded. Shimoga recorded the minimum temperature (11.9 °C) while a high temperature of 33.10 °C is recorded in Raichur.

Amongst the areas identified, Shimoga is the only place, which receives excess of 1500 mm rainfall. In other places, *i.e.*, Mysore, Belgaum, Tungabhadra Dam and Raichur, the rainfall ranges from 570.4 to 705.8 mm, which is quite inadequate to meet the water requirement of oil palm.

j) Kerala

The climate of Kerala is classified as humid tropical to temperate. The minimum and maximum temperatures range between 19°C in the month of January and 38°C in March.

Kerala gets rainfall from both North East as well as South West monsoons. Rainfall ranges from 1875 mm along seacoast to 3000 mm in the south. Palode receives 3410.4 mm rainfall. Most rainfall occurs between June and December. January, February and March are the dry months.

Relative humidity exceeds 80 per cent in most of the months except during April and May.

In Kerala the areas adjacent to Oil Palm India Ltd., plantations were identified for growing oil palm.

k) Maharashtra

The Konkan coast of Maharashtra comprising of Ratnagiri and Vengurla were identified for growing oil palm.

The characteristic feature of this coastal belt are warm and humid climate. The mean daily temperature is above 20°C throughout the year. The maximum temperature ranges from 29°C in August to 33.5°C in May. The minimum temperature ranges from 17.4 to 25.2°C. February is the coolest month, with May being the hottest.

Rainfall is mostly from South-West monsoon. The total rainfall ranges from 3800 to 5000 mm in Ghat region. The total rainfall received in Mulde is 3616.5 mm, with maximum rainfall in the month of July. January to April is dry period.

Relative humidity may be as high as 90-98 per cent during rainy season, which may come down to 60 per cent at many places after monsoon.

l) Meghalaya

The climate of Meghalaya varies with the altitude. The climate of Khasi and Jaintia Hills is uniquely pleasant. It is neither too warm in summer nor too cold in winter, but over the plains of Garo Hills, the climate is warm and humid, except in winter. True to its name, the Meghalaya sky seldom remains free of clouds. The average annual rainfall is about 1,150 cm.

Flood affected areas are mostly on the low altitude areas, bordering Assam and the India-Bangladesh international border. Flash floods have become a regular feature in these areas, due to massive deforestation, unchecked jhum cultivation. The flood water carries huge amount of hill sand, stone, logs and trees, which are deposited in agricultural fields due to inundation of banks in the foot hills, thus causing immense damage to crops.

The key to the health of the farm sector lies in the health of the forest cover in the state. Every peak, every square inch of the upper range of the hills need to be under mixed forest cover to protect the soil from leaching and erosion to help regulate and decrease the fury of streams and rivulets during the monsoon season.

Vegetation also helps to retain soil moisture and ooze it out during the lean winter months to balance vegetative stress caused by mono cropping in the valley; to bestow various other advantages to help maintain the fragile eco-balance. This ensures continuous cultivation of crops in the farm sector.

m) Mizoram

Mizoram, as a whole, gets an average rainfall of about 3,000 mm with Aizawal town having 2,380 mm and Lunglei 3,178 mm. During rains the climate in the lower hills is quite humid. A special feature of the climate here is the occurrence of violent storms during March-April which come from the north-west and sweep over the hills in the entire state. Rainfall is generally evenly distributed. The crops seldom suffer from drought

Temperature in the state varies from about 12 °C in winter (November-to February) to about 30 °C in summer. There is generally no rain or very little rain during the winter months. In April, storms occur and the summer starts. In April and May temperature goes up to 30 °C. Heavy rains start in June and continue up to August. September and October are the autumn months when the rains cease and the temperature is usually between 19 °C and 25 °C. The state has three district zones as shown below:

- Zone I: Temperate sub-alpine zone: Eastern and south Mizoram bordering adjoining Burma, Major portion of Chambai, Saitul areas ofAizwal district. S. Vanlaiphai areas of Lunglei district and Saiha Tuipang areas of Chfimtuipui district.
- Zone II : Sub-tropical Central Aizawl, Ngopa areas of Champhai and Lunglei sub-division of Southern Mizoram.
- Zone III: Humid Mild-Tropical Zone: Kolasib area of North Mizoram, Mamit areas of Western part of Mizoram bordering Tripura, Demagiri (Lunglei sub-div) and Chawngte sub-division of Southern Mizoram bordering Bangladesh.

n) Nagaland

Nagaland has a largely monsoon climate with high humidity levels. Annual rainfall averages around 1,800-2,500 mm, concentrated in the months of May to September. Temperatures range from 21 °C) to 40 °C).

o) Odisha

In Odisha, the climatic conditions are most ideally suitable to grow oil palm except a few areas, where cold temperatures and water logged areas exist.

Odisha has equitable climate, which is neither too hot nor too cold. The average minimum temperature ranges between 15.8 and 26.5°C, while the maximum temperature ranges from 28.7 to 37.1 °C. April and May are the hottest months.

The average rainfall is 1500 mm. In different zones, rainfall ranges from 1212 to 1562 mm. The rainfall received is from South-West monsoon.

The maximum relative humidity ranges from 84 to 92 per cent.

p) Tamil Nadu

In Tamil Nadu, Cuddalore, Karur, Nagapatnam, Thiruvallur, Vellore, Villupuram Thanjavur, Thiruvarur, Tiruchirappalli Pudukkottai, Perambalur, Tirunelveli, Kanyakumari, Madurai and Theni districts were identified for growing Oil Palm.

The climate of Tamil Nadu can be described as humid to sub-humid, sub-tropical climate. The annual maximum temperatures vary from 27.7 °C in Thanjavur to 41.3 °C in, Thoothukudi closely followed by Madurai (40 °C). The maximum temperatures do not exceed 34 °C in Kanyakumari. The annual minimum temperatures range from 15.0-26.8 °C, with Madurai recording the lowest. Thanjavur, Tiruchirappalli and Thoothukudi record temperatures of 27.4, 26.8 and 26.7 °C respectively. In Kanyakumari, temperatures never fall below 24 °C. May is the month of hot temperature, except in coastal areas, where June is hotter.

Most of the areas receive rainfall through North-East monsoon. Chengalpattu receives the maximum rainfall (1211 mm) followed by Thanjavur (1115 mm), while the least was in Thoothukudi (662.2 mm) and Kanyakumari (762 mm). The districts of Madurai, Tirunelveli and Tiruchirappalli receive around 850 mm of rainfall.

Relative humidity drops down to 39 per cent in the month of March in Tirunelveli and Thoothukudi districts. Maximum RH (90-95 per cent) is during the month of November in all the districts.

The annual potential evapo-transpiration ranges from 1441 to 2090 mm in all the districts.

q) Tripura

The climate in Tripura is sub-tropical, with warm and humid environment. The maximum temperature ranges from 27.34 °C in January to 35 °C in April. The minimum temperature is 9 °C in January and shoots up to 25 °C in August. January to March is the cold period, in which the temperatures are between 8 and 16 °C.

The total rainfall is 2,484 mm from South West monsoon and is fairly well distributed.

Relative humidity ranges from 62.4 to 85.5 per cent. It exceeds 80 per cent during June to October.

In most of the places, the rainfall is quite inadequate, but these areas were identified as potential areas because of the irrigation potential available through irrigation projects, bore wells, barrages *etc.*

As far as maximum temperature is concerned, the crop has withstood a high temperature of about 48 C prevailing in many parts of the East Coast for the last seven years. The minimum temperature of 11 C for a short while also did not have any deleterious effect on the crop. The abundant sunlight and favourable humidity are the favourable factors for the oil palm in these identified areas.

r) West Bengal

The climate of West Bengal is classified as tropical monsoon type. The maximum temperature goes up to 32.6 C in Cooch Behar and Jalpaiguri and upto 36.3 C in 24-Paraganas, during April/May months. The mean minimum temperatures are low in Cooch Behar and Jalpaiguri. 24-Paraganas has the highest minimum temperature (15.8 C) in the month of January.

The average rainfall is 1750 mm. It varies from 1200 mm in South Western region to 4000 mm in Northern region. Most of the rainfall is through South-West monsoon. Cooch Behar and Jalpaiguri receive double the rainfall than that of 24-Paraganas (North and South). January to March is the dry period.

Relative humidity is above 80 per cent only for three months (June-August) in Cooch Behar and 24-Paraganas (North and South). But in Jalpaiguri, it is above 80 per cent except in March and April.

5.1.7. Mitigating Effects of Climate Change

Palm oil is the most productive and efficient source of edible oil with a yield ten times that of soybean oil and five times that of rapeseed oil. The high land productivity oil palm is the key to mitigate climate change. This will minimise land use change and reduce potential GHG emissions from land use change. A changing climate in oil palm plantation would affect in many ways, with either benefits or negative consequences dominating in different agro-climatic regions. However,the factors that prevail regionally may change over time, as gradual and possibly abrupt climate change may develop in this century. Rising atmospheric CO_2 concentration, higher temperature, changing pattern of precipitation and altered frequencies of extreme events will have significant effects on oil palm production with associated consequences for water resources and pest/disease distributions.

Among the measures that are adopted by the oil palm industry to ensure sustainable development include zero burning in plantation development, integrated pest management, the use of oil palm waste as organic fertilisers, and other good agricultural practices. This will ensure long-term sustainability of the industry as well as ensuring minimal negative effects to the environment including GHG emissions.

The measures include among others, replanting of old and non-productive oil palm trees with high yield planting materials such as clonal materials as well as implementing other measures to achieve 26.2 tonnes of fresh fruit bunches per hectare per year (currently 19.2)and oil extraction rate of 23 per cent (currently 20.49) by 2020. These are to ensure that the increased palm oil production to cater the increase in demand for both food and non-food sectors are met through increased yields rather than expansion of new areas which may entail increased GHG emissions from land use change.

The anaerobic digestion of palm oil mill effluent (POME) at the mills has been identified as having the most harmful impact to the environment in Malaysia in terms of global warming and climate change as it generates methane (65 per cent) and carbon dioxide (35 per cent). In 2009, methane from POME in the country contributed an estimated 17.0 MTonnes of carbon dioxide equivalent to the environment. Thus, palm oil millers can play an important role in reducing greenhouse gas emission and mitigating global warming by installing the biogas capture process. EPP (Entry Point Project) No. 5 aims to promote and encourage all palm oil mills in the country to install biogas capture facilities by 2020.

The second thrust area in Malaysia for downstream expansion and sustainability includes three EPPs which are aimed at generating income from downstream value-added activities. These include:

- Developing oleo derivatives;
- Commercialising second generation biofuels; and
- Expediting growth in food and health based downstream segments.

These activities are important to ensure sustainable development of the other industries by using renewable sources from palm oil and its products. The use of palm oil as a feedstock for oleochemicals reduces the dependency on petroleum-based products and thus conserves depleting resources. Oleochemical products are also biodegradable, environmental friendly and as such, will help to mitigate climate change.

Climate change mitigation for a smart production of oil palm can be aimed through conservative agriculture, prevention of erosion through biodiversity conservation, developing thermo sensitive high oil yielding tenera hybrids, cropping/farming system and thereby residue management through bio-mass recycling, soil water conservation and mulching, bio-mass energy production, value addition and by product utilisation.

5.2. Soils

5.2.1. Topography

Soils with gentle slope are preferred, but undulating lands can also be used. If hilly areas are used there may be the need to construct terraces which greatly increased the initial planting investment and for this reason most plantations are established on lands with slopes below 21 per cent. Completely flat areas may need special drainage work.

Any oil palm project starts with general soil survey to determine and quantify the limitation and cost for correction: the need to construct the drainage system alleviate soil compaction, fertility and in general any situation that may adversely affect the yields in future. Oil palm can grow in a wide range of environments and soil conditions. However, soil and climatic conditions which are conducive for good vegetative growth and high yields of oil palm are:

- ★ Deep soils which permit root penetration to at least 75 to 100 cm. and absence of impediments such as compact impervious layer and stones/ laterites.
- ★ Fine sandy clay loam to clay loam with good moisture holding capacity.
- ★ Moderately well-drained soils. Oil palm can. tolerate short period of impeded drainage and flooding.
- ★ Soil pH within the range of 4.0-6.0. Oil palm, however, is generally quite tolerant to acid soil condition with proper maintenance of water-table.
- ★ Soils of high fertility status. Oil palm being a high nutrient demanding crop requires generous manuring to give maximum productivity.
- ★ Minimum rainfall of 1800mm distributed uniform throughout the year.
- ★ In situation of insufficient moisture during dry period, irrigation is necessary.
- ★ Optimum temperature for growth is 18-36°C throughout the year. However, where temperature extreme occurs, mean annual minimum not less than 20°C, while mean annual maximum not more than 30°C.
- ★ Nature of terrain gently undulating.

The evaluation of land characteristics for oil palm are given Table 5.3.

Table 5.3: Evaluation of Climate and Land Characteristic for Oil Palm

Sl. No.	*Characteristics*	*Not Limiting*	*Minor Limitation*	*Moderate Limitation*	*Serious Limitation*	*Very Serious Limitation*
A.	**CLIMATE**					
	Annual rainfall (mm)	>2000	1700-2000	1450-1700	1250-1450	<1250
	Dry season (months)	-	<1	1-2	2-3	>3
	Mean annual (max) temperature °C	>29	27-29	24-27	22-24	<22
	Mean annual (min) temperature °C	>20	18-20	16-18	14-16	<14
	Mean annual (°C) temperature	>25	22-25	20-22	18-20	<18
B.	**TOPOGRAPHY**					
	Slope	0-1	4-12	12-23	23-38	>38
	Water drainage	Moderate to imperfect	Well and somewhat excessive	Poor (aeric) (easily drained)	Poor (typic) difficult to drain	Very poor
	Flooding	Not flooded	Not flooded	Minor	Moderate	severe
C.	**PHYSICAL SOIL CONDITIONS**					
	Texture/Structure	Cs.SC.CL SiCs.SiCl	Co, L, SCo	SCL	SL,Lsf	LSCo.S.

Contd.

Sl. No.	*Characteristics*	*Not Limiting*	*Minor Limitation*	*Moderate Limitation*	*Serious Limitation*	*Very Serious Limitation*
	Consistency	Friable to mod. firm	Firm	Firm to very firm	Very firm	Compact
	Depth(cm)	>100	75-100	50-75	25-50	<25
	Depth to top sulfuric horizon (cm) Laterite/ rock fragment nodules	Nil	Nil	-	Up to 25 per cent nodules or fragmental laterite	>25 per cent and massive laterite
D.	**SOIL FERTILITY CONDITION**					
	pH	4-6	3.5-4.0	-	3.2-3.5	<3.2
	Weathering stage (effective CEC)	>16	20-3	<20	-	-
	Base saturation (per cent) A horizon	>35	20-35	<20	-	-
	Organic Carbon (per cent) A horizon	>15	<1.5	-	-	-
	Salinity (millimohs) -50 cm	0-1	1-2	2-3	3-4	>4
	Nutrient status	Low fertilizer requirement	Mod. Fertilizer requirement	Mod. to high fertilizer requirement	High to very high fertilizer requirement	Very high fertilizer requirement

The soil suitability classes for growing oil palm are given in given in Table 5.4.

Table 5.4: Soil Suitability Class

Class I	*Highly Suitable*	*Soils with No Limitations*
Class II	Suitable	Soils with only 3 to 4 minor limitations
Class III	Moderately suitable	Soils with minor and no more than 2 or 3 moderate limitations
Class IV	Marginally suitable	Soils with more than 2 to 3 moderate limitations and one serious limitation
Class V	Unsuitable	Soils with more than one very serious limitations

5.2.2 Soils for Oil Palm Cultivation

Oil Palm has a relatively shallow root system with most of the active roots found in the upper 30 cm of the soil. To maintain adequate nutrient supply to the palm, the nutrient concentration in the soil must be higher because of coarse and inefficient root system. Therefore, oil palm is very sensitive to the soil environment.

The elevation and slope of an area proposed for oil palm cultivation are important land characterization that determines its suitability. Oil palm is not recommended for planting in areas with an elevation above 200 m. The crop is best planted on slopes of less than 12°, and should not be planted on land with slopes of more than 20°.

Table 5.5: Detias of Soil Suitability Classes

Soil Characteristics	*Suitability Class*	*Highly Suitable (S1)*		*Moderately Suitable (S2)*	*Marginally Suitable (S3)*	*Unsuitable (N)*
	Degree of Limitation	*Not Limiting*	*Minor Limitation*	*Moderate Limitation*	*Serious Limitation*	*Very Severe Limitation*
Topography						
Slope per cent		0-4	4-12	12-23	23-38	> 38
Slope°		0-2	2-6	6-12	12-20	> 20
Wetness						
Drainage class		Moderately well drained	Well to some what excessive	Excessive or some what poorly drained	Poorly drained	Very poorly drained
Flooding		Not flooded	Not flooded	Minor flooding	Moderate flooding	Severe flooding
Physical Soil Condition						
Texture/structure		Cs, Sc, Cl	Co, L, Soc, Sicl	Scl, Co, Sics	-	-
Depth to root restricting layer (cm)		> 100	75-100	50-75	25-50	< 50
Depth to acid sulphate layer (cm)		> 100	-	75-100	50-75	< 50
Soil fertility condition						
Weathering stage (effective CEC) C mol (+)/100 gram		> 24	16-24	< 16	-	-
Base saturation per cent A horizon		> 50	35-50	< 35		
Organic carbon per cent A horizon		1.5 – 2.0	2.0 or < 1.5	-	-	-
Salinity (m mhos) 50 cm depth		0-1	1-2	2-3	3-4	> 4
Micronutrients		-	-	Deficiency	Toxicity	-

Source: Paramanandan, 2000, Malaysia.

Oil palm generally requires a well-aerated soil with sufficient moisture throughout its growth especially when the palms are young and the root system is yet to develop fully. Too much moisture due to flooding is also detrimental to the growth of oil palm as root growth and nutrient uptake can be seriously retarded. Thus, too little or too much moisture will adversely affect the growth and performance of oil palm.

Well-structured clay, sandy clay, clay loam and silty clay loam soils which are capable of retaining sufficient moisture and nutrients are ideal for oil palm cultivation.

Although most of the oil palm roots occur in the upper 30 cm of the soil surface, ideally oil palm prefers a soil that is free of any physical impediment such as presence of coherent rock, dense thick layer of lateritic gravel, stones, pebbles or chemical impediment such as presence of an acid sulfate layer or the organic soil materials to a depth of 100 cm, for free root penetration.

Soils having CEC >15 C mols kg^{-1} are considered as ideal for oil palm cultivation. Oil palm does not tolerate salinity (pH >8.5) and acidity (pH<3.5) and hence, soils with saline and acidic conditions within 5m of soil surface can be detrimental to oil palm growth (Table 5.5).

Soils which are unfavorable to the oil palm and must be avoided are as follows:

i) Poorly drained soils
ii) Laterite soils, *i.e.*, soils containing concretionary ironstone, usually in the form of gravel or thick bands of subsoil.
iii) Very sandy coastal soils
iv) Deep and peaty soils (more than 2.5 metres) and peat soils with underlying sand.

5.2.3. Soils Conditions in Major Oil Palm Growing Countries

The soil conditions of major oil palm growing countries around the world (Table 5.6) indicate that oil palm is grown from Alluvial soils to sandy loam soils.

Table 5.6: Soils of the Major Oil Palm Growing Countries in the World

Country	Soils
Brazil	Vertisols
Cameroon	Ferrallitic soils
Columbia	Alluvial soils
Congo	Hygro Kaolinite ferralsols or Ultisols. Sandy soils in Southern Congo.
Costa Rica	Alluvial calcareous soils and Alluvial volcanic soils. Also grown in clay soils.
Ecuador	Volcanic ash (Andosols) soils
Honduras	Rich Alluvial soils, marshy valley soils
India	Deep Red sandy loams, Alluvial silt loams, Laterite soils
Indonesia	Brown alluvial soils, grayish brown andosols and typical reddish or yellowish brown podsolic soils. Mostly Alluvial soils with varying water tables and sandy soils of Sumatra. Also grown in peat, ultisols and tertiary sandy soils

Country	Soils
Ivory Coast	Fertile acid sandy soils
Malaysia	Inland soils derived from acid Igneous Rocks. Sandy loam Kedah series to heavy silt clays of Batu series of sedimentary and metamorphic rocks. Coastal soils include Selangor (marine clay deposits) to Briah series (Alluvium deposits)
Nigeria	Acid sandy soils with PH ranging from 5-6 to 4-4.5. Also grown in Alluvial soils, clayey sand and sandy clay soils.
Sierra Leone	Soils derived from tertiary sand stone deposits and similar to acid sands if Nigeria

Source: Corley and Tinker (2003), The Oil Palm (Fourth Edition), Blackwell Science Ltd., U.K.

5.2.4. Soils of Oil Palm Growing States in India

In India, oil palm is grown in varied type of soils. The Table 5.7 gives a picture of the soil types in different states of the country.

Table 5.7: Soils of different Oil Palm Growing States in India

State	Soil Type
Andaman and Nicobar Islands	Peat Soils
Andhra Pradesh	Red sandy, Ted loamy, Red loam with clay base, Deltaic alluvium pH – 6.5 to 9.0
Arunachal Pradesh	Forest and mountainous soil and red yellow soils, Alluvial soils
Assam	Tea growing soils and river alluvium
Bihar	Major alluvial soilsm
Chhattisgarh	Red, yellow and laterite soils
Goa	Red loam and laterite soil
Gujarat	Black soils and alluvial soils are pre-dominant. Some are coastal alluvium and laterite soils
Karnataka	Deep and comprises of red and brown sandy loams and black clayey soil to the extent of 20 and 80 per cent respectively, with pH - 6.5 to 9.0.Redsandyloam with 7.0 to 8.3 pH. River alluvium and humus type soils.
	Heavy and black soils with 7.5 to 8.5 pH. Upper Krishna project and Thungabadra - 60 per cent black Cotton soils with 7.8 to 9.5 pH. 40 per cent is red soil with 7.4 to 8.5 pH.
Kerala	Lateritic and forest soils
Maharashtra	Lateritic and red loam gravelly soils
Meghalaya	Alluvial and forest soils
Mizoram	Sandy loam with acidic pH.
Odisha	Red Yellow soils, Coastal and river alluvial and brown forest soil, laterite soils.
Tamil Nadu	Laterite/gravelly, River Alluvium and red
	sandy soils, Red soils, red loamy soils and clay loam
Tripura	Deep sandy loam and river alluvial and Red Yellow, Lateritic soils.
West Bengal	Sandy loam with acidic pH.

In most of the places oil palm is being cultivated in plain lands except in Kerala and Tripura. While gentle slope can be acceptable, it is better to avoid steep slopes. Also one has to ensure that at least 1.0 to 1.5 m soil depth with good irrigation and drainage facilities are available.

The pH of the soil ranges from 5.5 to 8.5 and EC ranges from 1.0 to 2.0. Most of the upland soils are red loam, red sandy loam and also red laterite soils (Table 5.7). Black soils with adequate drainage facilities also have been used. Care should also be taken to prevent severe crack formation during summer. Marshy soils with adequate drainage facilities can also be used for raising oil palm. Proper soil management is absolutely essential to get better yield.

The plantations raised in all these soils have grown well wherever irrigation and drainage facilities are optimum.

Even in Kuttanadu (Kerala) soils where there is standing water oil palm was grown successfully by planting the seedlings on a raised mound. However, farming mounds and bringing the soil from long distance are costly affairs more over periodical maintenance of mounds also needs soil from outside.

Oil palm is grown in varied type of soils in India. Soil health management is very important. If it is water logged soils adequate drainage facilities are to be created first. Mound planting and widening the basins every year maintaining the drainage channel will help to establish good plantation. If soils are deep black clay it is necessary to add light textured soils in the palm basin before planting. Application of farm yard manure, vermicompost, or growing green manure crop and incorporation should be continued in the initially years. In light textured soils, soil organic matter buildup is necessary. This can be done by growing green manure crops and incorporation, of farm yard manure, vermicompost, tank silt *etc.*

In India oil palm growing is successful under various types of soil with adequate care and management of soil with moisture and nutrients. Organic residue management and maintaining soil health are very essential to get good yield.

Chapter 6

Origin, Habitat, Distribution and Botany of Oil Palm

This chapter deals with the origin, habitat, distribution and botany of oil palm.

6.1. Origin

Oil palm (*Elaeis guineensis* Jacq.) originated in the Guinea coast of West Africa and semi wild and wild palm groves still exist in the tropical West African countries. During the 15th century, oil palm was introduced to Brazil and other tropical countries by the Portuguese, but did not flourish in these countries till 19th century. The Dutch imported oil palm seeds from West Africa via Amsterdam during 1845 and four seedlings were planted in the Botanical garden, Bogor, Indonesia in 1846. The progenies of these palms were planted as ornamental palms in Deli and later came to be known as Deli dura. The oil palm seeds from here were taken to various parts of the world. During 1900, oil palm was planted in Sumatra Islands as ornamental palm in tobacco gardens. The commercial plantations of oil palm were established in 1910 and 1911. Indonesia and Malaysia took up commercial planting in 1915 and 1917, respectively.

6.2. Habitat

Oil palm grows naturally near rivers, where palms are subjected to less competition from flora, more light penetration and plentiful but not excessive moisture. The natural habitat for oil palm were in galleries; forestry's or forest outliers either near Raphia palm or in association with Elaeis and Raphia palms. It thrives well wherever man cleared a part of forest. It requires a relatively open area to grow and reproduce and thrives best when soil moisture is well maintained. It is natural to suppose, that the palms would find a place on the forest fringe near to rivers. The river habitat of this kind could be found right across Africa from Senegal to Angola and even Mozambique. Similar habitats have also been assumed by 'escapes' in Sumatra and Malaysia.

The fresh water swamps also have been suggested as a natural habitat for the oil palm. Palms growing in such swamps adjoining to rivers have been seen in several parts of Southern Nigeria. The palms are often found growing where there is standing, though not stagnant water for many months of the year. However, it should also be noted that oil palm will not tolerate permanently high water tables in impervious soils. Palm trees may grow up to 20 m and more in height. The trunks of young and mature trees are wrapped in fronds which give them a rather rough appearance. The older trees have smoother trunks apart from the scars left by the fronds which have withered and fallen off.

Oil palm tree will start bearing fruits after 30 months of field planting and will continue to be productive for the next 20 to 30 years; thus ensuring a consistent supply of oil. Each ripe bunch is commonly known as Fresh Fruit Bunch (FFB). In Malaysia, the oil palm trees planted are mainly of tenera variety, a hybrid between the dura and pisifera. The tenera variety yields about 4 to 5 metric ton of crude palm oil (CPO) 20 to 25 metric ton FFB/ha/year and about one ton of palm kernel oil. The oil palm is the most efficient oil-bearing crop in the world, requiring only 0.26 ha of land to produce one metric ton of oil while soybean, sunflower and rapeseed require 2.22, 2.0 0 and 1.52 ha, respectively, to produce the same quality oil.

6.3. Distribution

Oil palm is mainly distributed within 23.5°N to 23.5°S of the equator. Earlier it was grown in most of the African countries. Several countries with suitable climate like Brazil, Colombia, Ecuador, Panama, Costa Rica, The Solomon Islands, Papua New Guinea, Thailand, Philippines, Sri lanka, Myanmar and India, are also developing oil palm plantations (Figure 6.1).

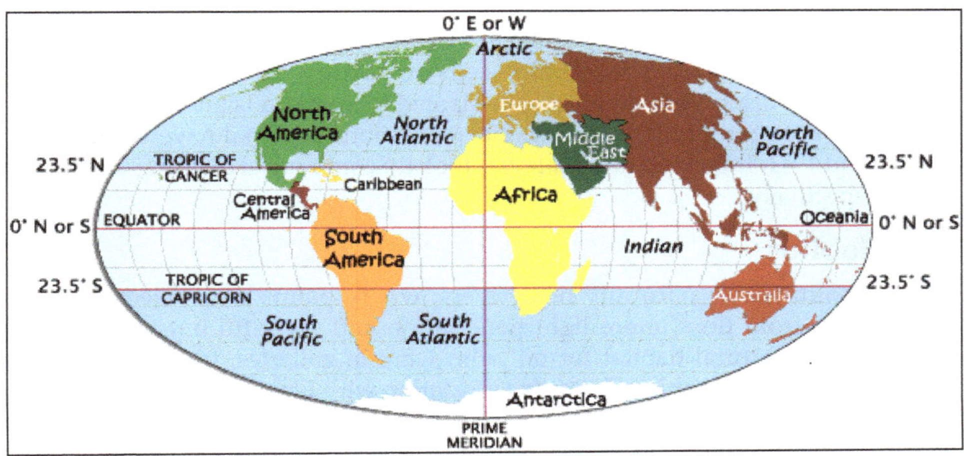

Figure 6.1: Oil Palm is grown in more than 33 Countries.

The centers of distribution of oil palm has been broadly classified into two areas :

i) The African oil palm, *Elaeis guineensis* is endemic to West and Central Africa. Extensive natural and semi wild palm groves are distributed along West Coast of Africa from Senegal to Angola (Zeven, 1967).

ii) *Elaeis olifera* is fund in Honduras, Nicaragua, and Costa Rica. Panama, Colombia, Venezuela, Surinam, Equador, and Brazil (Meunier, 1975; Ooi *et al.*, 1981; Escobar 1982; Raja Naidu, 1983).

6.4. Botany of Oil Palm

6.4.1. Taxonomy

According to the latest classification, the family Palmae has been divided into six sub-families. The sub-family Arecaidae is divided further into tribes and sub-tribes. The genera *Elaeis* and *Barcella* fall in the tribe Cocoineae and sub-tribe *Eleaideae*. The genus *Elaeis* consists of two species, *viz. Elaeis guineensis* and *Elaeis oleifera*. *Elaeis* is the word derived from the Greek word *'elaion'* which means 'oil', while the species name *'guineensis'* shows that Jacquin attributed it to its place of origin, the Guinea Coast.

The African palm (*E. guineensis*) originated from West Africa whereas *E. oleifera* is endemic to Central and South America and it hybridizes readily with *E.guineensis*. The main oil palm belt begins in Guinea and spreads South through Sierra Leone, Liberia, Ivory Coast, Ghana, Togo, Benin, Nigeria, Cameroon, Equatorial Guinea, Gabon, Zaire and Angola. From Central West Africa, the natural oil palm groove belt stretches through Zaire, Tanzania, Mozambique, and up to the island of Madagascar. It is believed that much of the spread to the East was due to dispersal by man. Natural distribution of *E. oleifera* is confined to Honduras, Nicaragua, Costa Rica, Panama, Colombia, Surinam, Brazil and Peru.

6.4.2. Seed

The oil palm seed is a nut consisting of a shell or endocarp, and one, two or three kernels. Due to the abortion of two of the three carpels in the tri carpellate ovary, the seed develops only one kernel in majority of the cases. Abnormal ovaries may contain up_to five kernels. The nut size varies according to the thickness of shell and size of kernel (1 to 13 g). The shell has fibres passing longitudinally through it and adhering to it and they are drawn into a tuft at the base. Each seed has three germ pores corresponding to the three carpels of the ovary. A plug of fibre is formed in each germ pore and these fibres join together to form plate-like structure at the base of the seed. The kernel lies inside the shell, which consists of layers of hard oily endosperm, which is grayish-white in colour. The kernel is surrounded by a dark brown testa covered with a network of fibres. Embryo is embedded in the endosperm facing one of the germ pores. The embryo is straight and about 3 mm in length. Its distal end lies opposite to the germ pore, which is separated from it by a thin layer of endosperm cells. The testa, the plate-like structure and layer of endosperm cells together form the operculum.

a) Seed Germination

The process of seed germination (Figure 6.2) involves the emergence of button shaped embryo in the beginning, which rapidly gains a plumular projection while radicle emerges from the end of the embryo. The plumule and radicle both emerge through a cylindrical ligule close to the seed. Within the seed, the haustorium develops rapidly. In three month's time, the spongy haustorium absorbs the endosperm fully and fills the cavity of the nut.

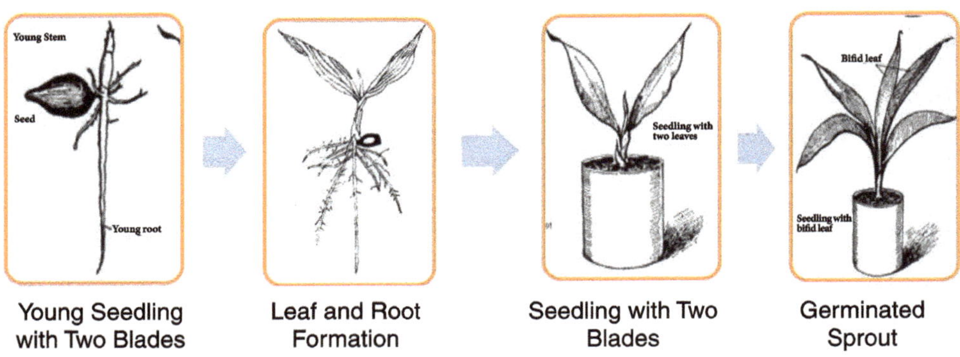

Figure 6.2: Stages of Germination.

Under natural conditions, germination is sporadic and may take several months, which is not desirable. If the seed is planted in sand or soil under natural conditions, irregular germination up to 50 per cent would occur after 3-6 months. To get uniform germination pre-heating technique is followed which has been dealt with in the chapter on Planting Material Production. A germination percentage of 90-95 per cent is obtained by this method.

b) Seedling

When radicle becomes 1 cm long, plumule emerges from the plumular projection. The first adventitious root emerges in a ring just above the radicle - hypocotyl junction and it gives rise to secondary roots before the emergence of first foliage leaf. Before the emergence of green leaf, two bladeless plumular sheaths are produced. The leaf is recognized by the presence of a lamina, which is produced after one month of germination. Thereafter, one leaf per month is produced until the seedlings becomes 6 months old. After 4 months, the base of the stem becomes swollen and the first true primary root emerges from it. The first few leaves are lanceolate. In later leaves, a split is formed resulting in bifurcated leaves and subsequently the leaflets become entirely separate.

6.4.3. Stem

Early growth of oil palm after the seedling stage, involves the formation of a wide stem base without inter nodal elongation, on which the stem column can rest firmly. The palm has one terminal growing point. Branched palm may occur due to the damage of the apical cells. The apical meristem lies in a basin-like depression

Figure 6.3: Leaflet Arrangement Along the Rachis in *E. guineensis* (a) *E. oleifera* and OxG Hybrids (b).

at the apex of the stem. The apex is conical in shape. It is buried in the crown of palm within a soft mass of young leaves and leaf bases commonly known as the 'Cabbage'. Stem-thickening is attained through a process called primary thickening which is brought about by the activity of meristem beneath successive leaf bases. Inter nodal elongation begins only after attaining the maximum diameter of the stem. During early years, the base of the stem assumes the shape of an inverted cone. Primary roots emerge in large numbers from this cone. As soon as internodes begin to elongate, a columnar stem with adhering leaf bases is formed. The leaf bases adhere to the stem for minimum of 12 years. The leaf bases start falling away from the base of the crown or the middle of the stem leading to the formation of smooth stem.

The leaves are produced at the apex in an orderly arrangement, which is roughly triangular. The arrangement gives rise to sets of spirals. In well-grown plants, two sets of spirals are visible, eight running one way and thirteen the other. Such arrangement is described as 8+13. If the leaf bases are numbered in the order, it becomes clear that in one way, every eighth leaf is seen to be in the same spiral, while in the other way every thirteenth leaf appears in the same (more nearly vertical) spiral. These spirals are in either direction, left hand or right hand, more or less on a 50:50 ratio. In practice total number of leaves can be clearly calculated by counting the number of spirals with 8 leaves and multiplying the figure by eight and then adding the number of leaves in the incomplete spiral. The average increase in height is from 0.3 to 0.6 m per year. The palm reaches a height of about 30 m. The width of the stem unclothed by leaf bases varies from 20-75 cm. The stem functions as a supporting, vascular and storage organ. The anatomy of the oil palm stem is typical to that of monocotyledons. In the cross section, there is a wide central cylinder with a very narrow cortex. In the periphery numerous vascular bundles with fibrous phloem sheaths are embedded in the sclerotic ground tissue. In the central zone the vascular bundles are found embedded in the parenchymatous ground tissue.

6.4.4. Leaf

In the adult palm, about 50 leaves could be observed in various stages of development.

Each leaf remains enclosed for about 2 years and then rapidly develops into a central spear leaf, which finally opens. The base of the leaf completely encircles the stem apex and in the adult leaf the base is persistent as a strong fibrous sheath. Mature leaf is simple, pinnate with linear leaflets on each side of the leaf stalk. The pinnate leaf is about 7 m long and consists of a petiole (approximately 150 cm long) and a rachis bearing 250-350 leaflets each may be up to 130 cm long. The leaflets are arranged on two lateral planes. In the species *E. guineensis*, leaflets are arranged in different angles and planes along both sides of the rachis. In *E. oleifera* and the Ox G hybrids, leaflets are inserted in one single plane (Figure 6.3). Variation in the plane angle as well as the angle of insertion of leaves and leaflets determine the general appearance of the tree, and facilitate the penetration of light into the foliage. Two kinds of spines are seen in the petiole region namely the fibre spines and the midrib spines. The fibre spines are seen on the lower portion, which is a modification of the leaf sheath, whereas the midrib spines are seen in the upper portion of the petiole. These are modified midribs of leaflets. The midrib is very rigid and the laminae sometimes tear backwards from tip. The number of leaves produced annually by a mature palm is about 25. Young palms may produce three or more leaves per month, but eventually leaf emission rate stabilizes at around two per month in adult plantations, when a palm may accumulate between 36 and 45 fully developed leaves, depending on pruning cycles and light competition.

6.4.5. Root System

Oil palm has a typical adventitious root system. The adventitious roots grow from a bole at the base of the stem and number several thousands in the adult palm. Primary roots are about 6 to 10 mm in diameter and 1 to 20 m long, branching into secondary roots (1 to 4 mm diameter), tertiaries (about 10 cm in length and 0.5-1.5 mm diameter) and quaternaries. The roots spread horizontally or descend at varying angles into the soil. The density of all classes of roots in the top 60 cm of soil usually decreases with distance from the palm. But in adult palm, total quantity of absorbing roots increases at least to a radius of 3.5 - 4.5 m. Absorption of nutrients takes place through the quaternaries and absorbing tips of primaries, secondary and tertiaries, bulk of which are seen in the top 15-30 cm of the soil

The anatomy of oil palm roots consists of an outer epidermis and lignified hypodermis surrounding a cortex in which well developed air lacunae are found. Within the cortex is the central stele surrounded by an endodermis, vascular strands of xylem and phloem and the pith. The quaternary roots are not lignified and perform the main function of absorption of water and nutrients. The roots of oil palm are characterized by the presence of pneumathodes. These appear on both underground and aerial roots. Such roots are supposed to ventilate the root system. The superficial root system makes this palm very sensitive to poor soil compaction and poor drainage. Besides this, superficial tillage may severely damage the root system particularly in young plantations.

6.4.6. Flowering

Oil palm comes to flowering 14-18 months after planting. It produces both male and female flowers separately on the same palm. Male and female phases do occur naturally in consequent cycles in a palm. Some individual trees may exhibit a phenomenon of producing more male inflorescence and less number of female inflorescence. This phenomenon causes no harm as long as the average annual production is satisfactory. Overall, there should be an average of 10-12 bunches per tree per year. Large number of male flowers occurs due to the following reasons resulting in low yield:

- Insufficient irrigation and/or irrigation at longer intervals.
- Non-application of recommended doses of fertilizers and other manures in appropriate quantities and time.
- Excessive pruning of fronds
- Ploughing deeply and close to the palms damaging the active feeding roots.

The various stages of flower opening to bunch formation are given in Figure 6.4.

a) Floral Biology

An inflorescence is initiated in the axil of each leaf although some of these may abort at an early stage. The inflorescence is enveloped in two inflated bracts or spathes, which open shortly before flowering. Initiated at the same time as the leaf (about 33 months before anthesis), the inflorescence bud will become sexually differentiated nine month later (22 to 24 months before anthesis). The process of sexual differentiation is not well understood but it is known to be under genetic control and highly influenced by environment. Favorable conditions induce higher female sex ratio. An individual palm will produce, during successive cycles, male and female inflorescences. The length of the male and female cycles varies widely according to genotype and environment. Cycles may sometimes be separated by 1 or 2 hermaphrodite inflorescence, but pure male and female cycles rarely overlap. In addition, usually only one inflorescence on a tree will be at anthesis at any one

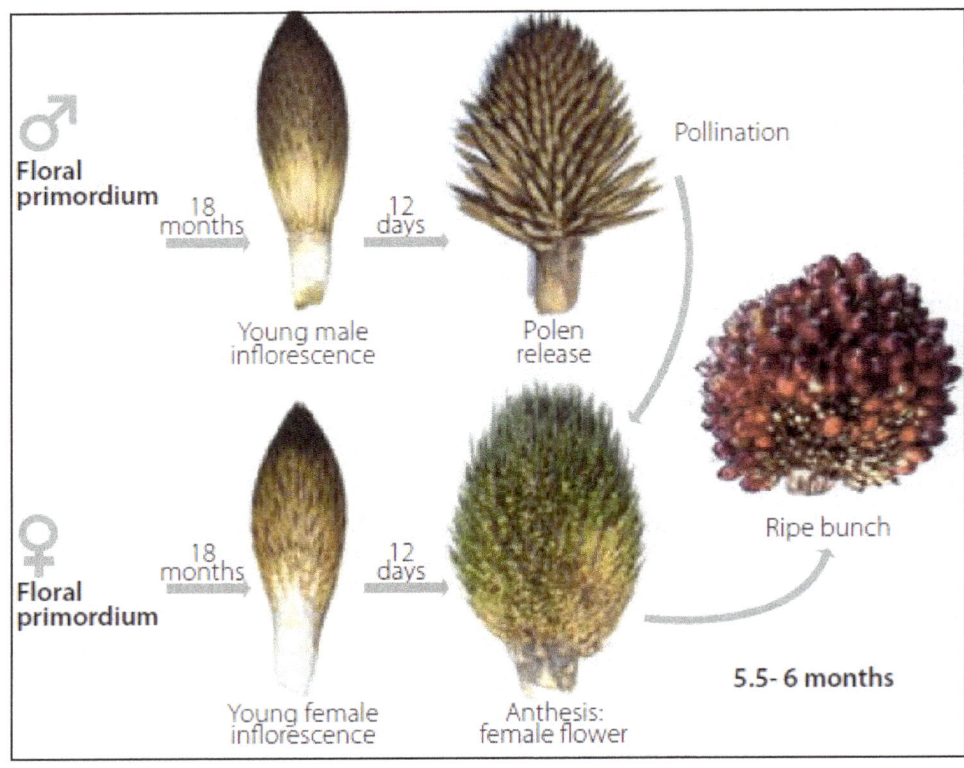

Figure 6.4: Flower Emergence to Ripe Bunch.

time. Thus although the oil palm is a monocious species, it is considered as strictly allogamous.

The male inflorescence (Figure 6.5a) is a peduncle approximately 40 cm long which bears 100-300 finger-like spike lets that are 10 to 30 cm long. Each spikelet arises from a central stalk bearing 600-1200 male flowers, which are yellow in colour having distinct aroma and mature from bottom to top. The flowers are arranged in a triangular bracket; each consists of a perianth of six mature segments, a tubular androecium with six or seven anthers and a rudimentary gynoecium. The anthers are bilobed and release pollen grains through the lateral splits. Each inflorescence produces 10-50 g of pollen, which may contain about seven billion grains.

Both are compound spadix and the pollinating beetles are foraging on the male flowers at the tips os spadix branches.

The female inflorescence (Figure 6.5b) reaches a length of 30 cm or more before opening. The female spikelets are thick and fleshy and develop in the axils of spiny brackets (Figure 6.6). The flowers are arranged spirally around the rachis of the spikelets. Each flower is housed in a shallow cavity and is subtended by a bract, which is drawn up into a spine. At the end of the spikelet also there is a spine. An

Figure 6.5: Inflorescences–a) Male and b) Female.

Figure 6.6: Female Inflorescence at different Stages of Receptivity.

average inflorescence from a mature palm may have more than 110 spikelets with over 4,000 flowers. Each female flower has a protection bract, two floral bracts and two whorls of three perianth each. All these bracts enclose a tricarpellary ovary. The two accompanying male flowers are abortive. At anthesis the trifid stigma curves outward. The stigmatic lobes when receptive are white to pale yellow in colour. Later a red stripe develops along those lobes as the flowers tum purplish indicating the end of receptivity.

Male and female inflorescences are produced in cycles. In young palms a great variety of hermaaphrodite inflorescence are produced. These consist of male, female and mixed spikelets in the same inflorescence. Young palms occasionally produce a peculiar type of inflorescence known as andromorphic. Here the male flowers are replaced by small solitary female flowers arranged in the same manner as that of the male flowers.

b) Pollination

The oil palm, hitherto thought to be wind pollinated, has been now proved to be an insect pollinated species. From West Africa, the original home of oil palm, eight species of pollinating weevils were reported. They belong to the order Coleoptera (Family - Curculionidae, subfamily- Derelominae). Weevils on oil palm in Cameroon include *Elaeidobius kamerunicus*. The only species found in South America is *E. subvittatus*. In India, *E. kamerunicus* was introduced in the oil palm plantations of Kerala and in Little Andamans during 1985. This weevil is dark brown in colour.

Adult weevils chew the anther filament. Eggs the flowers and larva feed on the spent flowers. Life cycle is completed within 11 to 13 days. Males live longer than females. The activity of the insects is in accordance with the receptivity of the male and female inflorescence. It was roughly estimated that 40 palms in a grove might be the minimum to sustain a sufficiently high continuous population of pollinators to pollinate all receptive female inflorescence. The weevils carry maximum pollen during the third day of anthesis. Antennae, rostrum, thorax, legs, *etc.*, are the main sites of pollen load. *E. kamerunicus* has a fairly good searching ability. It can survive in dry as well as in wet seasons. Its host range is limited to the genus Elaeis and even at this level, it cannot survive for more than two or three generations, on the most closely related species like *Elaeis oelifera*.

For introduction, male flowers having the weevils are cut from the palms and are transferred to the plantation where one wishes to introduce. It has to be made sure that they are not carrying any plant pathogens to other areas/countries while transferring.

6.4.7. Fruit

The fruit of oil palm is a sessile drupe varying in shape from spherical to ovoid or elongated. It varies in length from 2-5 cm and in weight from 3 to 30 g. The pericarp of the fruit consists of the outer exocarp or skin, oil-bearing mesocarp and the hard stony endocarp or shell. The endocarp together with the kernel inside forms the seed. The common type of fruit is deep violet to black at the apex and colourless at the base before ripening. Such fruits are called nigrescens. Another type of fruit is green before ripening and this is called virescens. Depending on the presence or absence of shell and other fruit characters, three fruit forms have been described in oil palm (Figure 6.7). They are:

- ★ **Dura:** The fruits are characterized by a very thick shell varying from 2 to 8 mm. This results in a low mesocarp percentage in fruits.

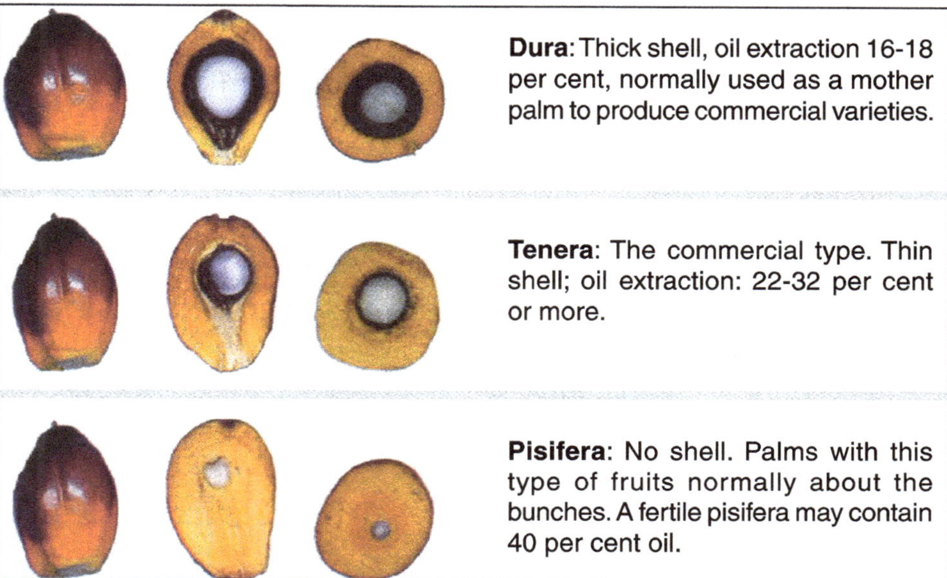

Figure 6.7: Description of Dura, Pisifera and Tenera Fruits.

* **Pisifera:** This fruit type is characterized by the absence of shell, the vestiges of the shell being represented by a ring of fibres around the kernel. They may contain embryo (fertile) or may be embryoless (sterile). The latter form is preferred as pollen parents.

Figure 6.8: Various Segments of Fruit Bunches.

- **Tenera:** The shell is thin (0.5 to 4 mm) with a characteristic fibre ring around the shell. Tenera is the hybrid between dura and pisifera and is the only commercially cultivated form.

The various segments of fruit bunch are shown in Figure 6.8.

Fruit bunch is an economically important component of oil palm tree which yields two types of oil namely crude palm oil and palm kernel oil and various byproducts for utility.

Chapter 7

Crop Improvement

Crop improvement is an important programme to improve the productivity, increase production and impart desirable qualities of biotic and abiotic stresses. Once the basic material of germplasm is assembled, established gene banks, evaluated and catalogued, these materials can be used for crop improvement.

7.1. Oil Palm Breeding

The ultimate goal of any plant breeding programme is the production of best varieties to satisfy the requirements of the consumers. In other words, a breeding programme is a dynamic and continuous release of superior planting material for optimization of production and quality of the end products. High extraction rates at no extra harvesting cost of the plantations, high early yields for early cash flow returns and yield stability for optimization of infrastructure should be aimed in designing breeding programme. Oil palm breeding is a lengthy process requiring large space, time and money. It takes some 20 years to develop progeny tested planting materials.

Plant breeding exploits the natural variability of the species for the benefit of the mankind. Breeders pursue the improvement of plant attributes through the application of genetic concepts during the process of selection for superior individuals, which will be finally reproduced in commercial scale.

Oil palm being a perennial crop, it is important that quality seeds are produced. This is possible only through strong breeding programme by selecting desirable and compatible parental palms, *viz.*, duras and pisiferas and the subsequent production of commercially cultivated tenera hybrids by crossing the two.

Oil palm breeding is a vast area involving genetic resources, methods and concepts of selection and breeding, varietal development, and hence it cannot be detailed in few pages. Therefore, this chapter is only a brief reference of the role of breeding in the development of high yielding oil palm planting material. It includes the concepts that every grower should know, in order to understand the importance of using certified seeds as a safeguard for their investments in the oil palm business.

7.1.1. Reproductive Biology

Oil-palm being monoecious, produces separate male and female inflorescences on the same palm. Although monoecious, it is functionally cross-pollinated due to the alternating cycle of male and female inflorescences. The major pollinating agent is the weevil *Elaeidobious kamerunicus*. The number of spikelets per inflorescence is similar in both genders but the number of male and female flowers per spikelet is highly variable. *i.e.* 700-1200 in males and 5-30 in females. The flowers are bisexual in origin, but in males, the stigmas are suppressed while in the female flowers, the stamens are undeveloped.

A large quantity of pollen (each 30g) is produced by an average size male inflorescence. Most pollen grains are shed in the first two days following anthesis and cease production within five days. At anthesis female flower is receptive for about 36 to 48 hours. Individual flowers within an inflorescence anthesis sporadically, with interval of up to four weeks between the initial and the last flushes. The second day of anthesis is most suitable for pollination when most of the flowers are receptive.

7.1.2. Genetics of *Dura*, *Tenera* and *Pisifera*

The oil palm is an exceptional example where selection progress is obtained through a single gene. This gene (Sh) controls shell thickness of the fruit:

Dura palms have kernels with a thick shell; pisifera palms have kernels with no shell and tenera palms have kernels with a thin shell.

- ★ Dominant homozygote (Sh+ Sh+) form the thick-shelled *dura* (D)
- ★ Recessive homozygote (Sh- Sh-) form the shell-less *pisifera* (P)
- ★ Heterozygote (Sh+ Sh-) form the thin-shelled *tenera* (T).

The average fruit and bunch composition of the *dura* and *tenera* is shown in Table 7.1.

Table 7.1: Average Fruit Composition of *Dura* and *Tenera*

Sl.No.	*Trait*	*Dura*	*Tenera*
1	Fruit to bunch (per cent)	60	60
2	Mesocarp to fruit (per cent)	60	80
3	Shell to fruit (per cent)	30	10
4	Kernel to fruit (per cent)	10	10
5	Oil to wet mesocarp (per cent)	50	50
6	Oil to bunch (per cent)	18	24

The *pisifera* is generally female sterile (no bunch production), although there are some fertile *pisifera* in the population. The tenera is generated by crossing the *dura* (female) with the *pisifera* (male), *i.e. dura x pisifera,* commonly referred as DxP crosses (Figure 7.1). The DxP progenies produce 30 per cent more oil compared to the *dura*. Seedlings found growing in plantations planted with DxP materials must not be used as planting materials because these are TxT offspring. A TxT cross segregates into *dura, tenera* and *pisifera* (Figure 7.2), consequently producing extremely low yield.

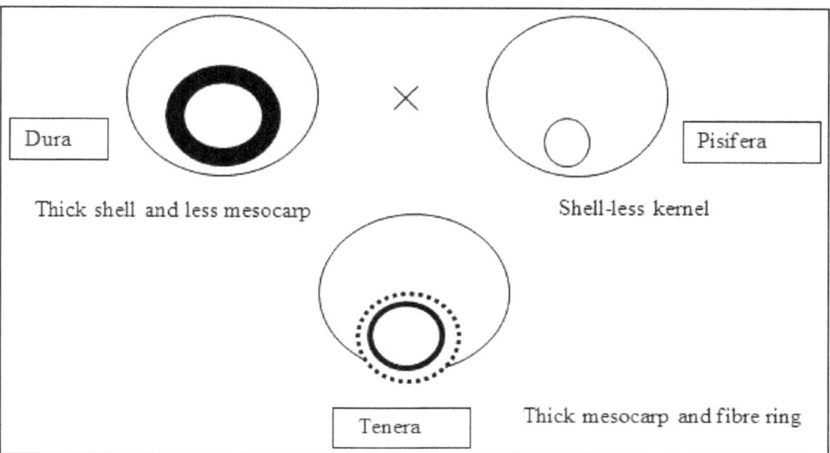

Figure 7.1: Fruit Composition of different Fruit Forms and Inheritance for Shell Thickness in D X P.

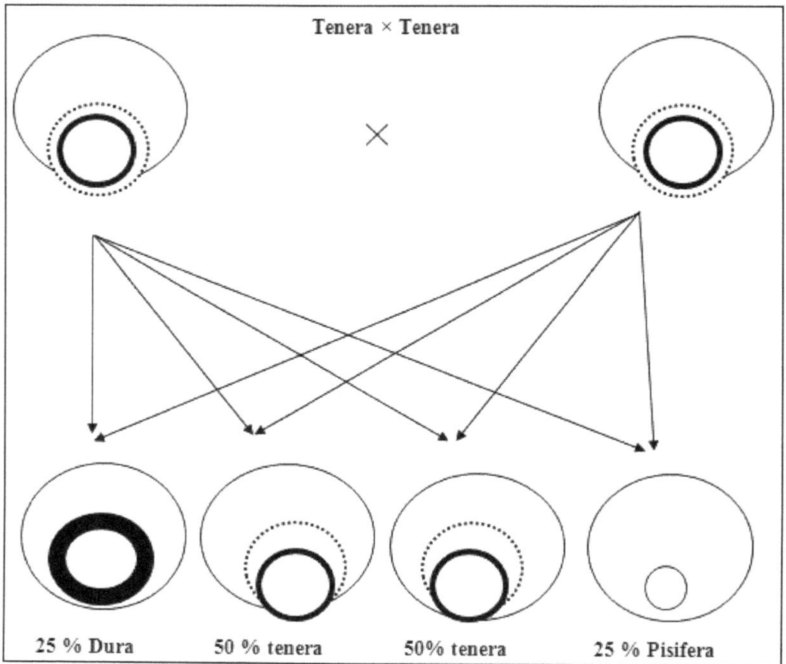

Figure 7.2: Inheritance and Segregation in Tenere x Tenera (T X T) Cross.

The segregation of different fruit-type palms according to the type of cross is illustrated in Table 7.2 without entering into details on the genetics of this segregation.

Table 7.2: Segregation of different Fruit Types

Possible Cross, Controlled or Natural Pollination	*Segregation of Palm Types*		
	Dura	*Tenera*	*Pisifera*
A *dura* crossed with a *pisifera*	none	100 per cent commercial seeds	none
A *dura* crossed with a *tenera*	50 per cent	50 per cent	none
A *tenera* crossed with a *pisifera* or vice versa	none	50 per cent	50 per cent (mostly sterile)
A *tenera* crossed with a *tenera*. Also open-pollinated seeds from a *tenera* commercial plantation	25 per cent	50 per cent	25 per cent (mostly sterile)

The fruit variation is the determinant factor for higher yields. The dura type has less palm oil production potential, because its thick shell reduces the layer of the pulp or mesocarp of their fruits where the palm oil is stored, while the tenera fruits present more pulp and consequently more oil. However, not necessarily all tenera palms will produce more oil; some superior duras may have better fruits than teneras with poor fruit quality. Finally, although the pisifera palm has more pulp in its fruits than the teneras and duras due to the absence of shell in their fruits, it is not planted commercially because normally it is sterile with no seed production.

7.1.3. Fruit Types, Production Components and Bunch Composition

Total amount of oil produced per bunch is determined by number of components as listed below that is generally measured by bunch analysis. The number of bunches harvested and their weights are recorded from individual palm at 7-10 day harvesting interval for 4-6 years. Bunch analysis for the determination of the oil content involves 3-5 bunches per palm. In bunch analysis, the ratios of fruit to bunch (FB), mesocarp to fruit (M/F), oil to wet mesocarp (O/WM), oil to bunch (OB), kernel to bunch (K/B) are estimated. Oil yield is the product of FFB and O/B (Table 7.3).

Palm oil, which is extracted from the fruit bunches constitutes the main useful product of the crop. Achieving high oil content in the fruits and high fresh fruit bunch (FFB) are the main targets of any oil palm breeding programme. Important variation exists between palms, related mainly to the differences in the ability to produce FFB and subsequently palm oil.

Table 7.3: Fruit Types, Production Components and Bunch Composition

F/B	= Fruit/Bunch (per cent)	= [FFWT + PFWT)/SWT] x [(BWT - STKWT)/BWT] × 100
FF/B	= Fertile Fruit/Bunch (per cent)	= [(FFWT/SWT)× ((BWT - STKWT)/BWT)] × 100
P/F	= Parthenocarpic/Fruit (per cent)	= [PFWT/(FFWT + PFWT)] × 100
M/F	= Mesocarp/Fruit (per cent)	= [(FSWT - FNWT)/FSWT] × 100
MC	= Moisture Content (per cent)	= 100 - [(DMWT/WMWT)/(WMWT)] ×100
O/DM	= Oil/Dry Mesocarp (per cent)	= [(DMWT - FWT)/(DMWT] × 100
O/WM	= Oil/Wet Mesocarp (per cent)	= [(DMWT)/(WMWT) × O/DM]/100
O/B	= Oil/Bunch (per cent)	= (F/B x M/F × O/WM)/10,000
O/F	= Oil/Fibre (per cent)	= [(DMWT - FWT)/(FWT)] × 100
K/F	= Kernel/Fruit (per cent)	= (KWT/FSWT) × 100
S/F	= Shell/fruit (per cent)	= [(FNWT – KWT)/FSWT] × 100
K/B	= Kernel/Bunch (per cent)	= (K/F × FF/B)/100
MNW	= Mean Nut Weight (g)	= FNWT/NOFNUT
MFW	= Mean Fruit Weight (g)	= FSWT/NOFNUT
P/B	= Parthernocarpic/Bunch (per cent)	= (P/F × F/B)/100
OY	= Oil yield (kg/p/yr)	= (FFB × O/B)/100
KY	= Kernel yield (kg/p/yr)	= (FFB × K/B)/100
TOT	= Total Oil (kg/p/yr)	= OY + (0.5 × KY)
TEP	= Total economic product (kg/p/yr)	= OY + (0.6 × KY)

BWT: Bunch Weight; SWT: Spikelet Weight; PFWT: Oil-bearing Parthenocarpic Fruit Weight; FSWT: Fruit Sub Sample Weight; NOFNUT FFB: No of Fresh Nut: Fresh Fruit Bunch; STKWT: Stalk Weight; FFWT: Fertile Fruit Weight; ESPKWT: *Empty Spikelet Weight FNWT KWT: Fresh Nut Weight: Kernel Weight

* ESPKWT = Empty Spikelet + infertile fruit (colorless and non-oil bearing).

7.2. Improvement Programme

Oil yield is the prime interest in oil palm cultivation. Kernel yield is a secondary product. The selection of these traits involves evaluation of the FFB yield and oil determination. In improvements programmes. oil palm breeders adopted the reciprocal recurrent selection (RRS) or its modification (modified reciprocal recurrent selection - MRS) as the main procedure (Figure 7.3). Under these schemes, the ***dura*** and ***pisifera*** populations are kept as distinctly separate base populations.

The following criteria should be aimed at for successful breeding: Improvement of base populations for their mutual combining ability, development of Breeding Populations of Restricted Origin (BPRO's) to accumulate favorable genes and introgression of selected BPRO material into proven populations.

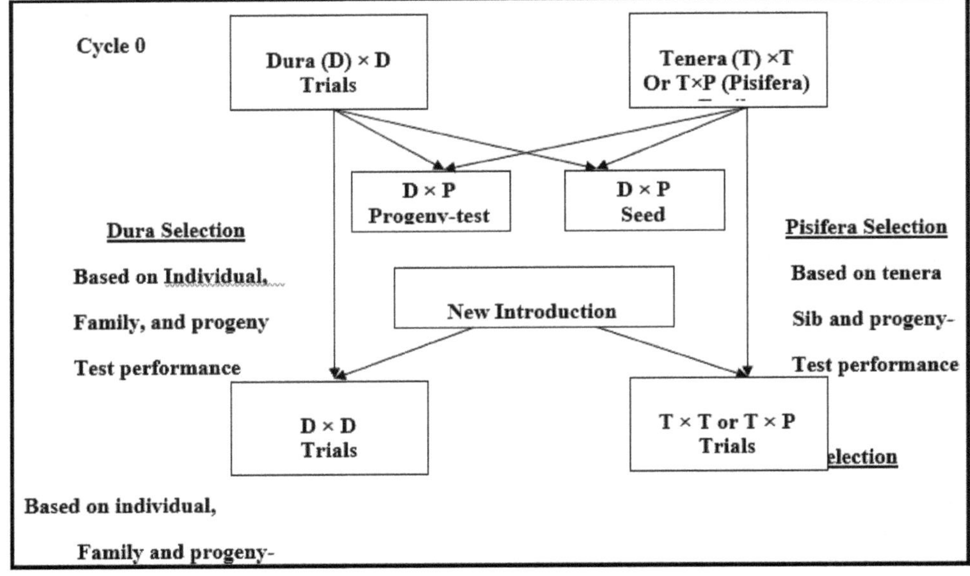

Figure 7.3: Tenera Production.

DxP Seed production involves the following:

- Selection of *dura* parent
- Selection of *pisifera* parent
- Progeny testing, to select *dura* and *pisifera* combination

The *dura* parents for use in seed production are selected on its own merits based on family and individual palm selection. On the other hand, due to the sterile nature of the *pisifera,* its yield potential is determined indirectly through progeny test with elite *duras.*

- The *duras* are selected for high bunch weight and low bunch number.
- The *teneras/pisiferas* are selected for high bunch number and low bunch weight
- *Pisiferas* with a good general combining ability (GCA) are selected based on the *tenera*

In progeny test experiment, DxP or *dura x tenera* (DxT) test, the *pisiferas* that give the best average performance of its half-sib progeny are selected.

7.2.1. Breeding Objectives

Like in any breeding programmes, sufficient genetic variability is a prerequisite for selection of economically important traits in oil palm also. However, the genetic base of oil palm breeding is extremely narrow and is heavily dependent on four Bogor palms for duras and limited teneras as the pisifera source. In this respect

PORIM had mounted numerous expeditions to Africa and South America to collect oil palm germplasm material.

In most crops the selection for yield is the main objective and the oil palm is no exception. In addition to oil yield, breeding for dwarfness and oil quality have assumed more importance. Oil palm breeding is carried out with the following main objectives:

(i) Improvement of oil yield
(ii) Development of dwarf palms
(iii Improvement of oil quality
(iv) Breeding for high bunch index (harvest index), ideo type palms
(v) Breeding for exploitation of GxE interaction and
(vi) Breeding for biotic and abiotic stresses.

The ideal palm breeding programme should aim towards the production of a more efficient and compact palm. Besides breeding for oil yield, new planting materials of specific traits are being developed. This is made possible with the large germplasm collection available in the research institutes.

7.3. Breeding Methods

The primary objective is to evolve genetically superior oil palm genotypes with higher yield. This can be achieved by manipulating any one or more of the factors contributing to yield, *viz.*, number of bunch, bunch weight, and number of fruits per bunch, mesocarp and oil content.

7.3.1. Selection

It is essential to use high-yielding mother palms for hybrid seed production. The characters considered for selection are number of female inflorescence, sex ratio, number of bunches, weight of fresh fruit bunches, percentage of fruit per bunch, fruit weight, percentage of mesocarp/fruit and percentage of oil in the mesocarp. Of these, number of bunches, weight of single fruit and percentage of mesocarp and kernel per fruit are given the maximum weightage because of high heritability.

For selection purposes yields of individual palms are recorded by weighing harvested bunches. Representative bunches are transported to a "bunch analysis laboratory" for determination of the various bunch and fruit components.

Studies done so far indicate that nursery selection is unlikely to result in substantial genetic improvements. The practical solution would, therefore, be to reduce the non-genetic variation to the extent possible and reject weak seedlings when appreciable variation still exists. Generally, there is a culling of 15 per cent each at germination and nursery stage.

7.3.2. Hybridization

Earlier breeding programmes were based on mass selection and family selection taking yield and yield attributes into consideration. There was a major change in the approach since the monogenic inheritance of shell thickness was elucidated.

Since only tenera hybrids are planted commercially the present day breeding programmes are designed to produce the best dura pisifera combinations. A simple procedure will be to test a large number of crosses to find out the best performer and produce seeds of these combinations by re-employing the 'proven parents' for many years till the parental palms are alive.

The present day breeding programmes are the modified version of reciprocal recurrent selection. This aims at identification of suitable dura mother palm and pisifera pollen parents which, when combined would give high yielding teneras. While duras can be assessed based on their own yield, the potential of pisiferas cannot be directly assessed. The worthiness of pisifera can be assessed only based on the performance of the progenies. Normally sterile pisiferas are preferred for hybrid seed production since fertility is normally related to shell thickness. Duras are selected based on a system of family and individual mass selection. Pisifera selection is based on the performance of DXP progenies. At every stage, the performance can be verified and it is possible to introgress genes from any desirable introduction.

The cycle of breeding can be continued bringing in more and more improvement with progressive cycles of breeding. Better results will be obtained if populations of wider origin are used. So also introgression of genes from introductions made from time to time should be advantageous.

7.3.3. Inter Specific Hybridization

American oil palm (*Elaeis oleifera)* is of interest to the breeders because of its dwarfness, oil fluidity and resistance to diseases particularly bud rot as well as vascular wilt and reduced susceptibility to the pest *Coelaenomenodera elaeidis*. Since only dura fruit form is available in *E. oleifera*, it is easily crossed with *E. guineensis* pisifera. The F1 hybrids are not directly exploited because they are low in fruit/bunch ratio, oil percentage and oil to bunch extraction. Undesirable characters of F1 hybrid have been improved through back crossing to *E. guineensis* parents, although it reduces unsaturated fatty acid (UFA) content and increases palm height. It is interesting that back-crossing the hybrid of *E. oleifera* from Surinam origin to *E. guineensis* parents produce shorter palms with compact canopy compared to DXP. Two important characters, *i.e.*, low height increment with compact canopy and high level of UFA content are among the traits sought for future planting material.

7.3.4. Molecular Breeding in Oil Palm

Until recently, plant breeders have depended on the conventional methods of hybridization and selection to insert new genes into plants. The conventional methods of hybridization and selection has several limitations such as the desirable genes may be linked to undesirable loci and are co-inherited. For example, the oil quality in *Elaeis oleifera* is desirable but linked to low yield trait. Moreover, sorting and selecting genetically suitable new genotype is an extremely slow process. The recombinant DNA technology provides a new avenue to overcome these problems. In addition, it is also possible to incorporate desirable genes from other sources such as micro-organisms, animals and plants. One of the aims of PORIM's oil palm genetic engineering programme is to develop transgenic oil palms to produce oil with iodine value of more than 72 and fatty acid composition with 8-13 per cent or less palmitic acid and 70-80 per cent or more oleic acid. The current palm has 44 per cent palmitic acid (C16:0) and 39 per cent oleic acid (C18:1).

The ability to alter the oil composition of oil palm requires (i) identification of genes of interest (ii) availability of method to introduce genes of interest into the plant cell (iii) understanding of the regulation and expression of the genes of interest and (iv) regeneration or multiplication of transgenic palms.

Molecular markers and genetic mapping are part of the intrusive 'new genetics' that is thrusting its way into all areas of modern biology. The location of markers and the relative distances from one another along each chromosome of an organism represents a genetic map. Genetic mapping provides a means of determining the location(s) of genes influencing a specific traits of interest and, as such, the information needed to implement marker assisted selection (MAS). Molecular breeding is well suited to a perennial crop like oil palm in which the economic products are not produced until several years after planting. The oil palm has a long selection cycle of 10–12 years. As such, achieving genetic gains through conventional breeding can be an extremely slow and tedious process. This fact is further confounded by the fact that oil palm requires large tracts of land for field planting as only about 140–160 palms are planted per hectare. The large land requirement directly translates to huge manpower requirement to manage and conduct breeding trials. This results in breeding trials often becoming very expensive and limiting in terms of the number of crosses that can be evaluated. The use of DNA markers for selection, in a crop like oil palm, could greatly reduce the number of breeding cycles. The use of DNA markers, especially by assessing their allelic variation at agriculturally important loci, can assist in making informed choices on the palms to be selected for crossing and/or subsequent field planting. These are decisions that will help to efficiently utilise the limited resources currently available in Malaysia with regards to land and labour shortage in conducting breeding trials. As such, planting materials can be produced faster and with greater precision. It is also argued that molecular breeding can assist the oil palm achieve its real potential oil yield, which is projected to be about 18 tons oil per hectare per year.

Projects have been designed at MPOB to develop tools and techniques for molecular breeding in oil palm. The first phase of the project was targeted towards the development of molecular markers. Since there was already an 'in-house' collection of ESTs, it was only natural that they be exploited as restriction fragment length polymorphism (RFLP) markers. RFLP markers are co-dominant, robust and easily transferable among different laboratories. The use of ESTs to reveal RFLP offers an added advantage that a map of known genes can be obtained. In the oil palm, the two single gene inherited traits (monogenic inheritance) of importance to plant breeders are shell thickness and fruit colour. A marker linked to the shell thickness character is of significant importance as it would allow breeders to distinguish the dura, tenera and pisifera types at the nursery stage. It would also allow breeders to eliminate dura contamination from the commercial tenera planting materials. As for fruit colour, the virescens (Vir) fruits undergo a more profound colour change on ripening and as such it makes it easier for harvesters to identify ripened bunches.

The identification of the Vir allele and its use in conjunction with the non-abscising genotypes will allow the identification of ripened bunches and hence reduce crop loss through fallen fruits. A progeny derived from the selfing of a tenera palm (Palm T128 from MPOB's Nigerian germplasm collection) and segregating these two monogenic characters was used for linkage map construction. Initially RFLP markers were used in the mapping effort. cDNAs derived from the young etiolated seedlings, mesocarp, kernel and inflorescence were used as probes. Since the oil palm is an out breeding species and a high degree of heterozygosity is expected in its genome, the codominant RFLP markers are expected to segregate in a 1:2:1 ratio in this selfed cross. An initial attempt at map construction using RFLP markers detected 11 linkage groups. The RFLP technique, however, suffers from the disadvantage of being labour-intensive and time consuming. Thus the AFLP technique is also being applied to the oil palm in an attempt to expedite the mapping programme. AFLP markers were scored as dominant and are expected to segregate in a 1:3 ratio in the selfed progeny. Work is also being carried out to develop markers based on simple sequence repetitive (SSR) DNA, or micro-satellites. These markers are also co-dominant and potentially useful for map saturation. Together with RFLP markers, SSR can serve as anchor probes for the development of a consensus map.

Extensive screening of the oil palm genetic collections is being carried out using RFLP and RAPD markers to study the level of variation within and between populations of *E. guineensis* and *E. oleifera*. These populations are screened at the seedling stage in order to optimize field planting of the collections based on the level of variation detected in individual populations.

Breeding for genotypes for higher yield at raised temperature may be a new challenge to oil palm breeders in the 21st century.

Chapter 8

Planting Material Production

Production of quality planting material is very essential for the success of oil palm cultivation. This is being done by setting up of hybrid seed gardens with proven parental palms of dura and pisifera source obtained from selected tenera x tenera materials. Potential dura palms are selected based on the yield and quality characters and used as mother palm source. Similarly ideal pisifera palms are selected, pollens are collected and crossed with duras mother palms to produce hybrid tenera seeds. Clonal propagation is the other efficient way of multiplying the planting material to a larger extent but still there are many shortfalls which need to be overcome. The detailed procedure for establishing seed gardens and production of quality planting materials is discussed in this chapter.

8.1. Hybrid Seed Production

With the understanding of the shell thickness inheritance, Tenera hybrid seeds are being produced by crossing elite dura and pisifera mother palms, after evaluating the performance in the field conditions by setting standards for yield and other bunch heritable characters. For a crop like oil palm with a low density and long standing in the field for more than 25 years, special attention should be paid for producing A grade seeds from highly reputable dura and pisifera source.

It is essential for a seed producer to have proven records for the following characteristics:

- ★ Very low dura contamination obtained from good field supervision and strict controlled pollination techniques.
- ★ Proven field performances in plantations and large-scale progeny trials.

- ★ Identified pedigree of all palms selected for seed production.
- ★ High yields of palm products obtained on plantation scale, with FFB yields in favorable environment over 25 MT/ha of FFB and extraction rates for oil and kernel above 25 per cent and 6 per cent respectively.
- ★ Precocity of the planting material limiting the immaturity periods to a maximum of one year.
- ★ Never use seeds collected from the plantation, as these seeds are T XT seeds giving 25 per cent pisifera and 25 per cent thick-shelled duras.

All over the world annually about 140 million seeds are produced which are not adequate to meet the requirement.

Hybrid seed production in India was started during 1982 using dura and pisifera of Thodupuzha (Kerala), which were probably planted during 1960 by Kerala Agricultural Department. Utilizing those palms, India has been producing 3 lakh sprouts annually from Palode. Indigenous large-scale production is being emphasized to meet the growing demand from the companies involved in the oil palm development.

In view of this, NRC for Oil Palm had been evaluating large germplasm materials collected over a period to have maximum D x P combinations suitable for different regions of identified areas.

Five seed gardens were raised, three in Andhra Pradesh, one each in Karnataka and Kerala. All these seed gardens can produce 10 million seeds though it is not really so which warrants careful management. Some more seed gardens have also come up. The details of seed gardens established and their production of hybrid seeds are presented in Table 8.1.

Under the present situation it is possible to go on slow progress in oil palm development. But if we want to go for a fast development of area expansion it will not be possible to meet the seed requirement locally and there will be need to import from other countries where very good compact and other planting materials are available.

8.1.1. Setting up of Seed Gardens

In setting up of seed gardens the following aspects have to be carefully planned.

a) Selection of *Dura* Parents

It is highly necessary to select suitable parents for obtaining hybrids of high yield potential. For selecting individual palms, it is necessary to analyze the yield components contributing towards bunch production and oil production. The important characters to be studied for the purpose are: number of bunches, weight of bunches and ratios like fruit to bunch, mesocarp to fruit, oil to mesocarp, fruit to oil and shell to fruit, *etc.* Selection standards for these characters could be fixed by considering total requirement of hybrid seeds and the general performance of the dura population over a long period of time.

Table 8.1: The Details of Oil Palm Seed Gardens in India

Location of the Seed Gardens	Year of Planting	Sprout Production (in Lakh)		
		Potential	Present level (2013-14)	Scope for Enhance-ment
A. Seed Gardens				
Indian Institute of Palm Research, Pedavegi (A.P.)	2000	6.0	3.1	2.9
Indian Institute of Oil Palm Research-Research Centre, Palode, (Kerala)	1982	8.0	3.1	4.9
M/s. Navabharath Private Ltd., Lakshmipuram (A.P.)***	1990	6.0	3.8	2.2
Department of Horticulture, Rajahmundry (A.P.)	1992	10.0	4.7	5.3
Oil Palm India Limited, Thodupuzha (Kerala)	1994	11.0	6.0	5.0
Department of Horticulture, Taraka, (Karnataka)	1994	8.0	3.7	4.3
Total		49.0	24.4	24.6
B. New Seed Gardens				
Morumpudi, Department of Horticulture, Rajahmundry (A.P.)	2012	5.0*	**New**	-
Gopannapalem, Department of Horticulture, Rajahmundry (A.P.)	2014	5.0**	**New**	-
Taraka (Taraka-II), Department of Horticulture, Taraka, Karnataka	2012	5.0*	**New**	-
Kabini, Department of Horticulture, Taraka, Karnataka	2012	8.0*	**New**	-
Total		23.0		

* Expected year of initiation of seed production-2020; ** Expected year of initiation of seed production-2020; ***Nava Bharat seed garden is not functioning now

Source: Status Report on Oil Palm, Oilseeds Division, DOAC and FW., Min. of Agriculture,GOI.

b) Selection of *Pisifera* Parents

Pisifera palms are characterized by the absence of shell in their fruits. In a pisifera bunch, different types of fruits are present. Pisifera palms are broadly classified into three classes based on fruit setting. They are fertile, semi-fertile and sterile palms.

Among the three classes of pisiferas, it has been observed that fertile pisiferas transmit thick shell to the progeny and hence are undesirable, while the sterile ones are capable of producing thin shelled tenera progeny and are hence desirable. So before selecting pisiferas, it is necessary to assess their potential by way of careful choice of palms based on fruit setting and progeny testing.

8.1.2 Pollen Collection and Pollination

a) Collection and Storage of *Pisifera* Pollen

Male inflorescence must be bagged five days before the flower opens for isolation. Late isolated inflorescence may have some proportion of early anthesis florets, which may attract the pollinating weevil and result in contamination. At anthesis, the inflorescence will be in the axil of 17-20th leaf from the spindle leaf. The spathes of the inflorescence are cut away by a level head chisel and the spikelets are thoroughly sprayed with a 40 per cent Formaldehyde solution diluted by adding 10 parts of water. This solution kills foreign pollen or insects adhering to the spikelets. Inside the bag also, spraying is done similarly by holding the bag downwards. Before carrying out any work connected with the pollen, the hand should be sterilized with alcohol.

The bags usually used for this purpose are made of terylene or cotton cloth having thick white polythene windows of convenient sizes on both sides to facilitate observation of the inflorescence. The polythene windows should have punch holes, closed with adhesive tapes for inserting pollination tubes. While bagging a thick band of cotton lint measuring 3″-15″ sprayed with Formalin and dusted with any insecticide is placed around the peduncle and the mouth of the bag is tied over the cotton lint with a strong thread. Three pieces of rat baits are placed on the subtending frond and neighboring fronds against rodent damage. Small metal tag is inserted between the spikelets to identify the isolated inflorescence. The bags, if necessary, may be tied at its upper corners to the nearby leaf bases. At the second or third day of opening of the male flowers, the inflorescence is cut and dried in shade along with the bag for five hours in the de-humidifier. The pollen could be collected in the bag by shaking the inflorescence thoroughly. After withdrawing the empty inflorescence, the pollen is dried at 35-40°C in an oven for 24 hours in the bag. The pollen may have to be sieved through a sieving chamber to remove all the extraneous particles and stored in a test-tube without any exposure to the outside atmosphere.

In seed production programs, it may be necessary to store the pollen for days together. Pollen may be stored satisfactorily after reducing its moisture to over 4 per cent by Calcium chloride in a desiccator at ambient temperature for about 6-8 weeks. This period can be prolonged by refrigeration at ±5°C or by storing in a complete vacuum. It has been observed that longer storage of pollen may attract the production of atrophied embryo in the seed, consequent low germination percentage and subsequent seedling abnormalities. Vacuum dried pollen if stored in ampules in deep freezers at -18 to -20°C, the storage period can be considerably extended and could be used as fresh pollen after re-hydration for 24 hours. Viability of pollen is often tested by germinating the pollen grains on a Maltose Agar medium.

b) Pollen Quality

To obtain good quality pollen, the following parameters may be taken into consideration.

i. Pollen with less than 7 per cent moisture content
ii. Pollen capable of producing more than 70 per cent of the germination are considered as of good quality.

c) Preparation of Female Inflorescence for Crossing

Female inflorescence should be bagged at least one week before the first flower is expected to open. The bagged inflorescence has to be observed daily for flower opening. If there are any pre-mature opening of flowers at the time of bagging the inflorescence has to be discarded. All other operations for the preparation of female inflorescence are same as those of the male inflorescence.

d) Pollination

Bagged female inflorescences are constantly watched for flower opening. In the first day of opening, pollination may be started. When pollination has to be carried out the pollen may be placed in a test-tube carrying a cork fitted with two L-shaped glass tubes stoppered with cotton wool. Before pollination, the surface of the bag should be sprayed with Formalin and the hands should be sterilised with Methylated spirit. Pollination is done after opening the glass tubes and inserting one of the tubes into the bag through the hole already provided in the window of the bag and blowing through the other tube. The glass tube is so manipulated as to ensure that pollen reaches all parts of the inflorescence. If the inflorescence is partly in bloom, second pollination will be carried out in the following day (about 2-2½ g of pollen + talc mixture is sufficient for one inflorescence – one part of pollen: 6 parts of talc). The adhesive tape is pasted again and then bag should be shaken well to help the pollen to reach all the female flowers. Pollination is normally carried out in the morning hours. Four weeks after pollination, the bag is removed and a cross plate bearing the pollinators code and date of pollination and cross identity is pierced into the young bunch stalk. The mother palms need special attention before pollination. Old leaf bases, spines and fibers of the leaves around the inflorescence have to be removed while bagging the female inflorescence. Space for the pollinator is made by making light cuts on the lower side of the leaf petiole of the subtending leaf so that it may be gently forced down to make access to the female inflorescence easier.

8.1.3. Preparation of Seed

The D x P crossed bunches ripen in about 5½ months after anthesis. On ripening (reddish fruit but not yet fully ripened) the bunch is harvested after covering it with a gunny bag to prevent any loss of loosened fruits and also to prevent damage to the fruits. Later the fruit bunch is taken out of the gunny bag and the spikelets with the fruits are cut and separated from the peduncle. Separation of the seed is done either manually or using mechanical depericarpers.

a) Depericarping

Removal of exocarp and mesocarp from the fruit is called depericarping. The spikelets with the fruits along with the loose fruits, if any, are kept as such for two to three days. By this time, it becomes easy to separate the fruit from the spikelet. Depericarping can be carried out by the following methods:

(i) Scraping off the mesocarp with knife
(ii) Retting in water
(iii) Rotary screen depericarper
(iv) Vertical drum type depericarper

The first method results in less damage to the seed but is impracticable for handling large quantities. In the second method, fruits are hand picked and kept in a gunny bag. Separate bags may be used to accommodate individual crosses. The gunny bags containing fruits are then immersed in running or stagnant water for retting and in about ten days time the mesocarp becomes very soft due to microbial action. Retting can also be done in running water but there is no difference in germination. Every day change of water is necessary if the gunny bags are immersed in stagnant water to avoid the foul smell on retting. Retted fruits are then depulped either by pounding the fruits with sand or by pressing under feet on a hard-floor along with proper watering. The de-pulped seeds are then washed repeatedly until clean seeds are obtained. Care should be taken to avoid trace of mesocarp or oil adhering to the seeds. With dura fruits clean product can be obtained whereas in tenera some quantity of fibre remains adhered to the smaller seeds. Mechanical depulpers are also employed in large seed production centres.

Mechanical depulpers like rotary screen depericarper and drum type depericarper are available in large-scale seed production centres. The Rotary screen depericarper consists of hexagonal cages revolving at 30 rpm. Water is fed to the cage to keep the fruit wet. The mesocarp is removed due to the continuous thrashing of fruits against the wall and the inside shaft while rotation. The fibre removed from the seed is hosed off by applying water at high pressure from the top pipe.

ASD, Costa Rica has fabricated an improved drum type depericarper for pulping the seeds very rapidly. It consists of a vertical drum with 2 inlets and 2 outlets. In the bottom a 1 HP motor is attached to drive a steel plate. Inlet is used to apply water continuously, so that it washes off and removes mesocarp and sends out through the lower outlet provision. Holes in the round plate and small gap between the round plate and wall facilitate easy flow of fibrous mesocarp to the lower outlet. Seeds that are trapped inside the round plate can be taken out through the upper outlet. Fruits are pulped by thrashing against drum walls and two-rod projection in the round plate.

It has been experienced at that it requires 4-7 minutes for pulping 25 kg of fruits fermented in water for 12 hours. There was no damage caused to the nuts of fermented fruits and the time taken for pulping was comparatively lesser than non-fermented fruits. In case of non-fermented fruits about 2 per cent of the nuts were damaged and consumed extra time for pulping. By applying water at high pressure fibrous mesocarp material could be pushed out very quickly.

Depulping is followed by seed treatment and drying of seeds to ensure moisture content suitable for storage.

b) Seed Treatment

Application of fungicide or insecticide to disinfect or disinfest the seed from pathogen or insects is called seed treatment. The pathogen can be present on seed coat or on the kernel surface or inside the seeds. Oil palm seed should be properly treated with fungicide (2 per cent Thiram or 0.1 per cent Bavistin) solution after extraction or before storage otherwise diseases like brown germ, *schizophyllum* may affect the seeds during germination.

8.1.4. Germination Techniques

For germinating oil palm seeds, many crude or refined heating techniques are used since many years. Out of these methods, the dry-heat treatment has been widely accepted mainly due to its advantage over the others.

This technique consists of heating the oil palm seeds inside the germinator at 39±1°C a condition in which the seeds are too dry for germination but at the same time, the seeds are fully live. This condition is achieved by soaking the seeds for 5 days with a daily change of water. The seeds are then dried for 24 hours in the shade and spread out in a single layer with periodical turning of the seeds, so as to attain 17-18 per cent moisture, which is found suitable for D X P seeds. The seeds are then placed in polythene bags of about 45 x 35 cm size at the rate of 500 to 700 seeds per bag. The open end of the bag is then secured by rubber bands so as to trap maximum air inside. Such seeds are kept inside the germinator in suitable racks and heated for 80 days continuously at 39±1°C. It is observed that for stored seeds, the duration of heating could be reduced to 40-60 days, depending on the period of storage.

After pre-heating, the seeds are removed from the germinator and they could either be stored or re-soaked outside in ambient temperature for a short period. For germinating the seeds, they are taken out of the polythene bags and soaked again in water for 5 days so as to regain the moisture content of 22 per cent. At this stage, the seeds may be treated with a fungicide as done for seed storage. Then the seeds are dried in shade for about 2 hours to evaporate the surface moisture. The seeds are then returned to the polythene bags and tied with rubber band with equal volume of air to that of the seeds inside. These bags are then kept in a cool place on racks of convenient sizes for germination. One week after placing, germination occurs and germinated seeds are picked out and the remaining seeds are put back into the bags and kept at ambient temperature in the room. After second week of germination, the remaining seeds are discarded since late germinates are known to give rise to poor, weak or abnormal seedlings. The germinated seeds are sorted to cull poor, abnormal, root retarded, weak plumule and brown germ (diseased) seedlings. There may be a delay of 10 days and germination starts rapidly and would complete in 10-15 days. If all the operations right from harvest to final drying are carried out correctly a germination percentage of 80 to 100 is not uncommon.

8.1.5. Packing and Dispatch

Pre-heated seeds or germinated seeds called sprouts can be transported to long distance after proper packing. Pre-heated seeds can be packed in polybags of convenient sizes and numbers but care should be taken to see that no moisture is lost from the seeds during transit. The moisture content can be adjusted at the destination and germination could be achieved.

The germinating seeds after the fourth day of germination can be dispatched to distant places if they are packed properly. Since they are very fragile, the sprouts are taken in poly bags of convenient sizes along with sufficient quantity of moistened sponge (PU foam) pieces and packed properly with provision for aeration. These sprout packets are then placed in cartons or wooden boxes depending on the type of transportation. Such sprouts can remain healthy inside the bags for about 10 days without any damage.

8.1.6. Seed Requirement Calculation

The planting density recommended for Dami DxP material is between 120 and 135 palms per hectare.

For each hectare, the seed requirement is calculated on the following basis:

- ★ Target Stand per hectare (example): 135 palms/hectare
- ★ Average culling losses: 25 per cent
- ★ Seed requirement: $\frac{135\times100}{75} = 180$ seeds

It is also recommended to include an extra 10 seeds per hectare to provide APM for 5 per cent replacements in the first year (see Nursery Preparation).

Total order per hectare = 180 + 10 = 190 seeds

A generous extra 5 per cent free allowance is to be added to each order to compensate for heavier losses than expected in transport and handling. Normally all seed suppliers will agree to give this margin when orders are placed with the seed suppliers.

8.2. Clonal Propagation

Currently oil palm planting materials are solely D P seeds. Being heterozygous, the progenies show a considerable level of variability. Oil palm is a perennial monocotyledonous plant and has a single terminal growing point. It does not produce suckers or offshoots like the date palm or some rattan species. Thus, it is impossible to propagate oil palm by vegetative method via conventional horticultural methods. With cloning, palms with desirable traits can be multiplied rapidly to increase production. Ortets are selected for oil yield and other characteristics such as low height increment or high iodine value (I.V). It is speculated that yield could be increased by more than 30 per cent using clones of the elite palms in the best family.

A number of tissue culture laboratories were established in various countries throughout the world especially in Malaysia, Indonesia, Costa Rica, IRHO, France, Thailand to propagate high yielding oil palms. The oil palm tissue culture process at the laboratory scale involves ortlet selection in the field, sampling of explants, callus initiation and multiplication, embryogenesis, shoot regeneration, rooting and hardening of ramets and field evaluation of clones. In 1985, the abnormal flowering behaviour in clonal palms was highlighted. Since then the technology refinement is going on and considerable success has been achieved to reduce the abnormalities. However, production cost of the individual plantlets is too much, and needs to be compared with the conventional sexual production. No doubt tissue culture technique is extremely useful to propagate parental - dura and pisifera - palms to produce biclonal seeds. Extensive research is being undertaken at various laboratories to overcome the abnormal problem especially using the molecular markers.

8.2.1. *In vitro* Propagation

The oil palm tissue culture process comprises several stages (Figure 8.1). Those are:

1. Sampling of explants from selected ortets,
2. Callus initiation in explants,
3. Embryoid formation (embryogenesis) in callus,
4. Embryoid maturation and multiplication,
5. Shoot regeneration,
6. Rooting, and
7. Plantlet transplanting to *ex vitro* conditions (Figure 8.1).

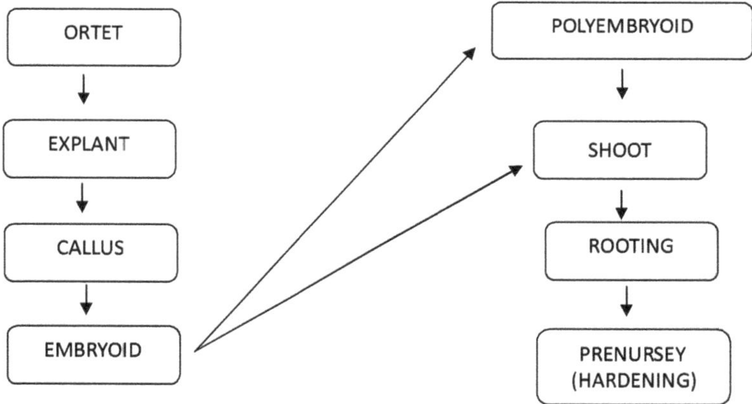

Figure 8.1: Oil Palm *In vitro* Process.

Ortet Selection for cloning is done with the best individuals within the best families, which have highly heritable traits. It is possible to initiate oil palm clones from young leaf, young inflorescence and root explants. Almost all ortets can produce callus, but the rate of explants producing calli differs significantly between clones. Callus initiates in the meristematic regions of the explants that are exposed to the culture media and also in floral and root primordia. An ideal ortet is one with the maximum number of desirable heritable traits, such as oil yield, low height increment, high oil quality and resistance to disease such as Ganoderma. Oil yield is derived from the oil to bunch (O/B) or oil extraction rate (OER) and yield (fresh fruit bunch weight, or FFB). Currently, ortets with at least 27 per cent of O/B and with a FFB yield of >200kg/palm/year are selected. New breeding materials with low height increment of only 20-25cm compared with 45-75cm of current commercial DxP planting materials are also selected for cloning. It is also worth considering selection for oil with high iodine value (I.V) because this trait appears to be heritable.

8.2.2. Types of Explants

It is possible to initiate oil palm clones from young leaf, young inflorescence and root explants. Both leaf and inflorescence explants do not require surface-sterilization since they are already surface-sterile. The young leaves are protected by the older leaf petioles while the inflorescences by two layers of spathes. Roots, on the other hand, are obtained from the 406 palm base. They have to be cleaned thoroughly and surface-sterilized, before they can be put to culture. There are advantages and disadvantages in using different explants.

8.2.3. Callus Induction

Almost all ortets can produce callus, but the rate of explants producing calli differs significantly between clones. They are more genotype dependent. Explants cultures are maintained on nutrient media such as Murashige and Skoog that had been modified to suit the cultures. For callus induction, it is common to incorporate auxins (plant growth regulators) such as α-Naphthaleneacetic acid (NAA) or 2,4-Dichlorophenoxyacetic acid (2,4-D) or both (Wong *et al.*, 1996). Adding cytokinins to the culture media supplemented with auxins seemed to suppress callusing. Callus initiates in the meristematic regions of the explants that are exposed to the culture media and also in floral and root primordia.

8.2.4. Embryogenesis

The rate of embryogenesis is very low, < 6 per cent depending on the genotype. Somatic embryos are organized structures produced from somatic cells (asexually). They can be discerned as pearly white opaque structures occurring sporadically in callus cultures, that have been incubated for 24 hr in the dark. Due to their asynchronous development, the embryoids may appear morphologically different. The common phenotypes obtained are nodular, torpedo-shaped, heart-shaped and globular.

8.2.5. Polyembryoid Cultures

Among the various types of embryoids, the nodular type is the most prolific. The embryoids expand and produce more compact clumpy, nodular structures called polyembryogenic (PE) cultures. The embryoid clumps are further subdivided into 1.5-2 cm diameters and subcultured repeatedly for longer periods. The PE cultures are maintained on hormone-free nutrient medium and subcultured at two monthly intervals. Currently the subculture cycles of PE have been limited up to 15 (about 30 months).

8.2.6. Shoot Regeneration and Rooting

For shoot induction, the PE cultures are subdivided into smaller clumps of about 0.5 cm diameter and cultured on basal medium for three months or more. 'Normal-looking' shoots are transferred to shoot development medium containing low level of NAA, while distinct abnormal-looking shoots are discarded. Shoots reaching a height of about 5cm are transferred individually to liquid rooting medium. Double layer technique involves solid medium with the liquid rooting medium. Using this technique, it is possible to increase the workers' productivity 18 times and to reduce the cost of rooting by about 94 per cent as compared to the previous rooting method. The average rooting success is about 85 per cent. Attempts at improving the rate of rooting are still going on.

8.2.7. Pre-nursery and Field Nursery

Individually rooted shoots are transferred directly to small polybags containing free draining sand and soil (1:1) mixture. Plantlets rooted by DL method are first hardened *in vitro* under shade for 5-7 days before transplanting. After hardening, the plantlets are separated under water followed by treatment with fungicide before they are transplanted individually into polybags as above. The plantlets are maintained under net with 80 per cent shade. One week after transplanting, a complete foliar fertilizer is sprayed on to the plantlets weekly. After 3-4 months, the plantlets are transferred to an open area or field nursery. In the nursery, a compound fertilizer is applied fortnightly. The plantlets are watered at least twice daily in pre- and field nurseries. Another method of hardening 409 plantlets is by planting them in sand beds. This method is specially developed to deliver plantlets to the various places as bare-rooted plantlets.

8.3. Planting Materials

There are four types of planting materials that are available namely, D XP hybride known as tenaras, Inter specific crosses *E. guineensis x E. olifera*, Semi and Bi clonal seeds and oil palm clones (Tissue culture plants). The bulk of the area is however, under tenera hybrid seeds.

8.3.1. Identification and Procurement

Initially Government of India appointed a four member Inter Ministerial Committee under the chairmanship of Dr. K.L. Chadha the then DDG to visit Costa Rica and later to Thailand, Malaysia, Indonesia, Papua New Guinea for identifying and recommending the best planting material suitable for India. Based on the recommendations the first consignment was received from ASD, Costa Rica. Subsequently planting materials were imported from IRHO, France, Papua New Guinea, Cote de'Ivoire *etc.* Recently Malaysia had also supplied one consignment. The seed gardens established at Thodupuzha and Palode could supply a small quantity of tenera seeds. Subsequently, four seed gardens were established in the states of Kerala, Andhra Pradesh and Karnataka with the parental plating materials obtained from Palode and Thodupuzha. One private sector seed garden was established at Lakshmipuram, West Godavari District of Andhra Pradesh with the same indigenous parental materials obtained from Palode. AII the seed gardens are producing tenera seeds. All the seed garden established with imported duras from Coasta Rica has also come to seed production. Presently 2.3 million seeds are produced locally and the rest imported from identified sources.

8.3.2. Commercial Cultivated Varieties

Tenera is the ruling hybrid grown all over the world. It is a cross between thick-shelled dura and shell-less pisifera. Tenera has thin shell, medium to high mesocarp content and high oil content.

Seed quality is of paramount importance for any Oil palm project. Only seeds from a recognized source should be used, which must be accompanied of both a phytosanitary certificate and another that guaranties its genetic purity. The origin of such seeds is necessarily a reputable breeding programme. It is certainly a false economy to obtain "cheap seeds" from an unknown source: The outcome can be disastrous when actual yields will be well below expectations in the original planning investment.

A breeding program only reproduces varieties resulting from the best combinations between selected mother palms and pollen source. These palms are selected for their ability to produce oil. Crosses are made artificially (controlled pollination) in order to guarantee genetic purity, so the whole resulting population (variety) will be teneras with a high potential to produce oil. The use of seeds from an unknown source may contain a high proportion of seeds that will originate palms of the dura and pisifera type, which will decrease the potential of the plantation to produce oil.

Commercial varieties are normally identified with a Ktwo – word name: the first indicates the origin of mother palm and the second indicates the source of pollen. As an example, the variety Deli x Ghana is produced by crossing Deli *duras* mother plant with pollen source originated from Ghana. In general the names used derivate from country, area, locality or institution that collected and selected the material. Some of the most common varieties available in the international market are the following (most characteristics indicated were observed in the south Pacific of Costa Rica in palms 8 years old).

i) Deli X AVROS

The mother palms deli originated from seeds planted as ornamentals along an avenue in Sumatra, Indonesia. The source of pollen comes from the breeding program of a private Dutch company based in Indonesia. This is an excellent variety but very sensitive to marginal conditions and poor management. The plant is vigorous, with fast stem growth rate > 70 cm/year, large bunches (>15kg) and fruits (>11g) and 27 per cent oil to bunch ratio.

ii) Deli x Ekona

The source of pollen originated in Lobe Cameroon. Stem growth moderate, medium bunches (<15kg) and fruits (<9 kg) and 28 per cent oil to bunch ratio and it presents moderate tolerance to drought, low temperatures and low solar radiations.

iii) Deli x Ghana

The pollen source came from the former "Nigerian institute For Oil Palm Research, (NIFOR)" and was later introduced to the experimental station in Kade, Ghana. Stem growth rate is slower (<60cm/year), leaves are short, orange petiols, medium size bunches(<15kg), and fruit (10g) and an excellent oil to bunch ratio (>28 per cent). Besides, this variety has good tolerance to low temperature and solar radiation.

iv) Deli x La Mé

The pollen came from Ivory Coast at the former "Institute de Recherches pour les Huiles et Oleagineux" (IRHO). Stem growth rate is slow (<60 cm/year), small bunches (<13 kg), with long thorns particularly toward its tip. Fruitlets are small (<9 g), some of which may develop over the surface. The inorescence peduncle is rather long. Oil to bunch is about 25 per cent. This variety also shows tolerance to drought in low lands.

v) Deli x Nigeria

The pollen source originated in Nigeria (NIFOR) and was further improved at the experimental station in Kade, Ghana. Stem growth rate is slow (<60 cm/year), medium size bunches (<15 kg), and fruits (10 g), and a high oil content: >28 per cent. It also shows good tolerance to drought. Within this variety two types of seeds can be obtained: those that will originate palms with black (nigrescens) bunches, or those that will produce both nigrescens and virescens bunches (green when unripe: orange when ripe).

vi) Deli x Yangambi

Pollen came from the Democratic Republic of Congo, and from here it was taken to Ivory Coast and other countries. These plants have a vigorous stem growth (>70 cm/year), medium size bunches (<15 kg), large fruitlets (>11 g) and high oil content (27 per cent). Its tolerance to drought is good at low lands.

vii) Tanzania x Ekona

This variety has a marked tolerance to low temperatures in uplands and also strives in under dry conditions. Its stem growth rate is moderate (60-70 cm/year), bunches are medium size (<15 kg), fruitlets small (<9 g) and good bunch oil content (26-28 per cent).

viii) Bamenda x Ekona

Excellent tolerance to low temperatures and drought in upper lands. Low stem growth rate (<60 cm/year), medium size bunches (<15 kg) and small fruitlets (< 9 g). Yield for this and the previous cold-tolerant varieties in marginal areas, and under very basic management has been 60 kg/palm/year at 4.5 years of age (12 liters of oil) increased to 150 kg/palm/year (20-30 l of oil) in six years in upper lands in East Africa.

8.3.3. Compact Varieties

Slow trunk growth and short leaves, as initially conceived when the original compact palm (OCP) was selected in 1970, is less evident after the second cycle of backcrossing to selected guineensis palms of different origins.

The selection of compact palms for seed production shall concentrate in populations F1, F_2, F_3, ... resulting from the recombination of elite palms of the second backcross cycle (BC_2), because of their reduced annual trunk increments and short leaves.

An alternative to produce high-density compact seed varieties shall result from selected BC_2F_1 mother palms crossed to proven 'guineensis' pisiferas, which indeed corresponds to a third backcrossing cycle (BC_3) type of palms. Another possibility is to cross selected Dura palms with pollen from Compact palms. These Compact seed varieties are be planted at 160-170 palms per hectare.

a) Compact x Ghana

This variety comes from the crossing of compact dura mother palms (Originating from the successive backcrossing of an *E. oleifera x E. guineenis* natural hybrid with exceptional characteristics to *E. guineesis* parental lines), with calabar parental lines originating from Nigeria (NIFOR) and introduced to Costa Rica from the Kade Experimental station in Ghana. The palms of this variety have considerably shorter leaves and trunks than standard *E. guineesis* varieties, so they can be planted at 170 palms/ha.

Important characteristics are slow trunk growth (45-50 cm/year), leaf length 6.6-6.9 m, medium bunch weight (18-22 kg), medium fruit size (9-11 g), oil in bunch excellent (28-30 per cent), moderate tolerance to drought, low temperatures and low sun light levels.

b) Deli x Compact

This variety comes from the crossing of compact pisifera palms (originating from the successive backcrossing of an *E.olelifera x E. guineensis* parental lines), with deli dura palms (Bogor, Java). The deli x compact palms have considerably shorter leaves and trunks than standard *E. guineensis* varieties so they can be planted at 170 palms/ha.

Important characteristics are slow growth (45-50 cm/year), leaf length 6.6-6.9 m, medium bunch (18-22 g), medium fruit size (9-11) and oil in the bunch (28-30 per cent). It has moderate tolerate to drought, low temperatures and low sunlight levels.

c) Compact x Nigeria

This variety comes from the crossing of compact dura mother palms (originating from the successive backgrossing of an *E.oleifera* x *E. guneensis* natural hybrid with exceptional characteristics to *E. guineensis* parental lines), with parental lines (pisifera) originating from materials introduced from the Kade experimental station in Ghana that were developed by NIFOR in Nigeria. The palms of this variety have considerably shorter leaves and trunks than the standard *E. guineenis* varieties, so they can be planted at 170 palms/ha. Another distinctive trait of these palms is that they produce *virescens* and *nigrescens* bunches.

Important characteristics are slow trunk growth (45-50 cm/year), leaf length 6.6-6.9 m, medium bunch (18-22 kg), medium fruit size (<9g) and oil in the bunch (28-30 per cent).

8.3.4. Other Promising Varieties ASD

Fran, Omega, Princes, Segio, Sunrise and Tornado are the other high density compact clones.

Yield of fresh fruit bunches in early stage (28-40 months) is 28-30 tons/ha/yr and at mature stage 32-45 MT/har/yr with oil content of 28-32 per cent in the bunch trunk growth rate is <40 cm/yr. It is suitable for high density planting 200 palms/ha.

a) Malaysia

i) Malaysian Palm Oil Board (MPOB)

MPOB has developed 12 varieties as shown below:

- ★ PS1: Dwarf Oil Palm Planting Materials-high yielding and dwarf charecteristics
- ★ PS2: High Iodine Value Oil Palms
- ★ PS3: Large Kernel Oil Palm Breeding Population
- ★ PS4: High Carotine *E. oleifera* Planting Material
- ★ PS5: Thin Shell Oil Palm Breeding Population
- ★ PS6: Large Fruit Oil Palm Breeding Population-duras.
- ★ PS7:High bunch index.
- ★ PS8: High Vitamin E Breeding materials
- ★ PS9: *Bactris gasipas* For Palm Heart
- ★ PS10: Long Stalk Oil Palm Bredding Materials
- ★ PS11: High Carotene *E.guineesis* Breeding Materials
- ★ PS12: High Oleic Acid Oil Palm Breeding Materials
- ★ PS 13:Low lipases

ii) FELDA Planting Materials

Felda DxP Yangambi is promising material with mean oil bunch ratio of 30.2 per cent (equalent to 25.8 per cent oil extraction rate or OER at the converstion rate of 0.855). The palms have high bunch number with medium sized bunches. Bunches are less spiky and fruit size is medium to large. Mesocarp to fruit contents is high. Nut size is moderate, while kernel to fruit content is moderate. Height increment is 52 cm/yr.

iii) United Plantation Berhad

DxP Progeny Performance indicates the first year harvest gives FFB yield of 16.6 tons/yr/ha and increases to 38.4 tons in the seventh year with 26.3 oil/bunched.

iv) Sime Darby Seeds and Agriculturals Services SDN BHD

- ★ Calix 600 is latest breakthrough from more than eight decades of breeding and selection. Bred from elite parental lines of Deli dura and AVROS pisifera,Calix 600 comes with highly advanced characteristics like supeior oil yield, high early yield, thin shell, thick mesocarp, uniformed growth, potential oil yield of nearly 10 MT/ha/yr, faster production of fresh fruit bunch (FFB), with oil-brearing mesocarp producing more oil per fruit and more efficient field operations.

- GH 500 Series DxP oil palms Planting Materials derived through crossing of Elite Deli Duras with 2nd generation BM 119Pisiferas. The commercial materials recorded potential yield of 47.65 MT/ha/yr 7 years after planting.
- Guthrie DxP oil planting Materials derived through hybridization of the 6th generation Dura parents with 7th introduction pisifera (URT) parents. Commercial planting Materials has recorded yield of more than 35 MT/ha/yr at prime age.
- HRU DxP oil palm planting Materials derived from the synthesis of HRU seeds are attributed to years of extensive evaluation from HRU oil palm breeding programme.
- Applied Agricultural Resources Sdn.Bhd. AA hybrid IS AA yields 22 per cent more oil than our previous DxP planting material (The Dumpy-AVROS). AA Dumpy characteristic is bred and fixed in AA Hybrid IS to reduce height increment and thus extend the economic life of the palms. Compactness of the palms allows a higher stand to be planted/ha. Through the utilization of clonal Dura mothers, AA hybrid IS palms are highly uniform both in the nursery and field. AA hybirda IS produces high bunch numbers and moderate size bunches. This gives it a better position to buffer the effect of stress on FFB production. Moderate size bunches translate into high oil extraction rates.

v) Sawit Kinabalu D x P Planting Materials

NPM Planting Materials Sdn Bhd, Jalan Sulaiman, Kuala Lumpur

DxP seeds are produced from Deli dura progenies from known pedigree and established performance and Avros Pisifera pollen of known pedigree and established performance. These hybrids can yield from 24 to 30 MT FFB/ha/yr, with oil to bunch percentage of 25.3 to 26.8 with kernel to bunch 4.8 to 6.7 per cent.

b) Indonesia

i) Sampoerna Agra PT Binasawit Makmur Sumatera salatan Indonaesia

Sawid Kinabalu DxP planting Materials- High oil extraction rate (OER) average height increment, High FFB yield due to high number of average sized bunches.

Important varieties

- Sriwijaya 1 (SJ-1 with Nigeria BSM 32, 33 and 41)
- Sriwijaya 2 (SJ-2 with Ghana BSM 15, 26, 43, 43 and 46)
- Sriwijaya 3 (SJ-3 with Ekona BSM 10, 29 and 56)
- Sriwijaya 4 (SJ-4 with Avros BSM 13)

- ★ Sriwijaya 5(SJ-5 with Dami BSM 17, 18)
- ★ Sriwijaya 6(SJ-6 with Yangambi BSM 28, 39)

c) Papua New Guinea

i) New Britain Palm Oil Limited

Dami Oil Palm D X P Seeds-very high yield, developed from Deli and Avros proven populations obtained from Malaysia in 1968. Precocious, start yielding from 24 months after planting (18 MT/ha) to more than 35 MT/ha/yr after 4years.

d) France: CIRAD–Tree Crops Department

i) Export Service, Montpelliar Cedex5

CIRAD DxP Oil Palm seeds Developed from CIRAD Deli x La Me materials. Yield under normal soil and climatic condition is 28 to 30 MT FFB/ha/yr,24 to 26 per cent actual CPO and 2 to 3 per cent actual PKO under mill extraction rate. Under water deficit of 200 to 250 mm/year the yield level will be 16 to 20 t/ha/year.

e) Thailand

i) Univanich Palm Oil Public Company Ltd., Thailand

Univanich Deli dura (D) and Yangambi pisifera (P) is a promising hybrid. Under irrigated condition it gives about 40 MT FFB and under non-irrigated conditions it yields about 30 MT FFB/ha/year. High oil to bunch ratio up to 29 per cent under irrigated condition and under non irrigated conditions up to 27 per cent has been recorded.

The new development in D XP crosses are-Golden Hope 500 series,FELDA D XP based on Deli x NPM X Yengambi (GMH43) and MPOB D X P crosses.

Estimated global production of Hybrid D X P seeds is 277 million of which the major producers are Indonesia (105 million) and Malaysia(101 million) with small quantities in other countries.

The best DXP hybrid seeds and Compact seeds are available for sales from producing countries. Inter specific hybrids, clonal, bi clonal and semi-clonal planting materials will be available in near future.

Chapter 9

Establishment of Nurseries

The objective of a nursery is to produce quality planting material which will set the basis for a uniform and healthy plantation. The production of good quality planting material completely depends upon careful attention at all stages. Direct planting of germinated seeds is not recommended because of the disadvantages like lack of uniform stand, large area to be managed which may lead to pest and disease incidence and above all proper selection of planting material may not be possible. Traditionally nurseries consisting of bags, baskets or beds used to be prevalent. Seedlings raised in such nurseries have to be packed in special containers to facilitate transport to long distances. Polybag nursery system was considered as most convenient, easy to use and easy to transport and hence, it is now prevalent in all oil palm growing countries.

Oil palm is propagated by seed, using Fl hybrid seed from controlled crosses that produce *tenera* types (dura *x* pisifera) (Figure 9.1). Seed is produced by companies specializing in oil palm breeding.

9.1. Nursery Establishment

9.1.1. Site Selection

The most important criteria about the selection of nursery site is availability of uninterrupted, unpolluted and adequate water supply for raising the seedlings. Area selected must be reasonably flat to ease the placement of the bags and should not be prone to flooding,water logging, shaded by adjoining crops or forest trees, and should be such that transport is easy.

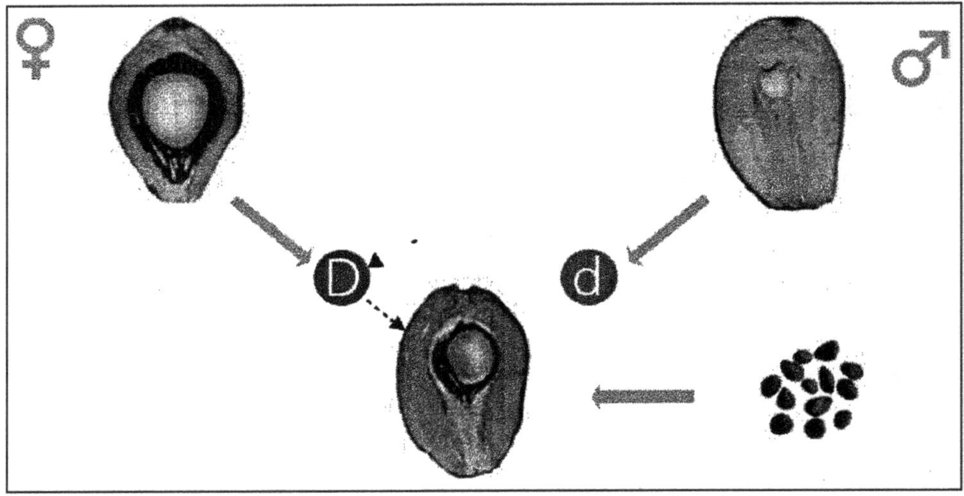

Figure 9.1: DxP Cross.

9.1.2. Choice of Soil

A loamy top soil with high fertility generally gives the best results. It retains water well and also has the correct drainage and aeration qualities important for good root development. A soil which holds together well when the seedling is removed from the pre-nursery bag for transfer to the main nursery is preferred. In India. A mixture of top soil, sand and well decomposed cattle manure in 1:1:1 ratio is commonly used for bag filling. Soil solarisation for a period of 45 days prior to filling helps to have a good soil free of diseases.

9.1.3. Fencing

Fencing is essential for the nursery to protect the seedlings from animal attack. Two major types of fencing are in use, *i.e.*, the conventional barbed wire fence and the electric or solar fence. Generally for protection against cattle and other small animals barbed wire fence is recommended but when there is a range of wild mammalian pests, electrified fence or solar fence is the best form of defence.

9.1.4. Filling of Bags

The bags are filled with potting mixture leaving one cm at the top of the bag. These are well-watered at least one week before they are needed. This gives the soil time to settle naturally and more soil can be added after a few days. After filling, the bags should be well-watered but not water-logged. They should be placed in shade to prevent drying and watered every two days until they are needed. While filling large polybags, normal practice is to quickly turn the bags inside out which has the effect of tucking in the comers and gives filled bag a level.

9.2. Types of Polybag Nurseries

There are two types of poly bag nurseries, namely, single stage and double stage.

9.2.1. Single Stage Nursery

Planting the sprouts directly into large polybags and raising the nursery upto the stage of transplanting is the concept of single stage nursery.

To decide whether to go for single or double stage nursery is a matter of choice. It is very difficult to recommend one system, since both systems have advantages and disadvantages.

a) Advantages of Single Stage Nursery System

i) Transplanting shock is totally absent in single stage nursery system, whereas in double stage nursery system shock is likely to be there at the time of transfer from small polybags to large ones.

ii) At a stretch layout, use of relatively lesser labour force, equipment and inputs over double stage is certainly considered as advantageous.

b) Disadvantages

i) It is necessary to have the full nursery infrastructure ready, large bags filled and irrigation of the full nursery are to be functional right from the initial seed delivery.

ii During the initial critical period, seedlings spread over large area require greater volume of water over pre-nursery system.

iii) Spreading over larger area makes it difficult to make critical observation and supervision, thus incurring more inputs.

iv) The system is unsuitable where availability of land is a constraint.

v) Culling in the last round results in very heavy monetary loss.

9.2.2. Double Stage Nursery

Raising seedlings in beds or small polybags (usually 250 gauge and 23 X 13 cm sized) up to three to five leaf stage and transplanting into bigger size bags (500 gauge or more thickness and preferably 40 x 45 cm in size).

a) Double Stage Nursery Procedure

i) Primary Nursery

Pre-nursery seedlings are grown in small and lesser thick polybag which facilitates easy transportation in large quantities from one place to another. Pre-nursery must be nearer to water source. In practice, beds with a length of 45.73m and 1.22m wide have been found to be convenient. These can accommodate 6,335 seedlings planted in bags of 15cm x 23cm size. Approximately, 13 rows can be held in one bed. Beds should be made parallel by putting flat wooden support along with pegs for anchorage. A path of 0.6m should be maintained between 2 beds to allow workers to carry out various operations (Figure 9.2).

a) Primary Nusery

b) Secondary Nursery

Figure 9.2: Double Stage Nursery.

ii) Secondary Nursery

Since the seedlings have to remain in polybags upto an age of 12-14 months before transplanting, it is necessary to transfer seedlings to larger polybags to provide for bigger root system and support. Normally black polythene bags of 500 gauge with dimensions of 40 x 45 cm size are used. In order to facilitate drainage the bags are provided with perforations in their lower half at an interval of 7.5 cm. The potting mixture used in primary nursery can be used in this case as well. Each bag can carry approximately 16 kg of the mixture. Eight weeks old pre nursery seedlings (five leaf stage) from smaller bags with their ball of soil are transplanted as such into the larger polybags.

Care should to be taken to avoid transplanting shock. Irrigation should be given immediately after transfer to bigger bags (Figures 9.2a and b).

iii) Advantages of Double Stage Nursery

- ☆ Main advantage is less irrigation requirement for very small section of nursery area, thus reducing maintenance cost and conserving water supply.
- ☆ Since all the seedlings are helds in a small section, it is easy to observe critically and cultural operation take less time.
- ☆ Culling is easy and can be done quickly in the first stage resulting in less wastage.

iv) Disadvantages

- ☆ Double operation is required which is labour intensive in the initial stages. Transplanting shock is inevitable in double stage nursery, especially, if transplanted at dry period. However, with proper care and supervision shock could be avoided.

b) Single Stage Nursery Procedure

A modification of polybag nursery is to directly plant seed sprouts in large poly bags with the advantage of eliminating the primary nursery and is known as single stage nursery. It is similar to secondary nursery of double stage except that the sprouts are directly planted in the large polybags.

i) Sizes of Polybags

Black polythene bags, UV sterilized (gauge 500 = 0.12mm) are to be used in the main nursery depending upon the age upto which the seedlings will be retained in the nursery. The size and thickness of polybags used in the main nursery in relation to the age of the seedlings are given in Table 9.1.

Table 9.1: Size and Thickness of Polybags Used in the Main Nursery

Field Planting/Age of Seedling	*Poly Bag Size (Lay flat)*	*Gauge Thickness (mm)*
9-12 months	38 x45 cm	500
12-18 months	45 x60 cm	800
18-24 months	60 x 75 cm	>800

Since most of the nurseries are keeping the seedlings upto 18 months it is necessary to use bigger size polybags of 45x60 cm. Presently only one size 45 cm bag is used for all ages of nursery upto two years or more. This is not advisable. The poly bag quality also should be good. If low quality recycled materials are used for making polybags, it will not withstand the plants even for one year and will get damaged.

So, again replacing the bag costs more. Instead of poly bags, plastic gunny bags are used for fertilizers. These get damaged quickly.

ii) Filling up of Poly Bags

The polybags should be filled with loamy top soil with good water retention but good drainage to avoid waterlogging. The best soils have sufficient structure to remain cohesive at time of transplanting about 16 kg of nursery soil is required for each bag. If only heavier soil is available it should be mixed with core sand to improve its structure. The cattle manure is composted and vermicompost can be mixed with the soil before filling.

iii) Soil Settling

After filling, the bags should be saturated with water every alternate day to ensure adequate soil settling but avoid waterlogging. The bags must be filled up again with the soil at least one week before planting, to 30mm below the top of the ploy bag. To maintain a good soil structure, the soil should be allowed to dry out excessively during settling.

iv) Handling of Seed Boxes

The seeds are normally dispatched in boxes each containing 3000 seeds. Inside the boxes the seeds are packed in bags of 200 seeds each from the same cross. On arrival put the boxes in air conditioned room and don't remove the seeds from the boxes until you are ready for planting. Inspect the seeds on arrival especially if any delay has occurred during transport. If the seeds look dry, open each bag and spray just enough water to humidify the seeds ensuring that excess water does not remain at the bottom of the bag. Observations on the damages and losses must be reported to the seed supplier as soon as possible.

9.3. Planting and After Care

9.3.1. Planting

To ensure successful nursery performance only seeds whose plumules (shoots) and radicles (roots) are fully differentiated should be selected for planting. The optimum stage of seed development for planting is when the radicle is between 5 and 10 mm long. At this stage, the embryo is big enough for the radicle to be identified easily but small enough to be planted without any risk of damage.

While planting, a hole should be made in the soil with a diameter larger than that of the seed to be planted and deep enough for it to be covered with 2 cm of soil. The seed should be placed in the hole gently but firmly ensuring that the axis of the shoot is vertical with the plumule pointing upward. After covering the seed with soil, it should be firmed down gently. Good contact with the soil is important but care should be taken to avoid pushing the sprout into the soil too hard and excessive firming after planting, as both actions could cause damage.

Normally planting of nurseries is initiated with the onset of rainy season. Shallow planting will expose the seeds to temperature or humidity fluctuations, increasing variability in seedling growth. Deep planting will produce etiolation and interfere with-the development of seedlings. Also, seeds planted upside down or sideways will produce seedling with twisted shoots to be culled out later. If planting in the main field is to be taken up during September to November, nurseries should be started between July to September of the previous year. Under our conditions, nurseries can be raised throughout the year in a staggered manner depending upon the availability of the sprouts as well as planting programme. Staggered sowing of nursery will help to avoid overage seedling.

9.3.2. Spacing

Spacing of polybags in the nursery should be done in such a manner that the palm can develop in a most natural way without competition for light from its neighbors (Figure 9.3). The spacing indicated in Table 9.2 will prove very suitable and is practical in the main nursery. If adequate spacing is not given, the seedlings will be lean and lanky.

Table 9.2: Spacing of Polybags in the Nursery

Field Planting/Age of Seedling (months)	*Triangular Polybag Spacing (m)*
09-11	0.75
11-13	0.90
13-18	1.25
18-21	1.50
21-24	1.80

Figure 9.3: Nursery Bags Arranged Over Polysheet.

9.3.3. Shading and Mulching

Since most of the oil palm plantations are coming up in harsher climates, shading may prove beneficial in raising a healthy nursery. In the initial stages, placing palm leaflets about 75 cm long and 7-8 cm wide forming two arches situated in a cross, facing East to West and North to South, respectively may serve the purpose. Shading can be done with coconut leaves, oil palm leaves, *etc.* (Figure 9.4). Netlon shades can also be provided which have the mesh providing 25 per cent shade The hades can be removed after 3 months. This shade also ensures uniform growth and vigour of the seedlings. After removing the shade, mulching can be done with palm shells, rice husk, finely shredded bunch refuse, saw dust, groundnut husk *etc.* which will help in moisture, preventing soil compaction and provide nutrition.

9.3.4. Weeding

Nursery bags should be kept weed free with weeding in regular rounds of manual weedings taking care to avoid damaging the bags. The nursery area also

Figure 9.4: Shading Nursery Bags with Palmyrah Leaves during the Summer.

be kept free of both broad leaves and grass weeds which host insect pests. Regular spraying between the bags with Gramaxin or Roundup should control most of the weed species. But if spray drips fall on the leaves scorching or death of seedling will occur. Better not to resort this method. Spreading polythene sheet on the ground will help to control weeds in the nursery area, particularly in sprinkler or perfo-irrigation systems.

9.3.5. Water Management in Nursery

Oil palm nursery requires regular water supply to ensure good growth and vigor of seedlings. Interruption in water supply even for a week would consequently affect the growth to a great extent. The following methods are employed for watering oil palm nurseries.

a) Use of Hose Pipe and Rose Can

This method is laborious and difficult particularly for larger area. However, to maintain a small nursery, a rose head fitted with 2.5 cm rubber or plastic pipe of required size (70 m) may be used to apply water. It is observed that negligence in doing work properly may lead to exposure of roots, rocking and leaning of seedlings subsequently resulting in stunted growth.

b) Sprinkler System

It is most commonly used system and is suitable for large-scale nurseries. One ha area may require about 15 sprinklers with 6 m coverage (Figure 9.5). However, it depends upon sprinkler nozzle coverage, water pressure, spacing *etc.* Section wise coverage is also possible by moving sprinkler from one section to another. The advantage in sprinkler system is less labour requirement and more uniform water supply while the disadvantage is that larger water droplet size washes away solid fertilizers in the bags and collect soil flakes on their surface. It is also found to encourage weed growth in the inter-space requiring regular weed control.

Figure 9.5: Sprinkler irrigation.

c) Drip Irrigation

In the drip system of irrigation, drippers will be provided for individual bags. It has the advantage of conserving water, maintaining uniform moisture to all plants and also avoid weed growth in the interspaces where the bags are kept (Figures 9.6a and b).

a) Drip lateral on ground and micro tube to discharge water

b) Drip lateral on the top of poly bags

Figure 9.6a and b: Drip Irrigation.

9.4. Fertilizer Application

Oil palm seedlings respond well to fertilizer application. The fertilizer schedule given in Table 9.3 is recommended up to 11 months.

Table 9.3: Fertilizer Application in Oil Palm Nurseries

Age of the Seedling (in months)	*Type of Fertilizer N-p-K*	*Amount (g/plant)*
2	18-46-0	3
3	18-46-0	5
4	18-46-0	8
5	18-46-0	12
6	15 -15-15+$MgSO_4$	20+o.5
7	15 -15 -15+Borax (1 1 per cent)	25+10
8	15 -15-15+ $MgSO_4$	25+10
9	15-15-15+ Borax	30+1.0
10	15 -15-15+$MgSO_4$	35+15
11	15 -15-15+ Borax	40+1.5

(ASP, Costa Rica)

The fertilizers have to be applied 6-8 cm away from seedlings during first application, 10-12 cm away during second and 15-20 cm away during third application. The fertilizer must be incorporated into soil by scratching the surface soil. Watering must be done immediately to avoid loss of fertilizer.

Neem cake solution in prepared by soaknig 1 kg neem cake in 40 litres of water for 2-3 days and the decoction is applied @250 ml per seedling once in 15 days. This helps to keep off the pests and diseases and thus to grow healthier seedlings. There is a possibility of reducing the usage of nitrogen fertilizer considerably. The fertilizer should not be applied when the foliage is wet. Fertilizer in contact with wet foliage results in severe leaf scorching which stresses the seedlings and delays its development.

9.5. Seed and Seedling Handling

Germinated seeds are dispatched as soon as the radicle and plumule can be clearly distinguished and after abnormal seeds have been discarded. By the time the seeds reach the destination, the radicle should be around 10 - 20 m long. This is the appropriate stage for planting.

- ★ Only bring the seed bag to the nursery on the day of planting. During planting period never expose seeds direct to sunlight and keep the seeds in boxes preferably in the shade covered with leaves.
- ★ Remove the seeds progressively from the bags which are ready to plant.
- ★ Avoid planting in very hot period and hot sunny days; Schedule planting to complete it on the same day

9.5.1. Bag Handling

Always lift filled bags with care by holding the bags around its middle. Never grasp by the top of the bag as this often causes tears and distortions, subsequently resulting in adequate watering and fertilization of the seedling. Organize the filling of bags close to the final location and avoid moving the bags once they are filled.

9.5.2. Hardening the Seedlings

In order to avoid root establishment in the ground, hardening has to be done periodically. It is nothing but tilting the bag to half at one time which will break the roots and after 15 days the other half is tilted so that the remaining roots also will be cut off from soil. Final hardening can be done one month before transport of seedlings.

9.5.3. Selection of Seedlings for Planting in the Main Field

Seedling with the following characters are selected in the nursery for the main field planting.

- ★ Seedlings with well-developed root system
- ★ Seedlings which have well bound roots with soil
- ★ Seedlings with fully opened frond spread
- ★ Unetiolated seedlings
- ★ Seedlings with good girth at collor region

a) Culling

Culling is one of the most important operations that is being followed in the nursery to ensure only best seedlings for the main field planting. Different types of abnormal seedlings can be seen in the nursery which have to be removed at every stage of growth of seedlings. It needs lot of experience.

The seedlings with the following characters should be culled out in pre-nursery are:

- ★ Narrow leaf
- ★ Crinkled leaf
- ★ Twisted leaf
- ★ Rolled leaf

The types of seedlings to be culled out in the main nursery include:

- ★ Runts - less vigorous and much smaller
- ★ Upright or sterile
- ★ Flat top
- ★ Limp form - Fronds hanging down instead of being erect
- ★ Seedlings with short wide internodes
- ★ Seedlings with narrow pinnae
- ★ Chimeras – seedlings with chlorotic or white leaf tissue

The various types of abnormal seedlings which need to be culled out are shown in Figure 9.7.

Figure 9.7: Various Symptoms of Abnormal Seedlings.

a) Culling of Seedlings at Preliminary Nursery

b) Culling of Seedlings at Main Nursery

Contd...

Figure 9.7–*Contd....*

b) Splitting and Planting Doubletons

Since oil palm seeds have tricarpellary ovary, occurrence of seedlings that produce two or three shoots is not an unusual thing and they are generally referred to as doubletons. Doubleton separation should be done in the pre-nursery stage using sharp knife or chisel. Two seedlings should be separated in the middle, in such a way that soil and root clods should divide equally since we find active root growth in both sides of the bag and roots should not get damaged. Separated seedlings should be transplanted in the fresh bags by adopting usual transplanting method. The transplanted seedlings must be provided extra care by giving extended water supply.

c) Over-aged Seedlings

The seedlings are generally maintained in the nursery up to an age of 12-15 months and then planted in the main field. But when the planting is delayed and seedlings are further retained in the nursery, those seedlings are termed as "over-aged seedlings". After 12 months in the nursery, pruning is required if the seedlings are overgrown to maintain the height to above 1.5 m from the ground. One or two weeks before planting pruning facilitates handling and helps to minimize the transpiration and transplanting shock. Pruning is also required if planting is delayed in order to avoid etiolation in nursery. At that point of time the mid canopy leaves are also trimmed to ease the movement between rows of seedlings. The common measures that can be followed for maintenance of over aged seedlings are to use a large size polybag and spread with wider spacing than usual. An appropriate spacing which would provide a leaf area index value of around 8 is required for normal growth of seedlings. The bag size and spacing may be proportionately increased according to the requirement of seedling's age as mentioned earlier.

d) Precautions to be taken to Avoid Over-Aged Seedlings

Since it is always better to plant seedlings of normal age, necessary precautions indicated below may be taken to avoid over-aged seedlings in the nursery:

i) Correct assessment of area expansion for the year and also the peak period of planting in that particular locality.

ii) Placing the seed indent according to the requirement in staggered periods so that instead of one time raising of nursery, it can be done at quarterly intervals.

Timely raising and distribution of seedlings will reduce the cost of production of seedlings, manpower as well as space.

9.6. Advanced Nursery Techniques for Primary Nursery

After years of intensive research in plantations on raising oil palm seedlings in the pre nursery G Planter had introduced propagation trays HY PLUG SERIES with unique and revolutionary features available to the oil palm industries for the first time with open base for excellent drainage to establish the root ball. This HYPLUG is having side vent to facilitate air root pruning and thereby promoting robust root system.

Open-design allows excellent drainage/aeration as well as better establishment of root ball. It is injection molded for durability and easy handling, made from recycled material and recyclable, side vents in all cells to enable "air root pruning" which seedling will grow a biologically homogeneous and robust root system, Ribs to direct roots towards holes to give maximum air root pruning. Plants are easy to extract from the trays due to the HYPLUG Tray series with a number of unique and revolutionary features which are space saving (accommodate up to 169 seedlings/m2), durable, cost effective by reducing transportation, weeding and transplanting cost and are user friendly by reducing filling time and easy to extract from cells.

These are also environment friendly eliminating poly bag waste and also last through repeated handlings for more than 20 cycles.

Now-a-days getting soils for filling the nursery bags is costly and also has lot of restrictions due to government regulations.

9.6.1. Filling the Planting Medium into the HYPLUG Trayz

Prepare sufficient planting medium in accordance to your agronomic recommendation. Normal planting medium such as mixture of top soil, sand, compost and POME are suitable for this planting method. Other high quality medium such as coco peat and peat soil are also acceptable.

a) Steps to fill up tray

- Place 24 or 32 cells tray on the flat floor.
- Fill up the tray with medium and level it.
- Lift up the tray to the height of 30-60 cm and drop it on the floor. This is to compact the planting medium. Repeat step 2 and 3 to make sure the medium is compact.
- Level the medium and it is ready for sowing.

b) Sowing the Seed

Make a 5 cm deep hole and place the seed in it. Cover the seed properly and apply sufficient water.

c) Translating

Hyplug tray is designed to raise pre-nursery seedlings for 3 months as standard agronomic practices and then seedlings shall be transferred to poly-bags in the main nursery. Steps for transplanting into the poly-bag are as follows,

- Extract the young seeding from the hyplug tray.
- Drill a hole in the planting medium in the poly-bag.
- Put the seeding in the planting hole and slightly compact the medium

d) Transporting

The beauty of hyplug tray is that these are light weight and easy to transport. With a proper stacking and arrangement, it is possible to transport up to 1400 seeding in a 3-MT lorry.

These HY PLUG trays are now available in the market (Figure 9.8). More than two companies are manufacturing these in Malaysia. It appears some distributors are also available in India. Different specifications are also available (Figure 9.9) and nursery lay out plan is given in Figure 9.10.

9.7. Management of Flood Affected Oil Palm Nursery

Due to calamities like cyclonic rains, floods, *etc.* water logging in the nursery site may happen which is inevitable.

Figure 9.8: HYPLUG Trayz.

Product	Dimension (mm)	Volume/ Cell (ml)	No. of Cell/m²
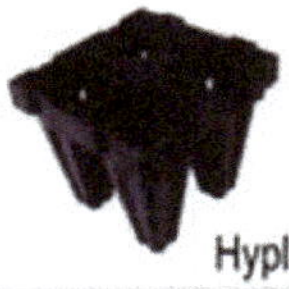 Hyplug-VS	69 x 60 x 120	225	169
 Hyplug-VC	58 x 58 x 120	205	169
 Hyplug-S	48 x 48 x 110	185	169
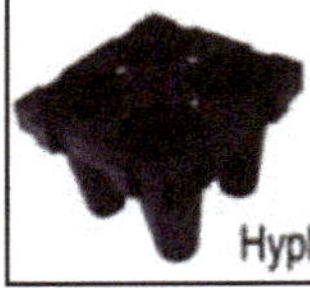 Hyplug-VR	58 (D) x 100	155	169

Figure 9.9: Different Specifications of HYPLUG Trays.

- Drain the water from the nursery site using any type of device like diesel pump or conventional methods on priority basis.
- From bunds around the nursery site to prevent the inflow of water. For farming the bunds sand bags can be used.
- Drain the water standing inside the poly bags and add soil to bags if severely eroded.

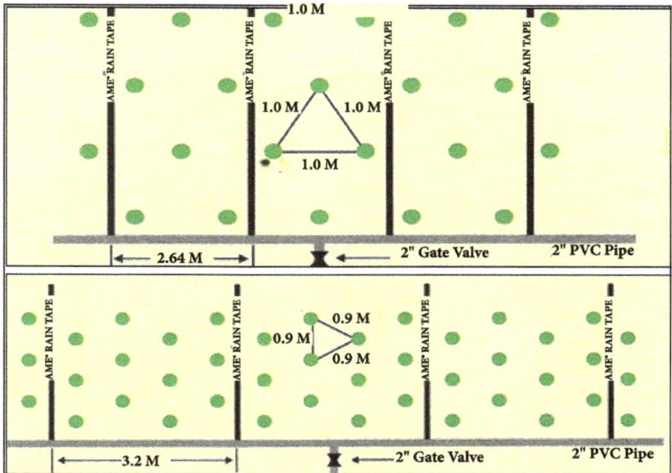

Figure 9.10: Nursery Layout Plan.

- Apply bavistin (1.0 per cent) and monocrotophos (0.036 per cent) sprays at weekly intervals to prevent budrot infection. Drench the plants with dithane M-45 to control leaf spot disease.
- Add neem cake to the bags.
- Apply 20 g of super phosphate and 25 g of Muriate of potash to the individual bag at monthly interval.
- Observe the seedlings carefully for any spread of pest or disease.

Chapter 10

Plantation Establishment and Management

Establishing oil palm plantations in non-traditional areas like India need proper care and management right from selecting areas which have adequate water resources for providing irrigation to the palms for the entire productive life span of about 25 years. Plantation management comprises of field planting and maintaining the plantations with proper input management and after care. Plantation establishment and successful management for getting fairly good yield has also been included in this chapter.

10.1. Plantation Establishment

For successful plantation establishment one has to take care of every aspect of planting procedure starting from field preparation, spacing, lay out, pitting, planting, after care *etc.*

10.1.1. Field Preparation

Field preparation varies with the type of land chosen for planting oil palm. In forest lands, the major activity will be clearing of forests including cutting the existing vegetation, collecting logs and burning the residue. In slope lands, circular platform with a diameter of 3-4 m and 7-8° slope back to the hill side is required. Terracing has to be undertaken in lands where the slope is more than 20°. In low lying areas, where water stagnation is a problem, mounds should be formed for planting. Drainage channels have to be provided in advance where drainage is a problem to avoid stagnation of water in upper layers of soil.

In India mostly plain lands are used for oil palm cultivation which may not require much land preparation. However, when the paddy growing lands are to be planted with oil palm care should be taken to avoid water stagnation. Provision of deep drains and also planting on mounds can be done. Later these mounds can be widened within 2 or 3 years. If contiguous areas are to be taken up with oil palm deep drains alone will be adequate.

10.1.2. Spacing

The optimum planting density for oil palm is the one that gives maximum production from unit area. When density is less than the optimum level, though individual palm yield increases to a certain extent, the total production per unit area will be lesser. Population above optimum level will also lead to reduced production of individual palms due to competition for water, nutrients and sunlight thus reducing the total yield per unit area. Although closer spaced palms gave higher yields initially, as the palms grow, mutual shading affects the yield adversely. Under more favorable conditions a population of 127-135 palms per hectare has been found to be optimum whereas under less favourable conditions as in India, higher densities of 138-150 palms per hectare are recommended. For efficient utilization of solar energy, the rows are to be oriented in the North-South direction. Equilateral triangular system of planting with 9 m spacing between palms will allow each plant to occupy the centre of a hexagon thus allowing better use of the area.

10.1.3. Lay Out

Different methods of marking planting points are discussed below:

i) In flat or gently sloping lands the stakes are placed at the base line 'B' taken in E-W direction at every 46.8 m (space for 7 rows = 7.8 x 6)). Palm rows A_1 and A_2 in N-S direction is marked by putting stakes. The stakes are placed for remaining palm rows in between these two rows by use of a chain on which marks are made at 9 m spacing. The first mark of this chain is placed on first stake of line A_1 and the seventh mark of the chain on the other end is placed on fourth stake of line A_2 (Figure 10.1). Then the stakes are placed on all the marks on chain thus filling the entire area of the block.

ii) Another method is to mark a base line A in N-S direction in one end of a block and put the stakes at 9m spacing on this line (Figure 10.1). From this base line another line B_1 in E-W direction is made with the corner stake on left hand corner remaining the same and stakes are placed at 7.8 m in this line. Similarly, another line B_2 is made after 54 m (6 x 9) from line A. Now a chain is used on which marks are made at every 4.5 m with two different colours so that the distance between similar colored marks is 9m. Thus, the marks on even numbered positions have one color and odd numbered positions have another colour.

   The first (odd) mark is placed on the second stake of B_1 and last mark on second stake of B_2. Then palm positions are marked by placing the stakes on the other colours (even). The chain is removed and placed on 3rd stake of B_1 and B_2 and the positions are marked by putting stakes on other coloured mark (odd) and continued till the block is completed.

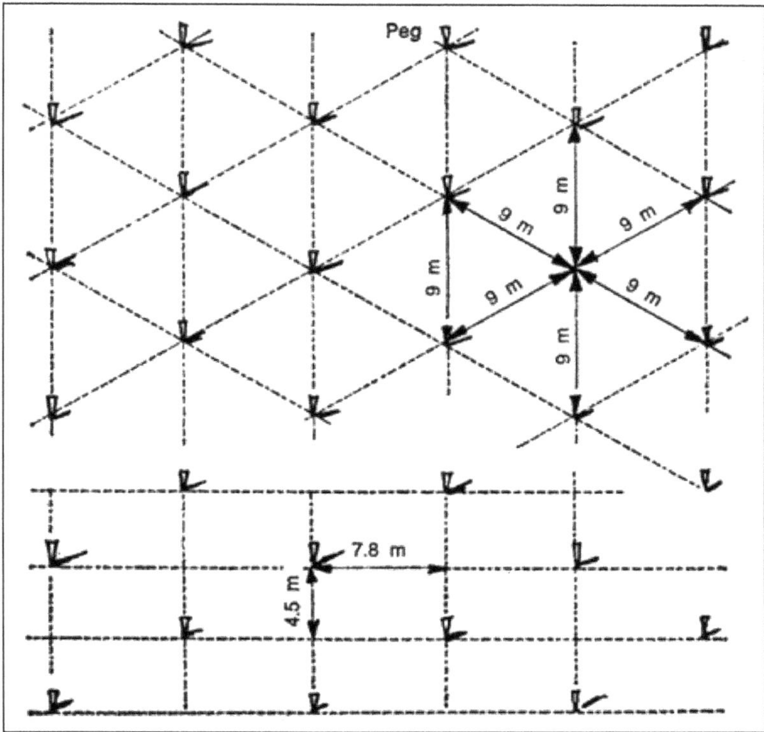

Figure 10.1: Equilateral Triangular System of Planting – Method 'A'

10.1.4. Pitting

The planting pit size should be 60 cm x 60 cm x 60 cm. Pits should be made where the pegs are marked making the peg as the centre of the pit. Pits can be made either manually using a spade or mechanically by using a post hole digger. The post hole digger can be operated by attaching to a tractor. Pits should be taken prior to planting and allowed to season. Any grubs, larvae *etc.* present in the soil will be exposed to hot sun and die. In deep black soils or saline soils it is advisable to take 1m x 1m x 1m pit and fill about 0.40 with a mixture of well decomposed FYM, red earth and top soil. This ensures better establishment (Figure 10.2).

10.1.5. Age of Seedlings

Seedlings of 10-14 months age are ideal for transplanting in the main field. At this stage, a well-developed tenera seedling will have a height of 1-1.3 m from base, good girth at collar and will have more than 13 functional leaves. Seedlings upto 24 months age can also be used for planting but proper care must be taken in the initial stages like daily irrigation, staking *etc.* Being a monocot crop, the establishment is not a problem when aged seedlings beyond 24 months are planted. However, in case of planting such aged seedling, the older leaves will dry off and new leaves will emerge. This is mainly due to transportation and transplanting shocks. In case

a) Tractor mounted Pot hole digger

b) Pit made using pot hole digger

c) A view of the field with pits

Peg

Soil from the top in one heap

Soil from the bottom in another heap

d) Separate heap of top and bottom soil

Figure 10.2: Preparation of Planting Pits.

of *in situ* plantations and regular hardening of seedlings, this problem will not be much. Above all, regular irrigation should be done and the first irrigation should be given immediately after planting.

10.1.6. Time of Planting

Planting in the main field can be done in any season but the best period is during the onset of monsoon so that the seedlings can establish well under favourable conditions. In case of planting during summer season adequate irrigation, mulching and growing cover crops like Sunhemp, *Sesbania, Desmanthes, Desmodium* etc in the basin will help to overcome the summer heat,maintain soil temperature, prevent weed growth and provide optimum micro- climate to the plant. It will also fix atmospheric nitrogen and make the same available to the plant.

10.1.7. Planting Method

Quality seedlings should be transported from the nursery to the planting site only at the time of planting. Seedling lifted and transported to the main field should be planted immediately. Seedling from the nursery should to be lifted once the planting pits are ready (Figure 10.3). Delay of 2 or 3 days in planting has to be

a) Preparing pit for planting

b) Filling pit and poly bag removal

c) Planting seedling

d) Planting, pressing all around and irrigation

Figure 10.3: Method of Planting.

avoided to ensure better establishment and to avoid transplanting shock. Heaping the seedling in the field for many days before planting should be strictly avoided. A mixture of 250g DAP or 250g Rock phosphate and 50g Phorate is applied at the base of the pit and mixed with the soil. The polythene bag is removed from the seedling by making a longitudinal cut and the seedling is put into the pit along with the soil. The gap in the pit is filled with soil and pressed firmly leaving the top portion so that seedling bole will be 2.5 cm below the ground level. Originally, it was suggested to plant at ground level. Since anchorage was not good and palms were likely to fall down during heavy wind slightly deep planting was suggested. Care should be taken to see that the soil does not get accumulated at the crown region, which may lead to rotting. As and when the soil accumulates it should be removed gently without any damage to the collar region.

Immediately after planting, a basin of 1m radius should be formed (Figure 10.4) and copious irrigation should be given. In case there is strong wind, staking may be done and the plant can be loosely tied to the stake at the base. The irrigation channels should not run along the palm rows. It should be on the sides as shown in Figure 10.5.

Figure 10.4: Making One Meter Wide Basins after Planting.

Figure 10.5: Forming Irrigation Channels-Correct Method.

In case of drip irrigation (Figure 10.6) the layout can be made before planting so that the water will be available adequately right from day one after planting.

Figure 10.6: Drip Irrigation.

10.1.8. Special Planting Technique

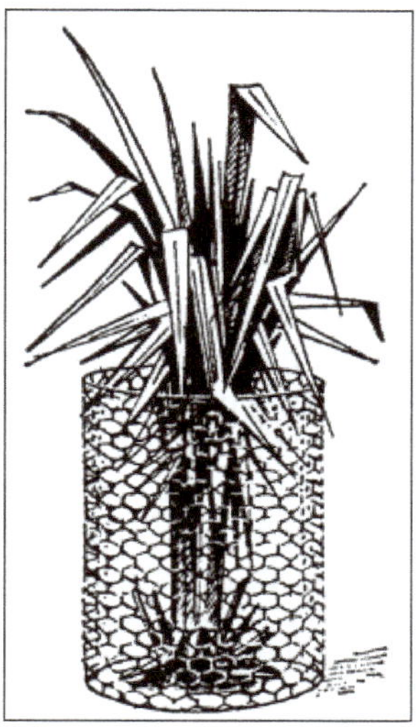

Figure 10.7: Wire Net around the Seedling.

In case of low lying wet land soils planting should be done on raised mounds to avoid water logging and poor aeration. Drainage facilities should also be created before taking up planting. Drainage channels should be dug in between two rows if severe water logging is there. If not drainage channels have to be taken one after three rows depending on the magnitude of water logging. These raised mounds have to be widened every year by one meter width.

10.1.9. Putting Wire Netting around Seedlings

Certain wild animals may eat the young oil palm seedlings. To protect the seedlings, surround them with wire netting (Figure 10.7). Leave the wire netting in place for about 18 months. After putting wire net in place, spread the mulch 20 centimeters thick around the seedlings. This mulch prevents the soil from drying out, and prevents weeds from growing. Use dry herbage, and spread it 15 to 20 cm thick at a distance of 30 to 40 cm around the crown.

10.1.10. Growing Green Manure Crops in Palm Basins

Quick growing sun hemp, *dhaincha*, desmanthes, desmodium, fodder cow pea *etc.* can be grown in the palm basins (Figure 10.8).This will help in maintaining favourable micro climatic conditions around the rhizosphere of oil palm seedlings, prevent weed growth, fix atmospheric nitrogen by forming root nodules in green manure crops and add biomass when it is pulled and ploughed back to soil. The crop will not be affected by sun burn or heavy winds. The sowing of green manure can be done during the first year of planting preferably before summer *i.e* in February/ March and incorporated in the month of June.

a) Sowing *dhaincha* seeds in basins

b) Sowing sun hemp seeds in the basins

Figure 10.8: Growing Green Manure Crop in the Basin of the Seedling.

10.2. After Cultivation

After planting the seedlings in the field, adequate care and management has to be taken right from planting to establish a good plantation and taking the yield within two years and six months.

10.2.1. Pruning of Leaves

Oil palm needs a few after cultivation activities like pruning of leaves, ablation, weeding in the basin and management. Oil palm leaves should not be pruned till it reaches harvesting stage. Maximum number of green leaves should be retained on the palm and as a regular practice all dead and diseased leaves should be pruned. Sometimes during harvesting it may become necessary to prune 1 or 2 leaves so as to gain access to bunches. Severe pruning will adversely affect both growth and yield of palm, cause abortion of female flowers and also reduce the size of the leaves. Palms aged 4-7 years should retain 6-7 leaves per spiral (48-56), those aged 8-14 years 5-6 leaves per spiral (40-49) and those above 15 years should have 4-5 leaves per spiral (32-40). Leaf pruning is carried out by giving a clean cut to the petiole using sharp chisels so that leaf base that is retained on the palm is as short as possible for otherwise it may catch loose fruits, allow growth of epiphytes and the leaf axils form a potential site for pathogens. Pruning is preferably carried out at the end of the rainy season. It is also better to carry it out during the low crop season when labourers are also available. Pruning is confined to only lower senile leaves during initial harvests but when canopy closes in later years, leaves are cut so as to retain two whorls of fronds below the ripe bunch.

10.2.2. Basin Management

Basins should be formed immediately after planting. During first year basins of 1 meter radius are to be taken around the palm ensuring that soil does not accumulate at the collar region. Sowing sun hemp/*dhaincha* in the basin after planting oil palm seedling during summer will help to protect the seedling from hot sun, provide optimum micro climate, prevent weed growth and fix atmospheric nitrogen and biomass which can later be incorporated in the basin itself at the time of flowering of green manure crop. Basins must be widened to 2m radius during second year and 3m radius basins must be maintained from third year onwards. Basin area of oil palm represents its active root zone and hence, it must be kept clean and weed free to avoid competition for nutrients and water. Soil inside the basins should not be disturbed by ploughing or any other inter-cultivation activities which will cut the absorbing roots (Figure 10.9).

10.2.3. Ablation

Ablation is the removal of male and female flowers produced in the early stages of plantation. The bunches produced in the initial years of planting will be very small and have low oil content. Ablation enables the plant to gain adequate stem girth, vigour and develop adequate root system. It improves the drought resistance capabilities of young palms. Flowering starts from 14th to 18th month after planting. Ablation should be started immediately after the appearance of inflorescence on the

a) Power tiller for farming basins

b) Power tiller working in adult plantation

c) Basins formed and manure mixed

d) Basins formed and irrigated

Figure 10.9: Use of Power Tiller for Basin Weeding, Mixing Manure, and Forming Basins.

palms. Such flowers can be removed easily by hand pulling at fortnightly/monthly intervals. Ablation can be extended up to 2½ to 3 years of crop depending upon the plant growth and vigour. Regular harvesting can be commenced after 3 to 3½ years of planting upon the growth and vigour for palms. Sometimes harvest can be taken up after two and half years of planting with proper management right from planting.

10.2.4. Mulching

a) Mulching with Organic Wastes

Mulching of oil palm basins is essential to conserve moisture as well as to control weeds. Mulching can be done with dried leaves, male flowers, coconut husk, empty bunches brought from factory, *etc.* In adult plantations all the cut leaves can be heaped in between two rows of oil palm which can act as a mulch. These mulching materials in addition to conserving moisture, maintain soil temperature, add organic matter and nutrients mainly potassium, and improve physical and biological properties of soil.

b) Plastic Mulches

Plastic mulches are used presently for control of weeds in the palm basins. About one meter radius mulch can be used for the first year and two meter radius mulch can be used for adult palm. FELDA has developed an eco-friendly mulch seen in Figure 10.10. The other plastic mulches commonly available in the market is used for adult palm.

c) Oil Palm Leaves Used as Mulch

In summer, sun hemp can be grown on the palm basin bunds especially in young plantations. This protects the palm from hot winds, provides good micro climate, conserves the moisture and adds organic matter when cut and put in the basins after summer.

In the light soils with less fertility it is suggested to grow one crop of sun hemp *Dhaincha* or *Sesbania* in a year and incorporate it in the soil during the juvenile period of the crop. This helps to increase the organic matter content which in turn increases water holding capacity and improves physical, chemical and biological properties of the soil. This is also a low cost method for adding organic matter to the soil.

d) Empty Bunches Used as Mulch

The empty bunches available after processing are being wasted in many mills and also cause environmental problems and occupy lot of space. These can be profitably used as mulch around the palm basins.

10.2.5. Release of Pollinating Weevils

Since oil palm is highly cross pollinated it is necessary to create conducive atmosphere for pollination. Studies have indicated that wind pollination is inadequate and results in bunch failure. However, after the release of pollinating weevil, *Elaeidobius kamerunicus* the problem of bunch failure has been drastically reduced and yields increased. Hence, it is suggested to release the pollinating weevil from 2 and 3 years after planting depending upon the crop condition. In a well-maintained garden, if the vigour of the palm is good, the weevil can be released at the beginning of third years. If the crop is weak with inadequate growth then the release should be made after 3 years.

Introduction of the weevil in India increased the fruit set from 36.8 per cent to 56 per cent resulting in 40 per cent increase in FFB weight and 11 per cent increase in fruit/bunch ratio. The maximum attainable pollination potential was as much as 70 per cent with 57 per cent increase in FFB weight. Introduction of *E.kamerunicus* into the oil palm plantations at Little Andamans increased FFB weight from 5 to 12 kg.

These weevils are multiplied in the male flowers and carry the pollen to the female flowers. It is very easy to multiply these under natural conditions. Spraying of insecticides will harm the weevil and so care has to be taken not to spray indiscriminately. Severe summer with high temperature, lack of adequate number

a) FELDA Mulch

b) Plastic mulch for adult palm

c) Plastic mulch for young palm

d) Leaf mulch for young palm

e) Dried grasses as mulch

Figure 10.10: Different Types of Mulches.

of male flowers, *etc.* may cause reduction in population. Under such circumstances, it is better to reintroduce the weevil and also induce male flower production in the palms by withholding irrigation to a few palms and pruning leaves, *etc.*

10.3. Weed Management

Weeds are unwanted plants in the cropping ecosystem. In oil palm plantations weeds can be grouped into grasses, sedges, broadleaf weeds, ferns and epiphytes. Weeds affect crop in several ways such as:

- ★ Reducing crop growth and yields through competition for nutrients, sunlight and space
- ★ Necessitating the use of control measures which are costly and at times cause injury to crop plants.
- ★ Interfering with harvesting, crop recovery and other agricultural operations.
- ★ Lowering the quality of crop produce
- ★ Harboring pests and diseases

Weed management serves to counteract these negative effects caused by weeds and favours the oil palm crop to achieve high yields at the lowest possible production cost. Efficient weed management has to strike a compromise between competition, crop loss, cost of control measures and the need to reduce soil erosion.

Weeds in oil palm plantations can be managed by applying several methods as manual or mechanical control, cultural control, and chemical control and also by integrated production system of using livestock in oil palm plantation.

10.3.1. Methods of Weed Management

Weed management in the early stage is very essential. Once the palm grows up and covers the ground exposure area it will be less. Weed management can be done by manual, mechanical, cultural, chemical and biological methods. Growing leguminous cover crops is one of the effective way to prevent weed growth.

a) Legume Covers

One sound method of weed management is the planting of legume covers in the entire plantation. Besides the many agronomic advantages, the planting of legume covers necessitates the complete removal of weeds at initial stage. Pure legume covers when fully established suppress weed growth. The legume species selected are usually a mixture of *Pueraria phaseoloides, Centrosema pubescens, Calapagonium muconoides, C. caeruleum* and *Mucuna cochinensis*. In most cases, pure legume covers or atleast more than 80 per cent coverage is preferred. In addition to weed management, oil palm grown in association with legumes also produced higher FFB yield (Table 10.1).

Table 10.1: Nutrient Content of Important Green Manure and Green Leaf Manure Crops

Crops	*Nutrient Content (Per cent on dry weight basis)*		
	N	*P_2O_5*	*K_2O*
	Green manuring		
Sesbania aculeata	3.3	0.7	1.3
Crotalaria juncea	2.6	0.6	2.0
Sesabania speciosa	2.7	0.5	2.2
Tephrosia purpurea	2.4	0.3	0.8
Phaseolus trilobus	2.1	0.5	-
	Green leaf manure		
Pongamia glabra	3.2	0.3	1.3
Glyricidia maculata	2.9	0.5	2.8
Azadirachta indica	2.8	0.3	0.4
Calotropis gigantea	2.1	0.7	3.6

Figure 10.11: Cover Crops in Oil Palm Plantation.

b) Natural Ground Covers

A suitable and less competitive natural ground cover of selected species of indigenous plants can provide the functions of good ground covers and reduce the competition from weed growth especially the noxious types. Natural ground covers should consist of selected species of indigenous plants or a mixture of *C. caeruleum* + indigenous plants. Amongst the acceptable species of indigenous plants are *Nephrolepis biserrata, Axonopus compressus, Ottochola nodosa* and *Paspalum conjugatum.*

In plantations where legume and natural ground covers are maintained, four types of weeding operations will be necessary as listed below

i) Clean weeding of palm circle to prevent competition from weeds and to facilitate loose fruit collection.

ii) Strip weeding to provide access for harvesting and other field operations.

iii) Selective spot weeding to remove noxious weeds from the legume covers or natural ground covers.

iv) Periodic control of legume covers or natural ground covers, if growth is too vigorous.

c) Manual Weeding

Manual weeding is often used in perennial plantation crops like oil palm as dictated by crop stage, cheap labour supply and type of weeds. But manual weeding using hand tools like spade can affect yield due to over scrapping of top soil and damage to feeder roots. Close cutting of the weeds with mechanical slasher is possible for strips, but difficult and not practiced for circle weeding. The low weed stands will regrow rapidly and compete strongly for nutrients. Moreover, hand weeding is highly time-consuming and in high rainfall tropical climate frequent weeding rounds are required to give the much needed accessibility for field operations.

d) Chemical Weeding

Chemical weeding can play an important role in efficient weed management. It has many advantages like being more cost effective over manual weeding,can be practiced over a large area with reduced labour input, wide range of weedicides to cater for the special needs of many target weeds and for various stages of crop cultivation. For the control of some tough perennial grass weeds, chemical weeding is the only proven cost effective weed control technique. Herbicides such as 2,4-D, 2,4,5T, halogenatedaliphatic acids, dalapan and TCA are found to produce abnormalities in oil palm

Efficient weed management in oil palm plantations at present and in the near future will depend very much on research and development of chemical weeding. The transfer of research findings for commercial adoption into routine spraying practices will result in effective weed management at lower cost.

The research conducted at NRC for Oil Palm indicates that spraying of systemic herbicide Glyphosate @10 ml in 1.5 l. of water per palm basins gives complete weed control. After 60 to 70 days another spray of the same chemical is to be done only on the weeds in the basins. Since the number of weeds will be less the chemical requirement also will be less spraying of this herbicide as and when the weeds are seen, at 3-4 leaves stage or at 60-90 days interval may help to reduce the weeds substantially. No side effect was noticed in the crop if the herbicide spray is done only on the weeds.

e) Plastic Mulching

Use of LLDPE material which is custom based for coconut, oil palm and rubber seedlings with the specification of 1.5m x1.5m and 2mx2m, black cover, UV stabilized is a eco-friendly cultural practice for weed control. It is very easy to install. Once a ground is clear of any debris or sharp stones pull existing weeds and rake the area flat and lay two sheets LLDP material per palm as a mulch. This mulch prevents light penetrating in side and weed growth. The common benefits of polythene mulches are prevention of weed growth at lower cost than normal practice, longer weeding interval, conservation of soil moisture, reduction in the risk of chemical scorching, easy to apply, cost effective and will lost long for two years. The fertilizer can be incorporated before laying out poly mulch.

f) Biological Control

Insects, spores or toxic factors of plant disease can also be used to attack and weaken the weed species. Fungal pathogens used for weed control, require repetitive application, and are referred to as biological herbicides or myco herbicides. Biological weed control in oil palm is very much under-developed. It must be encouraged in view of the long term sustainable benefits and environmental friendly measures.

g) Integrated Weed Management

Integrated weed management includes retention of appropriate inter row vegetation, establishment of leguminous cover crops, zero burning technique, mulching, mechanised application of herbicides, sheep grazing and biological control. The principle of IWM has been accepted but the details of the package have yet to be developed for testing and commercial adoption.

Since the main concern is to keep the basins free of weeds in order to arrest the wastage of irrigation water and fertilizers through weeds, the inter space may be maintained with cover crops or grass cover with mixed farming.

10.3.2. Bio-efficacy and Cost Effectiveness

For efficient weed management, information on bio-efficacy and cost effectiveness is very essential. This can result in efficient weed management with minimum or no crop injury and lowest production cost. Therefore, choice of herbicides, requires information on both technical and economic aspects. Herbicides used singly or in mixtures, have to be evaluated to compare their bio-efficacies and cost effectiveness. Sequential spraying of individual herbicides reduces herbicide usage compared to mixtures as it is based on the predominant weed present. There are advantages in spraying certain herbicides of *e.g.* Glyphosate to control grass weeds and Meta sulfuron-methyl to control broad leaved weeds and the mixtures of both to control mixed weed species. Glyphosate mixture in combination with one of the broad-leafed weedicides has superior bio-efficacy and cost effectiveness.

a) Crop Injury and Phytotoxicity

Herbicides are efficient in weed management only if they are used properly. The knowledge of crop tolerance and information on phytotoxicity are important to fully exploit the benefit of chemical weeding. Wrong choice of herbicide and careless or incorrect spraying technique can often lead to serious damage to the main crop. The benefits of weeding are nullified if weedicide phytotoxicity and crop injury occur. The choice and use of weedicides have to be guided by the results of experiments. Weedicides that can cause serious adverse effects should not be used at all.

10.3.3. Weed Succession/Weed Shifts

Successful eradication of certain weeds may result in the build up of another equally or more harmful weed. This may be due to the long term use of a particular herbicide leading to the colonisation of a resistant weed. By using suitable weedicide mixtures all the weeds present in the palm circle can be eradicated. But in the inter rows, natural ground covers and soft grasses have to be conserved by spraying only broad leaved weed killer, *e.g., Dicamba.*

10.4. Disaster Management

Oil Palm being a perennial monocot crop with shallow root system, *i.e.,* bulk of the roots being within one meter depth is likely to be uprooted due to cyclonic winds and rains,. Sometimes forest fire during summer time also affects the palms. In such cases one has to take immediate steps to rejuvenate the affected palms.A case study made at the farmers' fields at two locations after the 1998 super cyclone has been narrated here for the benefit of small holders. This management strategy is unique in the oil palm growing countries.

10.4.1. Management of Super Cyclone Affected Palms

The methodology followed for managing the three types of super cyclone affected palms *viz.,* totally uprooted, crown damaged and slanting palms are as under:

a) Management of Uprooted Palms

Uprooting of palm trees was considered the most serious damage observed. In this case there was complete uprooting resulting in falling of the palms on the ground. The root system was exposed to the extraneous environment subjected to further damage due to microbial infection and desiccation of root system. The management strategies adopted were- i) replanting of the uprooted palms, or ii) replanting with new seedling in place of fallen one.

b) Replanting of Uprooted Palms

In this case a pit of 1 cubic meter was dug close to the bole without disturbing the root system adhering to the soil, either manually or by using JCB proclaimer.It was noticed that digging a pit manually is not practical, as it requires more manpower and is also time consuming. Lifting and placing the fallen trees in pit with the help of tractor and ropes or locally available equipment like manually operated lever cranes *etc.* were impractical. Tractors with iron ropes as well as Proclaims were also used for lifting the palms.

c) Preparation of Pits and Planting the Uprooted Palms

To start with wider pits to accommodate the fallen palms bole were taken in the same place where the palms were uprooted. Power operated Proclaim is more economical to use than manually taking pits. Phorate granules were applied @25 g per palm to avoid termite and weevil problems. One kg rock phosphate and 2kg vermicompost or 20 kg farm yard manure were applied in the pit to accelerate the root growth. The crown of the fallen/uprooted palms were pruned leaving 6 to8 leaves from the growing point and tied. The palms were then lifted with the help of proclaim and replanted in the same pits. The pit was filled with soil around the bole and pressed hard to avoid air holes. After replanting the crown was sprayed with 0.2 per cent Bavistin followed by heavy irrigation to avoid sudden drying. All these operations were taken up before the roots and leaves dried in the sun. The planted palms should be erect without even slight bend. The soil around the palm should be firmly pressed without any air space. If need be bamboo pole support can be given. Irrigation should to be given regularly for better establishment.

d) After Care and Management

Application of NPK fertilizers at 20 per cent more than the recommended dose to be given once it is established. Leaves produced initially were drooped and dried. However, these palms gradually recovered after three years of replanting, with number of leaves, bunches, and FFB weights and have come to normal production.

10.4.2. Replanting with New Seedlings

In this case the uprooted palms were removed and new seedlings of 12 to 14 month age were planted. The normal oil palm cultivation and plant protection practices were followed. After planting of the seedling, the process of removal of inflorescences (ablation) was followed till third year.

10.4.3. Management of Crown Damaged Palms

In this case surgery was done at the crown region and the broken spindle and other broken leaves were removed. Bavistin spray of 0.2 per cent was applied to avoid any fungal infection and such palms were regularly irrigated and manured.

For crown damaged palms, only those leaves which are damaged should be cut and applied with fungicide spray. Care should be taken to retain maximum number of leaves for rapid recovery of the palms. Application of fertilizers 20 per cent more than the recommended dose (1440g N:720g P_2Os:1440 g K_2O) per palm per year should be made regularly for three years.

10.4.4. Management of Slanted Palms

The slanted palms were given support at the bole region by earthling up with soil and such palms were regularly irrigated and manured. Application of fertilizers 20 per cent more than the recommended dose(1440N:720P_2O:1440 K_2O/palm/year) were given.

Maximum effort should be made to make slanted palms which are tilted more than 25° straight rather than leaving them to grow naturally otherwise it may lead to breaking of palms after some years.

10.4.5. Performance of Palms after Three Years

The performance of palms of all the three types of damages were monitored regularly for four years and are being discussed as below:

a) At Gokavaram

The leaf production of normal, crown damaged, slanted, and replanted palms were 25.93, 25.15, 25.6 and 25.16, respectively while the sex ratios were 2.4, 2.1, 4.0 and 4.0. This clearly indicated that there is not much difference in leaf production and sex ratio in all the categories. The highest FFB yield was obtained by normal palms with an average total weight of 115 kg with the average production of 5 bunches/year followed by crown damaged palms with 114 kg FFB weight and average of 6 bunches/palm/year. Replanted and slanted palms produced 104 kg and 97 kg with average bunch number of 4.5 and 4.9/palm/years respectively. This indicated that the affected palms of all categories have reached the normal production.

b) At Lakshmipuram

All the four categories of palms *viz.*, normal, crown damaged, slanted, and replanted produced 27.4, 25.83, 26.76 and 26.93 MT FFB/ha, respectively. At this location also not much difference in leaf production rates of the affected palms was found. The highest sex ratio was obtained by normal palms in the rates of with 3.1 followed by replanted palms with 2.9, slanted palms with 2.2 and crown twisted palms with 2.1.The highest yield was obtained by normal palms with an average total weight of 155 kg with the average production of 7.4 bunches/year followed by Crown damaged palms with 133 kg and average of 5.7 bunches. Slanted and replanted palms produced 111 kg and 101.6 kg FFB with an average of 5.7 and 5.6 bunches/palm/year, respectively.

c) Performance of Replanted Up Rooted Palms vis-a-vis Newly Planted Seedlings

Exactly after three years of replanting, the uprooted palms are producing Fresh Fruit Bunches (FFB) weighing 25-30 kg, which is almost similar to that of normal palms. In order to get 20 MT of FFB in a newly planted seedling plot it will take six years, whereas in case of uprooted palms which have been replanted using the suggested technology, within three years of replanting a farmer can get 20 MT of FFB per ha. Though cost of cultivation is almost similar, replanted palms are producing FFB on an average 20 MT/ha. On the other hand, after three years of planting including the ablation period (removal of male and female inflorescences), the newly planted seedlings have just started production, which is not comparable to the replanted ones. Further, in the new seedlings being planted under the mature oil palms, the growth was not proper. A palm of four year old produced only 5 t/ha which fetched INR. 13750/ha, whereas the replanted palms after three years produced 20 MT FFB/ha fetch INR. 1,55000. Thus, replanting of the uprooted palms is suggested instead of going for planting with new young seedlings.

The cost of rejuvenation to get the same yield is comparatively lower than that of new plantation. Hence this technology of replanting of uprooted palms making it to yield within three years, is a sustainable, eco-friendly and low-cost technology which can be used to revive the palms uprooted by the natural calamities.

10.5. Oil Palm Cultivation in Wastelands

The excessive demand of land for both agricultural and non-agricultural uses has resulted in the development of vast stretches of different kinds of wastelands such as salt-affected land, waterlogged areas, gullied/ravenous lands *etc.* In the geographical classification of lands one can also see cultivable wastes, non-cultivable wastes and pasture lands to a sizable level. Besides there are sizable mine reject soils in the mining areas with deep reserve of water unutilized or underutilized. An ever increasing population places enormous demand on land resources which are indispensable for a country like India with 2.4 per cent of the worlds' geographical area supporting over 16 per cent of the world's population. Further, the country has 0.5 per cent of the worlds grazing lands but has over 18 per cent of world's cattle population. The tremendous pressure on land has led to conversion of forest lands into urban and industrial areas.

According to ICAR that Wastelands are lands which due to neglect or due to degradation are not being utilized to their full potential. These can result from inherent or imposed disabilities or both, such as location, environment, chemical and physical properties, and even suffer from management conditions. According to Integrated Wasteland Development Programme, Wasteland is a degraded land which can be brought under vegetative cover, with reasonable effort, and which is currently under utilised and land which is deteriorating for lack of appropriate water and soil management or on account of natural causes.

A few pictures of wastelands have been shown below (Figure 10.12a-g which are being used for raising oil palm in Nayagarh and Boudh districts of Odisha.

10.5.1. Wastelands of India

The situations of wastelands have been depicted in Figures 10.12a-g.

i) The state-wise wasteland distribution is given in Table 10.2 as estimated by NRSA as on 2003.

ii) **Wasteland status in India** estimated by various organisations in the country

Source	*Area (m.ha.)*
Ministry of Agriculture and the JNU, Deptt. Of Geography (1986)	175
National Land Use and Wasteland Development Council (First Meeting 1986)	123
Society for Promotion of Wasteland Development (1982)	145
Ministry of Rural Development and NRSA (2000)	64

Figure 10.12: Different Types of Wastelands

a) Wasteland with little slope

b) Wasteland plain

c) Wasteland in the foothills

d) Wasteland with shrubs

e) Wasteland with rocky patches

f. Clearing of the shrubs

g. Burning of the cut shrubs

Table 10.2: State-wise Wastelands of India–NRSA (Information as on year 2003)

State	*Wasteland (Area: In square km)*
Andhra Pradesh	45267.15
Arunachal Pradesh	18175.95
Assam	14034.08
Bihar	5443.68
Chhattisgarh	7584.15
Goa	531.29
Gujarat	20377.74
Haryana	3266.45
Himachal Pradesh	28336.80
J and K	70201.99
Jharkhand	11165.26
Karnataka	13536.58
Kerala	1788.80
Madhya Pradesh	57134.03
Maharashtra	49275.41
Manipur	13174.74
Meghalaya	3411.41
Mizoram	4469.88
Nagaland	3709.40
Odisha	18952.74
Punjab	1172.84
Rajasthan	101453.86
Sikkim	3808.21
Tripura	1322.97
Tamil Nadu	17303.29
Uttarakhand	16097.46
Uttar Pradesh	16984.16
West Bengal	4397.56
Union Territory	314.38
Total	**552692.26**

About 120839 sq km of area of J&K is remain unsurveyed. Total geographical area of India: 3287263 sq. km

Source: Wasteland Atlas of India-2005: NRSA.

10.5.2. Categories of Wasteland for Identification

a) Cultivable Wasteland

The land which is has potential for the development of vegetative cover and is not being used due to different constraints of varying degrees, such as erosion, water logging, salinity *etc.*

b) Uncultivable Wasteland

The land that cannot be developed for vegetative cover, for instance the barren rocky areas and snow covered glacier areas.

10.5.3. Oil Palm in the Wastelands of Odisha

The wastelands not cultivated for long number of years with shrubs, small trees, rocky patches, heavy clayey soils belonging to private farmers have been purchased by private companies and utilised for oil palm cultivation after clearing bushes, levelling partially after creating irrigation sources like bore wells, water harvest structures, check dam in nalas, intake wells in rivers *etc.* The details of two locations mainly with black soils with boulders and rocks utilised for oil palm cultivation are given in Table 10.3.

10.5.4. Preparation of Wastelands for Oil Palm Cultivation

Land clearing has to be made following the norms of Forest Department leaving the tall trees and clearing the bushes and other small plants. After clearing pits are made using pot hole digger with 9 x9x9 m spacing. Wherever the soils are not having depth the pit size can be increased to 3′ x 3′ x 3′ and some times more than that using proclaim which normally helps to have wider pits.(Fig. 10.13-10.16)

Figure 10.13: Planting Pits and Mixing Sand.

Figure 10.14: Planting Process and Mixing Soil.

Table 10.3: Oil Palm in the Wastelands of Badaliapada and Biraprathapur, Boudh Districts of Odisha

Name of Captive Plantation	*Area in Acres*	*Planting Date*	*Seedlings Age at Planting in Month*	*No. of Pruning at Nursery Stage*	*First Harvest Date*	*Fertilizer*	*Fertigation Frequency*	*Fertilizer Quantity per Month*	*Fertilizer Quantity per Annual*
Birapratapur (All Ghana X Compacta)	37.48	Block-1 11-07-11	23	5	Sept.'14 (3 yr, 2 months)	First we have given fertilizer Urea, SSP, MOP, MgSO4 and Boron. Three months once manually applied. Then we have given through fertigation, urea, Super, MOP monthly. Double dose of fertilizer and micronutrient applied six months once. Later instead of super, DAP applied monthly once and six months once double dose of fertilizer and micronutrient applied.	Monthly	1st yea 60-40-55	840-560-770
	0.73	Block-2 11-08-11	40	7				2nd year 120-80-120	1680-1120-1540
	38.21	Block-3 28-07-11	29	5				3rd year 175-110-166	2450-1540-2324
		Block-4 1-10-11							
							Now	185-110-220	2500-1500-3000
								10 Monthly doses in fertigation and twice manual application of double dose.	

Note: Seedling were pruned at the top for 5 to 8 times to arrest the over growth since the land was not ready. After planting pot watering was given for 4 to 6 months. After laying drip irrigation fertigation was also given as indicated. Firstly good rainfall from second fortnight of June to October. During summer irrigating adequate quantity is always problem due to inadequate availability of water.

Figure 10.14: Forming Wider Pits in the Stony Patches and Planting.

Figure 10.15: Forming Basin after Planting and Mulching in the Basin.

Figure 10.16: A View of Initial Plantation at Two Locations.

10.5.5. Initial Performance of Oil Palm in Captive Plantations in Wastelands

Even though pot watering was done for some months with thick mulches in the basins, the growth was fairly good in black soils where water holding capacity is more. Subsequent laying out of drip and monthly fertigation with one additional soil application annually and also adding vermicompost at 3 to 5 kg/palm/annum helped to boost the growth. It came to flowering in 18 months and ablation was done up to three years. The plantation had come to yielding stage. The water shortage reflected in the low yield to start with. If only adequate water of 150 to 200 l/palm/day could be given at the bearing stage the bearing would have been good and it should be possible to get 20 to 30 MT FFB/ha/year. Nature of growth is shown in Figures 10.18 abd 10.19 and they yielding palms at the fifth year is given in Figure 10.20.

10.5.6. Irrigation Sources Developed

All possible sources of tapping ground water resources, rain water conservation by check dams, setting up water harvesting structures, pumping water from rivers and nalas, creating storage structures,digging bore wells, open wells *etc.* were done. This helped to maintain the crop in a fairly good condition (Figures 10.21 and 10.22).

Figure 10.17: Growing Dhaincha in Palm Basin.

Figure 10.18: Oil Palm Growth at 26 Months after Planting.

Figure 10.19: Oil Palm in Rocky Patches of Birpradapur, Boudh District.

Figure 10.20: Five Year Old Palms in Yielding Stage.

Figure 10.21: River Water Pumping for Irrigation through Drip.

Figure 10.22: Water Harvesting Structures Developed for Irrigating Oil Palm.

Chapter 11

Water Management

Oil palm requires sufficient irrigation as it is a fast growing crop with high productivity and high biomass production. Continuous soil moisture availability encourages vigorous growth and increases yield of Oil Palm. Adequate supplies of water, good soil depth and water holding capacity contribute to water availability. However, insufficient irrigation will reduce the rate of leaf production, affects flower initiation, sex differentiation and low sex ratio due to production of more male inflorescence, inflorescence abortion and yield reduction. Water deficit increases, metabolic activity affecting stomata which will remain closed and inhibit the development of opening of spear leaf.

11.1. Water Requirement and Water Use Efficiency

The general belief is that oil palm needs more water than other irrigated crops like Paddy (Rice), sugarcane, banana etc. and would drain water sources over a period of time. Comparing the water requirements of many irrigated crops proves otherwise. Water requirements for different crops are given in the Tables 11.1 and 11.2 for easy understanding.

Two important points to be noted from Tables 11.1 and 11.2 are:

- ★ Cost of cultivation for oil palm is comparatively less than banana and sugarcane
- ★ Net income is comparable to the cost of cultivation and it is a perennial crop.

Table 11.1: Water Requirements and Irrigation Schedule of Major Crops in Tamil Nadu

Crop	*Water Requirement (mm)*	*Irrigation Schedule (IW/CPE ratio)*	*Irrigation Interval (Days)*
Rice	1200-1500	5 cm depreciation of previously ponded water	4
Maize	500-600	0.80	10-12
Cotton	600-650	0.40	10-12
Sugarcane	1800-2200	o.75	7-8
Banana	2000-2200	0.80	7-8
Ground nut	450-550	0.60	10-12
Chillies	650	0.80	7-8

Source: Palanisami *et al.*, Water Management Technologies, TNAU, 2003.

Table 11.2: Economics and Water Use Efficiency of different Crops in Navsari, Gujarat

Crop	*Irri-gation*	*Av. Yield (t/ha)*	*Cost of Culti-vation (Rs/ha)*	*Gross income (Rs/ha)*	*Net income (Rs/ha)*	*Water Require-ment (mm)*	*Water in l/kg of produce*	*Remarks*
Paddy (summer)	Surface	4.0	28000	42000	14000	1200	3000	Price Rs.8/kg + 10000 Rs/ ha straw
Sugar-cane	Surface	100.0	80000	200000	120000	1600	160	Price: Rs.2/kg
	Drip	125.0	92000	240000	148000	960	77	
Banana	Surface	60.0	135000	300000	165000	1600	267	Price: Rs.5/kg
	Drip	80.0	134500	400000	265500	960	120	
Oil palm	Surface	17.5	45000	70000	25000	1300	743	Price: Rs.4/kg
	Drip	20.0	50000	80000	28000	1200	600	

* Yield obtained at Navsari.

Water requirement for oilpalm is comparable banana and sugarcane with less maintenance cost and more returns.

11.2. Quantity of Water

The quantity of water required is determined on the basis of evapo-transpiration, rainfall and soil water reserves.

Potential Evapo-transpiration (PET) = Pan evaporation x crop factor (0.7 for oil palm)

Water requirement varies with the quality of soil, water holding capacity, evapo-transpiration rate and the palm age.

Based on the palm age, the water requirement varies as follows:

Palm Age	***Water Required in litres/day/palm***
1st year	40
2nd year	80
3rd year	120
4th year onwards	170-250

Based on the environment, location specific climatic and soil conditions the water requirements study should be carried out and implemented. In light soils, heavy watering often leads to the loss of nutrients through leaching and hence it should be watered lightly at frequent intervals. Water holding capacity of the soil can be enhanced by the incorporation of organic by-products derived from oil palm system. Such an enhancement will naturally conserve soil moisture and thereby help in minimising the frequency of irrigation.

11.3. Water Deficiency and its Effects

In young palms the most evident symptom of water deficiency is the accumulation of spear leaves. Later, inward curling (rolling) of leaflets' starts and subsequently the tips of leaflets of the lower leaves become necrotic. In juvenile palms water deficiency causes reduced dry matter accumulation, poor vegetative growth, yellowing and bronzing of the tips of the intermediate leaves, rolling of leaflets of lower leaves. In mature palms, folding of premature leaf and unexpansion of spear leaf are the initial symptoms. Later leaflets roll down, decreasing the photosynthesis rate. Leaf production decreases wilting and drying of leaves may take place (Figure 11.1).

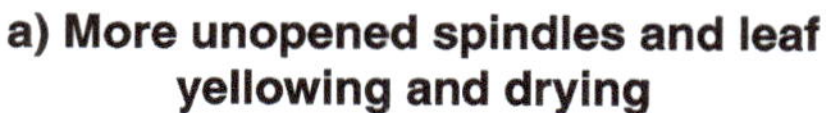

a) More unopened spindles and leaf yellowing and drying

b) Severe shortage of water-leaf drying

Figure 11.1: Water Deficiency Symptoms in Oil Palm Trees.

During dry months stomata closure occurs during the mid-day and results in the diffusion of CO_2 for photosynthesis. Water deficit experienced during any part of the year not only reduces the sex ratio and increases the number of male inflorescence length, spike let number and total number of flowers on the inflorescence, it also reduces the oil content of mesocarp. When the water deficit goes as high as 700 to 800 mm/annum, not only the growth and yield are affected but also the survival of oil palm is threatened.

Water being the major input required for oil palm, any deficit in terms of availability of water will affect the plant health and yield. Table 11.3 indicates the yield loss at different water deficit levels.

Table 11.3: Yield Loss at different Water Deficit Levels

Stage	*Water Deficit (mm/yr)*	*Symptoms*	*Yield Loss (per cent)*
1	Below 200	Not a serious problem	0-10
2	200-300	Non opening of immature and younger leaves, defective older leaves	10-20
3	300-400	Increased un-opening of younger leaves, defective leaves and drying of older leaves	20-30
4	400-500	Unopened immature leaves and dried leaflets	30-40
5	Above 500	Young leaves may not open, leaf bud cracks and breaks	More than 40

A number of experiments have been conducted in different countries regarding the effect of water deficit on FFB, bunch weight etc. and the results are given in Table 11.4.

The Table 11.4 clearly indicates the role of moisture in influencing the yield, bunch number and bunch weight. Higher the water deficit, higher is the yield reduction. Even though all these countries are growing oil palm under rain-fed conditions which is well distributed during a major period, still there are 4 to 5 months of dry period.

Under Indian conditions, almost all the states identified except Mizoram are having only seasonal rains and the length of dry period is more than six to seven months with less number of rainy days and more quantity at a time and hence only irrigated oil palm was suggested. The farmers are, however, using both micro irrigation as well as flow irrigation in basins and flooding the whole field. In a light textured soil the flooding not only leaches the nutrients down but also causes wastage of water. Micro irrigation with optimum quantity of water or fertigation is the best option for farmer to adopt.

It is clear that the major components of water loss are evapo-transpiration and surface run off which could be kept under control by adopting moisture conservation methods.

Potential evapo-transpiration varies with the climatic factors like solar radiation, wind speed and maximum and mean air temperature. In humid tropics suitable

Table 11.4: Details of Irrigation Experiments; Water Deficit, Age of Palms, Control and Irrigated Values and Percentage Increase for FFB, Bunch and Mean Bunch Weight

Location	*Water Deficit*[a] *(cm)*	*Age of Palms (years)*	*FFB yield (t/ha per year)*				*Bunch (No./palm per year)*			*Mean Bunch wt. (kg)*			*Ref.*
			Contr.	*Irrig.*	*Incr.*	*Per cent Incr.*	*Contr.*	*Irrig.*	*Per cent Incr.*	*Contr.*	*Irrig.*	*Per cent Incr.*	
Ivory Coast	660	4-8	6.4	14.5	8.1	125	8.7	15.8	82	5.5	6.8	24	1
Benin	582	8-11	16.1	23.4	7.3	47	7.2	9.5	33	15.7	17.4	11	2
Ivory Coast	275	3-5	6.5	11.5	5.0	76	10.6	17.1	61	4.3	4.7	9	3
Malaysia	?	13-16	30.7	34.1	3.4	11							4
Malaysia	82	5-10	24.7	25.3	0.6	2	13.6	14.1	3	13.1	13.0	-1	5
Malaysia	?	11-14	24.6	32.1	7.5	30	7.5	9.0	20	22.6	35.0	10	6
Malaysia	38	5-14	25.4	30.1	4.7	19							7
Ivory coast	572	5-10	16.6	22.3	5.7	34	10.6	14.3	35	10.9	11.0	1	8
Colombia	130	5-7	14.2	20.3	6.1	43							9
Colombia	366	4-8	14.2	19.3	5.1	36	9.5	12.5	32	10.5	10.8	3	10
Colombia	345	3-5	7.2	11.2	4.0	55	14.0	11.2	25	5.6	4.6	22	10
Malaysia	164	5-10	29.2	32.6	3.4	12							11
Thailand	214	8-14	18.7	24.5	5.8	31							12
Thailand	214	10-13	19.5	23.9	4.4	23							12
Malaysia	210	16-20	20.9	24.0	3.1	15	6.4	5.6	14	24.6	25.0	2	13
Ecuador	?	1	22.4	28.3	5.9	26	12,2	14.4	13	17.2	18.3	6	14

a: IRHO definition, maximum reached in one reason.

References: 1: Desmarest (1967), 2: de Taffin and Daniel (1976); 3: Ochs and Daniel (1976); 4: Chan (1979); 5: Corley and Hong (1982); 6: Chan *et al.* (1965), 7: Chuah and Lim (1989); 8: Prioux *et al.* (1992); 9: Corley (1992); 10: Unipalma (unpubl.); 11: Lim *et al.* (1994); 12: Palat *et al.* (2000); 13: Kee and Chew (1993); 14; Mite *et al.* (2000).

for oil palm cultivation, PET has been worked out to be 150 mm/month with <10 rainy days and 120 mm/month during months with >10 rainy days. Under any circumstances if the water stress is likely to cause 25 per cent reduction in the annual dry matter production, it might lead to 50 per cent reduction in FFB yields.

11.4. Methods of Irrigation

11.4.1. Basin or Flood Irrigation

When water is not a constraint, basin irrigation can be taken up. Required quantity of water can be given at weekly intervals or once in five days depending upon the soil condition. Irrigation channels must be prepared in such a way that the individual palms are connected separately by sub-channels. For light soils, frequent irrigation with less water should be given. If more water is given at a time leaching of nutrients will be more. In heavy soils irrigation interval can be longer.

11.4.2. Drip Irrigation

If irrigation water is limited and land is of undulated terrain, drip or micro sprinkler irrigation can be advantageous. When drip irrigation is given, care should be taken to avoid clogging and assure uniform discharge of water. Four drippers have to be placed for each palm. If each drippers discharge 8 l. of water per hour, 5 hours of irrigation per day is sufficient to discharge 160 l of water/day. Drip irrigation has not only increased the FFB yield from 15.5t/ha to 17.3t/ha but also improved the uptake of potassium.

Fifteen years of research on irrigation to oil palm in Thailand have shown that drip system of irrigation is the best for oil palm. While yield responses upto 10 MT FFB/ha over the other system, costs are comparable, management is relatively easy and irrigation can be comparatively profitable. For maximum response to irrigation, fertilizer inputs must be increased after ascertaining which nutrient is critical. Where lower rates of irrigation are used, additional fertilizer dose may not may not give response.

Even though drip is the best system,the quantity of water should be assessed based on the calculation. In some places farmers are using more water and more fertilizers in the light textured soils, thereby both water and nutrients are lost by percolation. Uniformity in discharge through drippers at all places and the quantity of water discharged should be checked periodically.

11.4.3. Micro-Sprinkler Irrigation

This system is a slight modification of drip system, where micro-sprinklers are used in place of drippers. These apply water in a circular pattern to produce more wetted area than drippers. In each palm basin 2-4 micro-sprinklers can be arranged. Different micro-sprinkler nozzles like 180° jet, 360° jet, rotary type *etc.* are available. Micro-sprinklers help in maintaining high humidity in the plantations.

The major setbacks encountered in the drip and sprinkler system are as follows:

(*i*) Physical damages caused to the pipelines often by intercultural operations and rats.

(*ii*) Water stagnation in the pipeline between two subsequent irrigation sessions often leads to the development of micro-organisms and algae on the inner surface thereby obstructing delivery.

11.5. Drainage Management

Drainage is a problem only in areas where there is adequate rainfall uniformly distributed throughout the year. Basically three types of drains (Table 11.5) are laid out in any oil palm plantation.

(*i*) Field drains are subsidiary drains which run parallel to the palm rows and their frequency often determines the intensity of drainage. In heavy coastal clays, they are dug for every fourth row while in a better structured clay, field drains at every sixth row will be sufficient for better drainage.

(*ii*) Collection drains are those which collect water from the field drains and are often cut at right angles to meet the field drains. These mostly run parallel to the collection roads.

(*iii*) Outlet drains are ones which collect water from the collection drains and run directly back to the water source.

Table 11.5: Dimensions of the Drains Varies with the Types

Type of Drain	*Width (m)*		*Depth (m)*
	Top	*Bottom*	
Field	1	0.3	1.1
Collection	2-2.5	0.6	1.25-1.75
Outlet	3-5	1.0	2-2.5

Gradual build up of soil and detritus at the bottom of the drains not only reduces the depth but also the efficiency of the drainage system. So periodical clearing of the drains is quite essential. Normally the field and collection drains are cleared in every 6 months while the outlet drains are cleared either annually or once in 3 years. Access to the palm rows for inspection can be obtained on both sides of a drain either directly or over a foot bridge.

11.6. Moisture Conservation Methods

In mature plantations, fronds falling from taller palms have a high kinetic energy sufficient to cause soil detachment, reduced infiltration and higher surface run-off resulting in soil erosion. To check erosion and conserve moisture the following measures are commonly adopted in oil palm plantations.

11.6.1. Cover Cropping

Raising of cover crops has been dealt in detail a separate chapter. Bimonthly slashing of the cover crops itself will serve as a mulch in moisture conservation.

11.6.2. Silt Pitting

It is normally practiced in terraced lands. Pits are dug at the rate of one for very four palms. The excavated soil (from the pit) is placed 0.5 m away from the top of the pits, to minimize the back wash of the soil into pits. Silt pits will get refined in 2-5 years depending on the soil type and rainfall intensity. However it is advisable to construct new ones rather than de-silting the older ones.

11.6.3. Mulching

Natural mulches not only reduce the water loss but also assist in water acceptance. Oil palm shell after kernel extraction acts as a best mulch and it is also prevents the formation of hard cap. Mulching with FFB led to better vegetative growth and subsequently high yields. But in estates with acute labour shortage located away from palm oil mills, FFB mulching is difficult and other alternate sources of mulching should be explored. Consequently palm trunk chips(PTC) was found to be advantageous than FFB which lost its physical structure after 6 months of mulching while PTC conserved moisture for a longer time as they retained their hard physical structure even after 15 months of mulching. Mulching cost is also 50 per cent lower than those of FFB mulching. Application of POME also improved the water holding capacity through increase in soil organic matter content. Depending upon the available water sources, various sizes and depths of water harvesting structures can be made as shown in Figure 11.2.

Figure 11.2: Farm Ponds

Plastic mulches can be used for mulching the palm basins at the young as well as adult stages for conserving moisture and preventing weed growth

11.7. Water Harvesting

Keeping in view the limited availability of water in irrigated oil palm cultivation it is all the more essential to store the rain water when we get rains and use it for irrigation. This is a time old practice in every village to save water and use it for various domestic purposes including drinking, washing, animal use *etc.* Nowadays it is termed as water harvesting, water shed management, farm ponds *etc.* These water harvest structures help in recharging the open and bore wells as well as farm ponds. The various sources of water harvesting structures are shown in Figure 11.3.

a) River Water Pumping

b) Constructing Sunk Well in Rivers

c) Small Farm Pond along *Nala* Side.

Figure 11.3: Water Harvesting Structures

Chapter 12

Nutrient Management

Oil palm is a gross nutrient feeder due to high growth rate, biomass production and yield and therefore, demands a balanced and adequate supply of macro, secondary and micro-nutrients for maintaining growth and yield. The adult palm produces two leaves and two inflorescences per month thereby making year round harvest possible. To get good bunches the availability of nutrient should be continuous. Thus, efficient integrated nutrient management is essential.

12.1. Nutrient Removal

Large quantities of nutrients are being removed by oil palm which is comparatively more than that of other plantation crops. The estimated uptake of nutrients is given in Table 12.1.

Table 12.1: Nutrients Removed by Various Plantation Crops

Crop	*Yield*	*Nutrients (kg/ha/year)*				
		N	*P*	*K*	*Mg*	*Ca*
Oil Palm	25 MT fruit bunches	93.5	11.0	92.7	19.3	20.3
Coconut	2400 kg copra	20.8	6.8	99.8	7.0	3.5
Cocoa	1125 kg dried leaves	25.5	5.0	50.0	6.3	3.2
Coffee	1125 kg made coffee	40.0	7.3	50.3	--	--
Tea	1350 kg dried tea leaves	62.5	4.5	28.3	3.0	5.5
Rubber	1928 kg dry rubber	19.1	3.8	15.5	2.6	--

Source: Ng. S.K. and Thamboo (1967).

The nutrient uptake of adult oil palm tree is given in Table 12.2.

Table 12.2: Estimates of Nutrient Uptake by Adult Oil Palm (kg/ha/year)

Components	*Nutrients (kg)*				
	N	*P*	*K*	*Mg*	*Ca*
Vegetative matter (net cumulative)	40.9	3.1	55.7	11.5	13.8
Pruned fronds	67.2	8.9	86.2	22.4	61.6
Fruit bunches (25 tons)	73.2	11.6	93.4	20.8	17.5
Male Inflorescence	11.2	2.4	16.1	6.6	4.4
Total	192.5	26.0	251.4	61.3	99.3

Source: Ng S.K. (1972).

It shows that very high quantities of N and K are being removed by an adult oilpalm every year. Leaves and fruit bunches remove the major part of the nutrients. This indicates the need to replenish the nutrients to the plantations. To ensure correct estimates of the nutrient intake, one has to carefully collect the soil and leaf samples and estimate the nutrient content.

12.2. Importance of Nutrients

For a balanced growth, vigour and reproduction of palm, all major, secondary and micronutrients are required. The functions of these nutrients and their deficiency symptoms are given below:

12.2.1. Major Nutrients

a) Nitrogen

In oil palm, characteristic yellowing symptoms are developed under 'N' deficiency conditions. Nitrogen is found to be essential for rapid growth and fruiting of the palm. It increases the leaf production rate, leaf area, net assimilation rate, number of bunches and bunch weight. Excessive application of nitrogen increases the production of male inflorescence and decreases female inflorescences thereby reducing the sex ratio.

b) Phosphorus

In oil palm seedlings, 'P' deficiency causes the older leaves to become dull and assume a pale olive green colour while in adult palms high incidence of premature desiccation of older leaves occurs. Phosphorus application increases the bunch production rate, bunch weight, number of female inflorescences and thereby the sex ratio. However, lack of response to P due to P fixation in soils is very common in the tropics.

c) Potassium

Potassium is mostly required for the production of more number of bunches by increasing the production of more female inflorescence. It also helps to increase the weight of bunch and total dry matters.

When potassium is deficient, growth as well as yield is retarded and it is translated from mature leaves to growing points. Under severe deficiency, the mature leaves become chlorotic and necrotic. Confident orange spotting is the main 'K' deficiency condition in oil palm in which chlorotic spots, changing from pale green through yellow to orange, develop and enlarge both between and across the leaflet, veins and fuse to form compound lesions of a bright orange colour. Necrosis within spots is common, but irregular. Mid-crown yellowing is another prominent 'K' deficiency condition of the palm in which leaves around the 10th position on the phyllotaxy become pale in colour followed by terminal necrosis. A narrow band along the midrib usually remains green. There is a tendency for later formed leaves to become short and the palm has an appearance with much premature withering.

Potassium removal is large compound to be normal exchangeable 'K' content in most striking and have been named as orange frond while the lowermost leaves are dead, those above them show a gradation of colouring from bright orange on the lower leaves to faint yellow on leaves of young and intermediate age. The youngest leaves do not show any discolouration. The most typical Mg deficiency symptoms is the shading effect in which the shaded portion of the leaflet will be dark green while the exposed portion of the same leaflet is chlorotic. Heavy rates of 'K' application induce Mg deficiency particularly on poor acid soils.

12.2.2. Secondary Nutrients

a) Calcium

Calcium is essential for proper development of meristematic tissue. In the absence of adequate Ca, terminal buds and apical tips of roots do not develop properly. Typically, most of the Ca in plant cells is found in the vacuoles (as Ca oxylate and Ca phosphate) and in the middle lamella of cell walls (as Ca pectate). In oil palm much of the Ca is located in the mesocarp of the fruit, where it exists as Ca oxylate crystals. Calcium plays an important role in the regulation of membrane permeability and particularly in the strengthening of cell walls. A typical symptom of Ca deficiency is the breakdown of cell walls and the subsequent breakdown of petioles and upper parts of stems. Plants that are deficient in Ca are highly susceptible to fungal infections, due to the weakness and breakdown of cell walls.

Calcium is usually the dominant cation (Ca^{2+}) in soils, even under acid conditions that favour leaching losses. It is a very common element in the earth's crust and is a component of such minerals as dolomite, calcite, apatite and gypsum to name a few. Because of its abundance in the biosphere, deficiencies of Ca in plants are rare. In many oxisols and ultisols of the tropics, Ca levels are very low and responses to additions of lime are often dramatic. Nevertheless, this is usually more a response to neutralization of aluminum and pH correction than to correct plant nutrient Ca levels. It is not uncommon to observe that Ca plays an important role in the quality of certain crops. Some research workers have suggested an ideal range of per cent saturation of 65-75 per cent as being optimum. However, most experiments show that optimum yield can be obtained over a much wider range of per cent Ca saturation.

Calcium also activates several enzymes, although it's role in this regard is not nearly as pervasive as that of Mg. In particular, Ca is an activator of membrane bound enzymes, such as ATPase's. Calcium present in the vacuoles of cells contributes to proper anion-cation balance. Specifically, it acts as a counter-ion for both inorganic and organic anions.

b) Magnesium

Magnesium is a component of the chlorophyll molecule and is thus essential for the process of photosynthesis. Many experiments have shown the effect of tissue Mg levels on photosynthetic activity. When leaf Mg drops below the critical level, photosynthetic activity drops rapidly and yields can be severely affected. Only about 15 per cent of the Mg present in plants is associated with chlorophyll. This element also plays many other roles in plant ribonucleic acid (RNA). There is a lengthy list of enzyme reactions for which Mg is essential. More than any other element, Mg acts as a general "enzyme activator" in plant cell.

c) Sulphur

Sulphur is a component of the amino acids cysteine and methionine and is thus essential for protein formation. Many experiments have demonstrated the close correlation of S content and protein levels in plants. Sulphur is essential for the formation of vitamins such as biotin and thiamin as well as glutathione and coenzyme A. The implications with respect to crop quality are thus obvious. In addition, S is essential for chlorophyll formation. Sulphur is a component of several of the aromatic compounds responsible for the characteristic odors of such plants as garlic, onion, mustard and cabbage. These crops have high requirements for S.

12.2.3. Micronutrients

Micronutrient elements, iron, manganese, copper and zinc are not generally found limiting in the nutrition of oil palm on acid soil conditions. Boron deficiency is occasionally found in young palms in the field in the form of reduction of leaf area in certain leaves producing incipient ' little leaf', advanced 'little leaf' with extreme reduction of leaf area and bunching and reduction in the number of leaflets and 'fish bone leaf'. Leaf malformation including 'hook leaf' and corrugated leaflets are some other associated symptoms. Soil applications of 50-200 g borax dehydrate, depending on age, and severity 10.4.

12.3. Fertilizer Recommendations

For irrigated oil palm in different agro-climatic conditions, the fertilizer schedule is being worked out. However, the fertilizer schedule in Table 12.3 is considered satisfactory for oil palm under rain fed conditions. The quantity of fertilizers to be applied to meet the above requirements is given in Table 12.4.

A balanced application of nutrients is essential for higher economic yields. If balanced application of macro, secondary and micro nutrients are not maintained, the imbalance creates antagonistic effects.

There is a strong interaction between N and P fertilizers and application of N fertilizer will be wasted unless adequate P fertilizer is also applied.

In K deficient soils, application of nitrogenous fertilizers containing calcium and also indiscriminate application of phosphorus is not advisable because of the antagonistic effect of calcium (present in all phosphoric fertilizers) on potassium. In the same way high doses of potassium lead to deficiency of magnesium or calcium in plants.

The suggestions are that optimum yield would be obtained when potassium: phosphorus ratio ranges between 2.5 or 3.5, nitrogen: potassium ratio 2.2 and nitrogen: phosphorus ratio 15.3 in mature plantations.

Fertilizer requirements and general fertilizer schedule for oil palm are given in Table 12.3-12.4.

Table 12.3: Fertilizer Requirement for Oil Palm

Age of the Palm	*Nutrients in Terms of Fertilizers (g/palm/year)*			
	'N' as Urea Ammonium Sulphate	*P_2O_5 as DAP SSP RP*	*K_2O as MOP*	*$MgSO_4$*
I Year	870 (or) 2000	- 1250 (or) 1000 or	667	125
	700 (or) 1610	435 - -	667	125
II Year	1740 (or) 4000	- 2500 (or) 2000 or	1333	250
	1400 (or) 3215	870 - -	1333	250
III Year onwards	2610 (or) 6000	- 3750 (or) 3000 or	2000	500
	210 0 (or) 4825	1305 - -	2000	500

Table 12.4: General Fertilizer Schedule for Oil Palm

Age of the Palm	*Nutrients (g/palm/yea*			
	N	*P*	*K_2O*	*$MgSO_4$*
I year	400	200	400	125
II Year	800	4000	800	250
III Year onwards	1200	600	1200	500

In gardens where 20-25 tons FFB/ha is obtained by 6th year, an additional 20 per cent of the recommended dose can be applied to maintain the productivity. Location specific fertilizer dose, depending upon crop growth, yield, soil response and availability have to be assessed from time to time.

Borax @ 100g/palm/year is recommended when the deficiency symptoms are noticed. This may be applied in two split doses as soil application.

The percentage of nutrients available in the selected fertilizers is given in Table 12.5.

Table 12.5: Percentage of Nutrients in Selected Fertilizers

	Analysis	*Nitrogen (N)*	*Phosphate (P_2O_5)*	*Potash (K_2O)*	*Sulphur (S)*
Nitrogen Fertilizers					
Urea	46-0-0	46	0	0	0
Ammonium Nitrate-(Granular)	34-0-0	34	0	0	
Ammonium Sulphate-Urea	34-0-0	34	0	0	11
Ammonium Sulphate	21-0-0	21	0	0	24
Anhydrous Ammonia (gas)	82-0-0	82	0	0	0
Urea-Ammonium Nitrate Solution	28-0-0	28	0	0	0
Phosphate Fertilizers					
Mono-Ammonium Phosphate	12-51-0	12	51	0	1.5
Mono-Ammonium Phosphate	11-55-0	11	55	0	0
Ammonium Polyphosphate Solution	10-34-0	10	34	0	0
Nitrogen Phosphates					
Ammonium Phosphate Sulphate	16-20-0	16	20	0	14
Ammonium Nitrate Phosphate	23-23-0	23	23	0	0
Ammonium Nitrate Phosphate	27-14-0	27	14	0	0
Urea Ammonium Phosphate	27-27-0	27	27	0	0
Urea Ammonium Phosphate	34-17-0	34	17	0	0
Potash Fertilizers					
Potash Chloride	0-0-60	0	0	60	0
Potassium sulphate	0-0-52-12	0	0	52	12
Sulphur Fertilizers					
Ammonium Sulphate	20-0-0-(24)	20	0	0	24
Gypsum (agricultural)	0-0-0	0	0	0	17
Elemental Sulphur*	0-0-0	0	0	0	90-99
Ammonium Thiosulphate Solution	12-0-0-(26)	12	0	0	26

The nutrient contents of common fertilizers and manures used in oil palm plantation are given in Table 12.6.

It is always better to be apply nutrients based on the leaf nutrient analysis. In adult plantations it would better to have the analysis once in two or three years.

12.4. Sources of Fertilizers

12.4.1. Inorganic Fertilizers

Fertilizers are applied in defined chemical form, either as single (or straight) chemicals or as specially formulated mixtures or compounds in granular form. The most frequently used nitrogen fertilizer in oil palm is Ammonium sulphate. On the basis of nitrogen content alone urea is the cheapest form of this nutrient. Rock phosphate is the cheapest and most widely used source of phosphorus. On soils of pH above 6, use of more soluble phosphates like Super phosphate, double super phosphate and Triple super phosphate becomes necessary. For oil palm Potassium chloride can be applied which contains 60 per cent K_2O and considerable amount of chlorine, to the extent that it probably fulfills the chlorine requirement of palms in most areas. There is no harm in applying compound fertilizers to oil palm as they are all supplying more than one nutrient. Nutrient content of important fertilizers and organic manures are given in Table 12.6.

Table 12.6: Nutrient Content of Common Fertilizers and Manures

Fertilizer	*Nutrient Content (per cent)*				
	N	*P_2O_5*	*K_2O*	*S*	*Others*
Urea	46	-	-	-	-
Ammonium Sulphate	21	-	-	24	-
Ammonium Chloride	26	-	-	-	Cl
Calcium Ammonium Nitrate	25	-	-	-	Ca
Single Super phosphate	-	16	-	12	Ca (20 per cent)
Potassium Chloride	-	-	60	-	Cl
Potassium Sulphate	-	-	50	18	
Di ammonium Phosphate	18	46	-	-	-
Rock Phosphate	-	18-20	-	-	Ca
Gypsum(Agric. Grade)	-	-	-	13	Ca (16-19 per cent)
Magnesium Sulphate	-	-	-	13	MgO (16 per cent)
Farmyard manure	0.5-1.0	0.15-0.2	0.5-0.6	-	-
Poultry manure	2.87	2.90	2.35	-	-
Castor Cakes	5.5-5.8	1.8	1.0	-	-
Neem Cake	5.2	1.0	1.4	-	-

12.4.2. Biofertilizers

Biofertilizers are microbial inoculants containing living cells of micro-organisms (bacteria, fungi, algae, *etc.*) which when applied on the seed or soil brings about substantial quantities of required mineral elements. These biofertilizers increase the availability of nutrients to plants, increase the fertilizer utility rate and thereby increases the production. The main groups of biofertilizers are:

i) Nitrogen fixing bacteria: They convert the atmospheric nitrogen into nitrate or ammonia forms and fix them in soil, *e.g.*, Rhizobium, Azolla, Azatobacter, Azospirillum.

ii) Rhizobium cultures are mixed with the green manure seeds before sowing for better nodule formations and fixation of nitrogen by green manure plants.

iii) Azatobactor and other organisms need lot of organic matter for their own multiplication and action. Addition of FYM/organic residues will help to get better benefit out of these organisms.

iv) Vermi-compost with vermi castings can be utilised along with the farm yard manure to compost the organic litter of oil palm added to the soil for quicker availability of nutrients.

v) Phosphate solubilising bacteria: They have the capacity to bring down the insoluble phosphate in soil to soluble form which can be utilized by plants. They increase the efficiency of rock phosphate and super phosphates, *e.g.*, Pencillium, Aspergillus.

Oil Palm being a crop which adds lot of organic matter to soil, these biofertilizers may be of greater use for conversion of organic matter into nutrient form and subsequent utilization by palm.

12.4.3. Organic Recycling

Oil palm depletes large amount of nutrients from soil as it is a perennial year round yielding crop, but at the same time adds lot of organic matter to the soil which improves the physico-chemical and biological properties of soil. Oil palm leaves, male inflorescence, empty bunches, fibres, shell, mill effluent can be used for organic recycling. A proper recycling of all these wastes can fulfill more than 50 per cent of nitrogen and potassium requirement and almost complete requirement of micronutrients by the palm.

Recycling of oil palm wastes and regular incorporation of green manures in the plantations definitely adds considerable quantities of nutrients required for oil palm and are able to reduce the dependence on synthetic fertilizers. This will help in the maintenance of the ecological balance and protection of fast deteriorating environment. The soil health is also maintained.

12.5. Methods of Application

Normally, fertilizers are applied by broadcasting in the basins of 3 m radius to enable rapid uptake by the palm feeding roots. It is important to determine the concentrations of feeding roots. In older palms the highest concentration of feeding roots will always occur in the weeded circle.

12.5.1. Basin Application

All the fertilizers should be evenly broadcasted over the clean weeded circle about 60-90 cm away from the base and incorporated into soil by forking. The area of application should be extended as the weeded palm circle becomes progressively enlarged as the palm gets older.

12.5.2. Inter Row Application

Maximum number of absorbent roots of 6 year old palms were found in the centre of interrow (3-5 m from the palms). The fertilizer could be spread in bands in the middle of the interrow or even over the whole interrow surface. However weeds have to be removed before fertilizer application. Under irrigated conditions weed management in the whole area is difficult, hence only basin application is ideal.

12.5.3. Fertigation

The soil application of fertilizers is more labour intensive and loss of nutrients through leaching, volatilisation *etc.* is high. Fertigation refers to application of fertilizers through trickle system of irrigation. This enables easy absorption of nutrients and help in efficient utilization of applied nutrients. The fertilizers reach the roots in solution form uniformly and increase their use efficiency. This system of fertilizer application needs completely soluble or liquid fertilizers.

12.6. Frequency of Fertilizer Application

The normal practice is that fertilizer application should be done twice a year for adult palms. Depending upon the labour availability, application of fertilizer can be done once in three or four months interval. However, more frequent applications during the early years are more effective. The reason for this is that a palm is capable of absorbing and utilizing only a certain maximum of applied nutrients according to its development stage. The amount of vegetative development increases so rapidly during the early years that uptake capacity changes quickly. This necessitates more frequent application of fertilizers to prevent interim stages of temporary nutrient deficit of the growing palm.

Application of fertilizers at recommended time and doses are mandatory up to 36 months. For the newly planted crop, the first dose of fertilizer may be given three months after planting. More split applications are recommended for very sandy soils where leaching occurs rapidly. It is also suggested that all applications of individual nutrients should be made within as short a period as possible if the risk of inducing nutrient imbalance is to be minimised. Normally, N, P, K, fertilizers are mixed and applied. In case of Calcium or Gypsum, it should be applied separately. Along with the second dose of fertilizer 50-100 kg FYM or 100 kg green manure and 5 kg neem cake per palm can be applied.

The nutrient availability chart given (Figures 12.1 and 12.2) charts have to be clearly understood and the soil physical, chemical and biological constraints if any identified and corrected for efficient utilization of applied nutrients.

The nutrient balance is to be calculated taking in the account of nutrient uptake for growth and production, nutrient losses by leaching and volatilization,nutrients immobilized on demand side and atmospheric nutrients, symbiotic nitrogen and soil nutrients supply side.

In case of acid soils lime application once in a year is recommended. In case of saline and alkaline soils application of gypsum is necessary before rainy season. Growing Dhaincha in the inter space and incorporating it 60 days after sowing will also help to reduce salinity and improving soil conditions. When green manure seed is sown it is desirable to broadcast 20 kg of gypsum.

Soil pH

Nutrients	4.0	4.5	5.0	5.5	6.0	6.5	7.0	7.5	8.0	8.5	9.0
Nitrogen	NA	SA	MA	MA	HA	HA	HA	HA	MA	MA	SA
Phosphorus	NA	NA	SA	MA	HA	HA	HA	MA	MA	SA	MA
Potassium	NA	SA	MA	MA	HA	HA	HA	HA	MA	MA	MA
Sulfur	NA	SA	MA	MA	HA	HA	HA	HA	MA	MA	MA
Calcium	NA	NA	SA	MA	MA	MA	HA	HA	HA	HA	MA
Magnesium	NA	SA	SA	HA	HA	HA	HA	HA	MA	MA	MA
Iron	HA	HA	HA	HA	HA	MA	MA	SA	SA	SA	SA
Manganese	HA	HA	HA	HA	HA	MA	MA	SA	SA	SA	SA
Boron	MA	MA	HA	HA	HA	HA	HA	MA	SA	SA	MA
Copper	NA	SA	MA	HA	HA	HA	HA	MA	SA	SA	SA
Zinc	HA	HA	HA	HA	HA	MA	MA	SA	SA	SA	SA
Molybdenum	NA	NA	SA	SA	MA	MA	HA	HA	HA	HA	HA

Acidic | -------Optimum pH Zone for Soil------- (5.5–7.5) | Basic

Key:
NA = Not Available
SA = Slightly Available
MA = Moderately Available
HA = Highly Available

Figure 12.1: Nutrient Availability Chart.

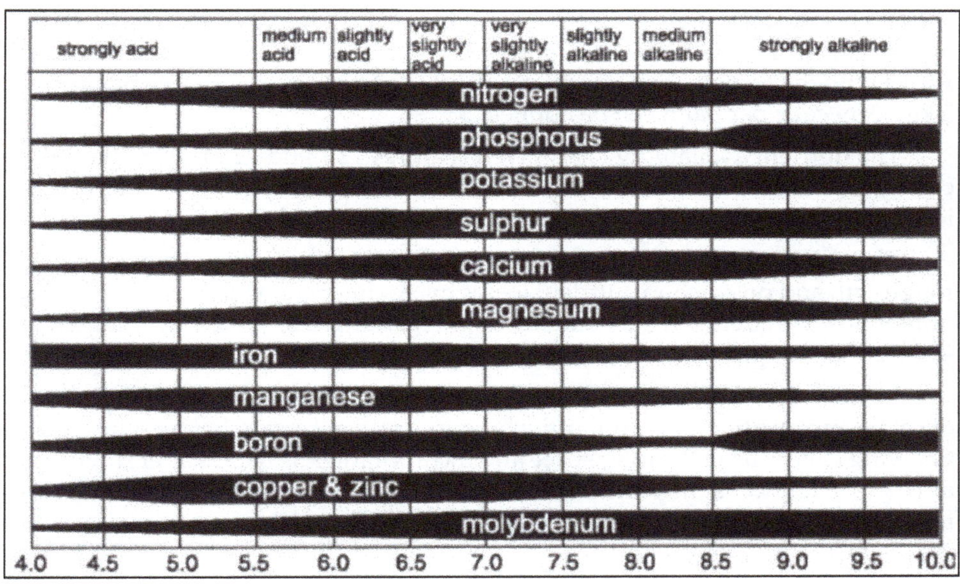

Figure 12.2: Trough Chart of Nutrient Availability.

12.7. Fertilizer Application Time

Nutrients applied as fertilizer if not absorbed by plants are subjected to losses as dissolved nutrients in surface run off and eroded sediments through volatilization, nitrification or leaching and fixation or adsorption. Timing of fertilizer application is crucial to minimise the excessive losses in surface run off.

The best response to fertilizers was obtained when the fertilizers were applied at the beginning of the rainy season rather than at its end. However application of fertilizers during heavy monsoon rains should be avoided to prevent leaching losses.

Very dry periods should be avoided for fertilizer application if adequate irrigation immediately after fertilizer application is uncertain. The palm will not be in a position to absorb nutrients from the soil without adequate moisture.

12.8. Soil and Leaf Analysis

12.8.1. Soil Analysis

a) Equipment Required

- ☆ Auger
- ☆ Spade
- ☆ 2 red buckets (labelled "Circle 0-20 cm" and "Circle 20-40 cm).
- ☆ 2 blue buckets (labeled "Stack 0-20 cm) and "Stack 20-40 cm).
- ☆ Water proof pens
- ☆ Recording forms
- ☆ Labels
- ☆ Plastic bags

b) Procedure

- ☆ At each Leaf Sampling Unit (LSU), a sample is taken from 0-20 cm depth and 20-40 cm depth, from the palm circle(on the path side) and the frond stack.
- ☆ Remove surface debris (dead frond material, loose fruits, weeds) before taking the soil sample.
- ☆ Place the samples in one of the four labeled buckets
- ☆ Bulk the samples for each zone and depth for each block.
- ☆ After thorough mixing, take two 0.3kg sub samples (one for analysis and store one for the each bulk sample.
- ☆ Air dry the samples in paper trays before dispatch or storage.

The results of soil analysis carried out at regular intervals throughout the life of a plantation may reveal important temporal changes in soil properties and prompt corrective action, involving adjustments to the amount and type of fertilizer applied. The soil analysis provides important background information but is not the main criterion for the preparation of yearly fertilizer recommendations for the reasons mentioned above, and because leaf analysis provides more useful information on how rapidly a soil can supply nutrients and water to palm roots.

12.8.2. Leaf Analysis

The basic procedure and sample preparation is described below:

a) Leaf Sampling Procedure

(*i*) Determine the phyllotaxis of the LSU palm. Identify leaf 1 (the first fully opened leaf). Follow the spiral around the trunk and determine the position of leaf 17.

(*ii*) Remove leaf 17 from the palm with a chisel or sickle leaving a long petiole base. Mark the petiole base with blue paint (to facilitate checking by supervisory staff). Cut the leaf into three sections of equal length. Cut and remove three upper and three lower rank leaflets from the centre of the middle leaf section (*i.e.* where the rachis ridge narrows to a point) with a clean sharpknife. Place the leaflets in a labeled plastic bag (record estate name, block number, date sampled, *etc.*). Move to the next LSU palm and repeat the operation.

(*iii*) Sampling is completed when a plastic bag containing the samples from each LSU palm has been prepared for each planted field.

(*iv*) Unless oven dried, samples should reach the analytical laboratory within 24 hours. If drying is carried out on the plantation, sample preparation is carried out within 24 hours.

b) Preparation of Leaf Samples for Dispatch

(*i*) Remove the distal and proximal portions of the leaf lets using a clean, sharp knife or guillotine, retaining the middle 20 cm section. Wipe each leaflet with lint free cloth using double deionized water. Remove the leaflets midrib and slice the leaf lamina into 2 cm pieces.

(*ii*) Place the cleaned leaflets pieces on trays and dry in a fan oven at 65-80 C for 24 hours. Leaflets may then be stored before they are sent to the laboratory for analysis.

(*iii*) If the plantation is equipped with a leaf grinder, samples are ground to pass a 20 mesh sieve. If samples are to be digested by 'dry ashing', samples should be ground with special leaf milling equipment instead of the conventional hammer mill since a very even particle size is required for successful 'dry ashing'.

(*iv*) Each dry samples ready for dispatch or storage should weigh about 0.2 kg and a duplicate of each sample is kept in the estate sample store in case of samples loss and for future reference.

c) Choice of Leaf Material

Since the pioneering work of Chapman and Gray (1949), the lamina from pinna removed from the midpoint of the 17th fully opened leaf has been adopted as the standard reference tissue for leaf analysis in mature palms. Leaf 17 is chosen to represent an intermediate position between the first fully opened leaf where cell expansion and maturation is under way and the oldest leaf where nutrients may have been found to show the greatest difference in the concentration of N, P and K. In addition, yield has been shown to be more closely correlated with nutrient concentration in leaf 17 than physiologically younger leaves.

Leaf analysis may be carried out on young palms (where leaf 9 is the reference tissue) but more usually, sampling is delayed until three years after field planting (when leaf 17 becomes the reference tissue).

Various shortcomings associated with the use of leaf 17 as the reference tissue have been highlighted (Foster, 1976) and palm nutritional status for K has been found to be more closely related to rachis K concentration than lamina K concentration in oil palm fertilizer experiments carried out on K deficient soils in Malaysia (Teoh and Chew, 1988). Some workers have found less variation in leaf K and Mg concentration in leaf 1 compared with leaf 17, and it has been suggested that the supply of the elements B, Mn and Fe is more precisely assessed by the analysis of leaf 3. Nevertheless, whilst there is increasing interest in rachis analysis, leaf 17 remains the standard tissue for leaf analysis in oil palm cultivation. The characteristics of leaf 17 may be summarized as follows:

i. Easily identified (lower part of the rachis makes a 45° angle to the vertical orientation of the trunk).
ii. Fully developed.
iii. Not senescent.
iv. Carries an inflorescence in the leaf axil (and is therefore linked to a nutrient sink).
v. Nutrient concentration shows better correlation with yield than that of other leaves.

d) Sampling Density

If Leaf Sampling Unit (LSU) palms are properly selected, a sampling density of 1 per cent is usually considered acceptable as long as the block size is at least 20 ha. When the planting density is 135 ha^{-1} this results in 27 LSU palms per 20 ha block. Abnormal palms (due to the effects of genotype, diseases or pests), unproductive (sterile) palms, and palms located near to rivers, roads and railways are not selected as LSU palms.

e) Frequency of Leaf Sampling

Leaf sampling is usually carried out once each year, and the minimum time interval between fertilizer application and leaf sampling is 2-3 months to avoid distorted results. Occasionally, non- routine leaf sampling may be required to investigate a particular problem (*e.g.* nutritional disorder in a particular field.

f) Sources of Variation in Leaf Nutrient Content

In addition to the effect of soil and fertilizers there are nine other sources of variation in leaf nutrient concentration, and the impact of all of these factors should be considered since they may account for some of the differences in palm nutrient status. This underlines the importance of training workers on how to select leaf 17 properly.

(*i*) Variation with leaflet rank

(*ii*) Effect of palm growth rate

(*iii*) Variation with palm age

(*iv*) Variation with leaf age

(*v*) Variation with climatic conditions

(*vi*) Diurnal variation in leaf concentration

(*vii*) Variation with planting material

(*viii*) Variation with fruiting cycle

(*ix*) Effect of changes in inter-palm competition.

12.8.3. Leaf Nutrient Analysis

Leaf nutrient analysis in oil palm for diagnostic purpose is useful, for it is a perennial crop which is relatively slow growing, can provide easily defined and standard leaf material for analysts and the responses to fertilizer application will take a longer time as compared to annual crops. The leaf nutrient analysis data can be compared with critical limits, and fertilizer recommendations can be made.

Critical level can be defined as 'the concentration of the element in the leaf (dry matter basis) above which a yield response from the element in the fertilizer is unlikely to occur. The detailed list of nutrients along with their critical levels are provided in Table 12.7.

Critical levels given are not universally useful since they could be modified by many other factors. Thus, it can be said that critical level is that at which the supply of the element is barely, but just above the point of limiting growth or yield. When critical level for element A has been reached, some other factor may then become limiting for growth or yield. This factor may be element B or it may be any environmental factor such as climatic conditions or a cultural factor, *e.g.*, drainage. Hence, relying solely on leaf nutrient levels of any particular nutrient ignoring other nutrient levels or other factors such as environment, cultural factors *etc.*, may be misleading.

In adult palms, leaf 17 is used for estimating the leaf nutrient levels for the reason that this leaf lays conveniently on the easily recognised spiral in the succession 1, 9, 17, 25 of the phyllotaxis and is in the middle of the foliage, the lower part of the rachis making an angle of about 45°. The leaf is fully developed, not yet senescent and carries an inflorescence in its axil whereas in younger palms leaf 9 is used as the index leaf. The seventeenth leaf is the most satisfactory for N and P determination and has no disadvantages for Ca, while first leaf gives less variation for Mg and more especially for K.

Table 12.7: Nutrient Concentration of Oil Palm Leaves (Deficiency, Optimum and Excess Levels - per cent)

Age of the Palm	*Nutrients*	*Deficiency*	*Optimum*	*Excess*
Young palms (below 6 years)	N	<2.5	2.6 -2.9	>3.1
	P	<0.15	0.16 - 0.19	>0.25
	K	<1.0	1.1 - 1.3	>1.8
	Mg	<0.20	0.3 - 0.45	>0.7
	Ca	<0.30	0.5 - 0.70	>1.0
	S	<0.20	0.25 - 0.40	>0.6
	Cl	<0.25	0.50 - 0.70	>1.0
	B (ppm)	<8	15	>35
	Zn (ppm)	<10	15-20	>50
Old palms (below 6 years)	N	<2.3	2.4 -2.8	>3.0
	P	<0.14	0.15 - 0.18	>0.25
	K	<0.75	0.9 - 1.2	>1.6
	Mg	<0.20	0.25 - 0.40	>0.7
	Ca	<0.25	0.5 - 0.75	>1.0
	S	<0.20	0.25 - 0.35	>0.6
	Cl	<0.25	0.50 - 0.70	>1.0
	B (ppm)	<8	15-25	>40
	Cu (ppm)	<3	5 - 8	>15
	Zn (ppm)	<10	12-18	>50

Source: Uon Uexkull and Fairhurst (1991) IPI Bulletin 12.

The optimal size of area to be sampled clearly depends on the terrain and soil variations and may be between 4 and 5 hectares. However, sampling of tenth palm in every tenth row, giving a one per cent sample is quite common. For this system the unit area should be at least 18 hectares, since not less than twenty five palms per unit should be sampled. If the plot size is small 7th palm in every row can be sampled.

For sampling, periods of drought or very heavy rain are avoided and 6 months or an absolute minimum of 3 months are elapsed between fertilizer application and leaf sampling. Two methods are commonly used for sampling leaf 17. In one, after the leaf has been cut down, it is cut into three equal parts and six leaflets are taken from each side of the middle section; of these twelve six will be upper rank and six lower rank leaflets. The leaflets are bulked from each sampling unit and placed in a polythene bag. Another method is to take leaflets from the entire length of the leaf on alternate sides of the rachis at intervals of about ten leaflets. Leaf samples should be prepared within 24 hours. The leaflets are first thoroughly cleaned with damp cotton wool. The midribs are then removed and the marginal 2mm of the laminae is cut off. Taking of whole lamina is preferable over the middle 20-30 cm. The laminae are cut into small pieces and dried for about 5 hours at 65-70°C in an air-drought oven. This does not remove all moisture but dries the samples sufficiently for storage or dispatch.

12.8.4. DRIS

Diagnosis and Recommendations Integrated System (DRIS) reflects the nutritional balance and indicates not only the nutrient more likely to be limited but also the order in which other nutrients are likely to become limiting (Table 12.8). DRIS has been found reliable in diagnosing nutrient requirements and scheduling the fertilizer application.

12.9. Nutrient Deficiencies

Work on deficiency symptoms dates from a classic experiment at Nkwelle in Nigeria in 1940 and the discovery by Hale in 1947 of the connection between leaf potassium content and bronzing or orange spotting of leaves. The knowledge about deficiency symptoms was obtained by three methods.

i. Leaf injection and spraying of chlorotic leaves.
ii. Correlation of symptoms with leaf nutrient contents and yield.
iii. Inducement of symptoms in sand culture.

12.9.1. Nitrogen

Nitrogen is essential for rapid growth and fruiting of the palm.

a) Factors affecting Availability

(*i*) Poor aeration - Water logging, prolonged inundation, inadequate drainage or a high water table in an area creates poor soil aeration thus leading to denitrification (conversion of NO_3 to elemental N) and deficiency.

(*ii*) Lateritic soils - Lateritic soils are poor in fertility and acidic in reaction. Under strongly acidic conditions the nitrification process (transformation NH_4 to NO_3) may be inhibited and reduces the availability of nitrogen to plants.

(*iii*) Soil depth - Deficiency symptoms appear as a result of planting on soil with variable depths of top soil or when planted close to large boulders just beneath the soil surface due to non-availability of the nutrients.

(*iv*) Grass weed competition - When the grasses proliferate in the inter-row space they induce nitrogen deficiency through competition for the nutrients.

(*v*) Insufficient N application.

(*vi*) Inhibition of N mineralization due to the effect of very low pH on soil microbiological activity.

(*vii*) Soil is shallow, terrain is hilly and palms are closer to the proximity of boulders.

(*viii*) There is acute shortage of available soil N.

(*ix*) Plants suffer from transplanting shock.

b) Symptoms

In seedlings, the most obvious visual symptom is the development of uniform pale green colour over the leaves. As the deficiency becomes severe, there is a gradual colour change to yellow. In older palms, chlorosis appears first on older fronds, later spreading to younger tissues. The tissues of the midrib become bright yellow or orange (Figure 12.3).

a) Nitrogen Deficiency **b) Magnesium Deficiency**

c) Copper Deficiency **d) Potassium Deficiency**

e) Boron Deficiency **f) N/K imbalance**

Figure 12.3: Nutrient Deficiency Symptoms.

c) Control Measures

(*i*) For seedlings (2-3 leaf stage), 2 per cent urea solution is used, repeated at weekly or fortnightly interval.

(*ii*) Eradication of grasses around the seedlings.

(*iii*) Improving drainage by using drains.

(*iv*) Application of recommended dose of N fertilizer and addition of organic manures like farm yard manure and neem cake will help to overcome the problem.

(*v*) When N fertilizers like urea are applied there should be adequate moisture in soil or irrigate immediately. Otherwise there will be volatilization loss of N.

12.9.2. Phosphorus

a) Factors Affecting Availability

In small seedlings, the oldest leaves become dull and assume a olive green colour. The chlorotic condition increases in severity, but the seedlings do not become fully yellow before necrosis of the tips occur. However clear symptoms of phosphorus deficiency does not appear under field conditions but the trunks of affected palms may show a a pronounced pyramid shape earence and shortened fronds. However, other plants in the oil palm ecosystem show symptoms indicative of soil P infertility.

b) Causes

(*i*) P deficient legume cover crops (*e.g. Pueraria phaseoloides*) produce abnormal small leaves and are difficult to establish. Sufficient P is important to realize the potential for biological N fixation from legume cover crops.

(*ii*) P deficiency occurs most commonly where soil contains very low concentration of available P (less than 15 mg P kg^{-1}, Bray II).

(*iii*) Inadequate application of P to high yielding palms.

(*iv*) Palms planted in areas where top soil has been eroded and run off.

c) Management

On acid soils, rock phosphate is to be used to provide maintenance requirements Apply 600 gm of phosphorous per palm per year in 2 to 3 equal split doses

12.9.3. Potassium

a) Causes

(*i*) Soils of granite origin

(*ii*) Peat soils

(*iii*) Soils with poor moisture retention ability

(*iv*) Heavy application of fertilizers *viz.*, Calcium ammonium nitrate, Rock phosphate, Kieserite.

b) Symptoms

The development of K deficiency symptoms occurs in the older fronds. The most common deficiency symptoms are

(*i*) **Confluent orange spotting:** Symptoms first appear as pale green rectangular spots which gradually become irregular and colour changes through olive green to bright orange with coalescence between adjacent spots until a rectangular mass develops on older pinnae of severely deficient palms. Necrosis develops in the centre of older spots. When heavy confluent orange spotting occurs, older fronds appear copper or bronze in colour from a distance.

(*ii*) **Mid crown yellowing:** The earliest symptom is a rather dull uniform Khaki or ochre chlorosis in one of the younger fronds situated in the upper part of the crown. The chlorosis spreads until the whole pinna is uniformly affected, with the pinna margin later becoming necrotic. The 'shading effect' characteristic of magnesium deficiency is also found in this symptom *i.e.* where one leaflets is covered by another and the covered portion does not show chlorosis.

(*iii*) **Orange blotch:** Large, elongated and diffuse olive green blotches appear on pinnae of older fronds, located about halfway along the length of pinnae and in pairs on each side of midrib. With increasing age, the discolouration is bright yellow and then orange, with well defined margins between them and healthy tissues. Ultimately all pinnae become chlorotic, which contain a mass of minute spots.

(*iv*) **Pre-mature desiccation:** The terminal results of above mentioned K deficiency symptoms are confluent orange spotting, mid-crown yellowing and orange blotch end in marginal or terminal die-back of the pinnae. Tissues become grey in colour, brittle and desiccated. Finally fragmentation occurs.

c) Control Measures

Application of 3-4 kg Potassium chloride/palm to be followed after 6 months based on leaf analysis data. This should be followed by leaf analysis and canopy assessment after application to confirm that deficiency have disappeared and leaf K concentration has increased.

12.9.4. Magnesium

a) Causes

(*i*) Heavy application of potash fertilizers.

(*ii*) Soils having very small concentrations of exchangeable Mg

(*iii*) Soils of granite origin

(*iv*) Peat soils and soils with topsoil removed.

(*v*) Acid-sulphate soils

(*vi*) Inadequate application of Mg to high yielding palms or palms growing on Mg deficient soils

b) Symptoms

Olive green and ochre coloured areas appear on the pinna of older leaves and merge gradually with the healthy green tissue. The yellow colour spreads down towards the frond midrib untill the whole pinnae is affected, becoming a deep orange colour in later stages and finally leaf gets desiccated. 'Shading effect' is the most diagnostic feature of Mg deficiency as the symptoms occurs only on leaflets exposed to light. However, newly emerged leaves will not show any deficiency symptoms.

c) Control Measures

a. Application of 2-4 kg/palm of Kieserite (Magnesium sulphate) will be required to remedy the deficiency in adult palms.

b. For seedlings (8 month old), 7g/palm Kieserite is applied. If normal Kieserite fails to cure deficiency, spraying of 2 per cent solution of Epsom salt at 3-4 intervals for a period of 2-3 weeks is required.

12.9.5. Boron

a) Causes

Boron deficiency occurs most commonly where:

(*i*) Palms have received large application of N, K and Ca fertilizers.

(*ii*) Soils contain very small concentrations of available 'B' (*e.g.* sandy soils, peat soils).

(*iii*) Soils with very low (<4.5) or high (>7.5) soil pH.

(*iv*) Increased removal in bunches, due to improved pollination by *Elaeidobius kamerunicus*, has not been balanced by the addition of 'B' fertilizer.

(*v*) Inadequate application of 'B' to high yielding palms or palms grown on 'B' deficient soils.

b) Symptoms

The first symptom of 'B' deficiency is the shortening of young leaves giving rise to the characteristics "flat-top" appearance (Figure 12.4).

Boron Deficient leaves are dark green, brittle and misshapen or wrinkled (Figure 12.4) and produce symptoms which have been described as "hooked leaf", "fish-bone leaf" and "blind leaf" which are easily identified in the field.

c) Control

Severely affected palms may require a corrective of 200g Sodium borate per palm, applied in the palm circle, close to the stem. The application of 'B' in the axils has been suggested but is not recommended as it may result in uneven distribution of 'B' in the palm and 'B' toxicity.

Figure 12.4: Boron Deficiency Symptoms.

12.10. Disorders

12.10.1. White Stripe

a) Causes

(*i*) N: K ratio in the leaves is more than 2.5.
(*ii*) Boron deficiency
(*iii*) Excessive N application

b) Symptoms

Narrow bands of chlorotic tissue appear along the length of the pinnae. In the early stages, the chlorosis produces a strip clearly delineated from the adjacent green tissue. Strip may be at any portion between the midrib and edges of lamina. In severe cases, there is change in morphology of crown, *i.e.* upright and obconic in shape.

c) Control Measures

Regular leaf analysis to check leaf N: K concentrations and ratios. Application of 2.5 - 4.5 kg Potassium chloride/palm together with temporary cessation of 'N' fertilizer application.

12.10.2. Plant Failure

a) Causes

Deficiency of potassium and magnesium.

b) Symptoms

Symptoms are markedly seen on the older fronds. In early stages, yellowish to orange coloured spots develop which enlarge both in number and size with increasing age. Adjacent spots coalesce, until the pinnae appear bright orange or bronze in colour. Ultimately, the entire frond dies and is typified by one or more whorls of semi erect dull grayish brown fronds.

12.10.3. Leaf Base Wilt

a) Causes

(*i*) High potassium content
(*ii*) Low magnesium levels
(*iii*) Variation in phosphorus contents
(*iv*) Susceptibility of dumpy varieties.

b) Symptoms

The lowermost fronds bend downwards at the point at which stalk is inserted on the stem and continue to bend until the fronds come to rest on the ground. When the disorder is severe, the palms become enclosed within a tent of fronds. Cracks later develop in the petiole. It is normally associated with period of bunch production. Affected palms normally recover spontaneously. Since it is not associated with loss of crops, no control measures are required.

12.10.4. Peat Yellows (Cu Deficiency)

Peat yellows noticed in peat soils of Malaysia and Indonesia and results from nutrient deficiencies and imbalances. The first symptom will be the development of pale green to yellow interveinal chlorotic streaking of pinna of youngest fully opened fronds. The streaks extend from the distal end to within 5-8 cm of proximate end of leaf pinnae. Vascular strands in leaf pinnae stand out in sharp contrast to the cholorotic streaks due to greater chlorophyll production in tissue adjacent to pinnae veins Later yellow speckling sometimes develop within the chlorotic streaks which accounts for the yellows in the disorders name. Affected fronds are shortened,turned pale orange in colour and moribund fronds become desiccated and die.

a) Causes

Peat yellows occurs most commonly where soil is deficient in potash, soil deficient in cu, large doses of Mg fertilizer have been applied, large application of N and large applications of P without sufficient K.

b) Management

Preventive measures have not been recommended for this disorder. Continuous leaf analysis will help to identify outbreaks at an early stage so that different treatment regimes can be tested out in goo d time.

12.10.5. Leaf Breaking

The main symptoms are bending of leaf petioles, drooping of leaves mostly on the outer whorl, twisting and fracture of the middle whorl, bending of the upper portion of the crown, scorching symptoms on the upper side of the leaf rachis (Fig. 12.5). The reason for the leaf breaking could be nutrient imbalance, a larger number of branches, more leaves per palm in the crown, exposure of leaf bases to high temperature and reduced vapour pressure.

View of leaf breaking in plantation | Close up view of single palm

Figure 12.5: Leaf Breaking Symptoms.

12.10.6. Bunch Failure

Bunch failure is common in young plantation but it is also seen in mature plantations the Figure 12.6 shows the symptoms of bunch failures.

Figure 12.6: Bunch Failure.

a) Symptoms

(*i*) The first symptom of bunch failure is the loss of the glossy appearance with characterizes healthy fruit.

(*ii*) Bunch failure is scattered irregularly or is confined to a few fruits over the bunch.

(*iii*) Aborted fruits became desiccated and shriveled but do not fall from the bunch failed bunches are rapidly invaded by a large number and a wide range of insects and microorganisms.

b) Causes

(*i*) Inadequate pollination/absence of pollinating weevil

(*ii*) Inadequate nutrient status and irrigation especially during at the time of high yields

c) Management

There is no recovery once bunch failure has started and hence all control measures must be aimed at avoiding those conditions favouring bunch failure. Most important is providing adequate nutrient status especially for palms newly in production, where bunch failure is a result of adequate pollination efforts should be taken to release the pollinating weevil *Elaeidobius Kamerunicus* and correct the situation.

12.10.7. Chimeras

The occurrences of chimeras are purely genetical. It is the asymmetrical pattern of yellowing of leaflets on some fronds in a palm. It is observed on rachis, leaflets and midrib with variation in pattern of distribution.

Nutrient deficiency symptoms of oil palm seedlings, probable cause and corrective action to be taken are given in Table 12.8.

12.11. Yield Gap Analysis

The highest theoretical oil yield of oil palm is 18 tons of crude palm oil/ha/yr. The calculated yield of a high yielding palm at Nava Bharat oil palm plantations at Lakshmipuram of West Godavari District was 12 tons/ha at the age of 8 years after planting. In small holders plantations many progressive farmers are getting more than 40 MT FFB/ha/yr. or 8 t of oil/ha/yr. Large number of farmers who are adopting the optimum package of practices are getting 20 MT FFB/ha. Many other farmers who are not adopting the packages fully are getting very low yields of 5 to 10 MT FFB/ha. In Malaysia and Indonesia many large scale plantations are giving 40 to 50 MT FFB/ha (8 to 10 t oil/ha/yr.)

The gap between actual achieved yield (Y) and the maximum yield potential (Y-max) can be apportioned into 3 parts – i) Yield Gap 1 arises from in efficiencies during development of a plantation until the end of the immature period; this limits the attainable yield (Y-a) in the mature stand compared to Y-max for a particular site. This yield gap 1 is very relevant to Indian oil palm growers. After taking up the oil palm plantation the land is leased for growing inter crops and those who are taking up the inter crop do not leave any space around the palm, even cut off leaves or tie the leaves of oil palm and do not irrigate and manure after harvest of intercrops; ii) yield gap(G2) arises from inaccurate assessment of nutrient requirements, which further reduces the attainable yield to the nutrient-limited yield (Y-n). This is also common and many farmers apply fertilizers rationally not a balanced fertilizer application. Iii)Yield Gap 3 (G3) arises from inefficiencies in the management of the mature stand, bringing Y-n down to the actual yield (Y). There are limited opportunities for plantations to correct Yield Gap 1 – the first during the

Table 12.8: Nutrient Deficiency Symptoms of Oil Palm Seedlings, Probable Cause and Corrective Action

Nutrient	*Deficiency Symptom*	*Probable Cause*	*Corrective Action*
Nitrogen (N)	Uniform paling or yellowing of the entire leaf area	Insufficient N fertilizer. Waterlogging, excessive water in the polybag. Insufficient watering causing N volatilization. Intensive solar radiation, removal of shade usually causes temporary N deficiency. Deficiency can also be caused by weed competition, particularly some grasses	Adjust the N fertilizer rate. Reduce watering and/or control and if necessary improve drainage system. Water after application of N fertilizer. Ensure that seedling shade is managed correctly.
Phosphate (P)	No specific symptoms. Poor root development and as a result, poor seedling growth	Nursery soil source is P deficient	P deficiency should not occur if nursery soil has been prepared correctly. Change source of soil for future bag filling
Potassium (K)	Development of small, olive green spots that later turn bright orange-yellow and transmit light. Affected seedlings have a 'flat top' appearance	Sandy soil, clay soil or soil derived from marine alluvium that either are low in K or fix K have been used to fill polybags. Seedlings held in nursery too long	K deficiency should not occur if nursery soil has been prepared correctly. Change source of soil for bag filling. Some orange spotting may be natural to some oil palm varieties and is not due to K deficiency
Magnesium (Mg)	Appearance of bright orange discolouring of older leaves. Shaded leaves show no deficiency symptoms	Bag filling soil contain insufficient Mg. May occur in sandy soils or soils that contains some organic matter. Intensive solar radiation. Excessive application of N and K fertilizer may induce Mg deficiency	Control the source of soil for bag filling. Use fertilizer containing Mg. Kieserite (magnesium sulfate) at 7 g/bag for 7 months old seedlings, increasing up to 15 g for older seedlings. Reduce application of N and K fertilizer. Ensure that the observed symptoms are not due to chimera genetic abnormalities
Copper (Cu)	Chlorotic speckles appear on the edge of the youngest leaves. Seedlings severely stunted. Leaflets may show desiccation and necrosis in the affected area. Cu deficiency is rare	Bag filling soil contain insufficient Cu and/or K. May occur in peat soils. Excessive application of N, P and Mg fertilizer may induce Cu deficiency	Do not use peat soils as bag filling soil. Seedlings can be treated with a 0.05 per cent solution of $CuSO_4$. This can be mixed with the foliar fertilizer applications for four consecutive applications
Boron (B)	The symptoms are deformity of leaf development such as 'hook leaf', 'wrinkle leaf' and 'whiskers'. Leafs are brittle and dark green. Palms develop a flat top appearance. B deficiency is quite common	In a few cases these symptoms may be due to genetic abnormalities	If leaf abnormalities is severe or growth poor, the affected seedlings should be rouged. If the symptoms are less severe borax (11 per cent B) can be used: 40 g/16 liters of water to be used in 400 plants. The spray is directed to the soil since boron can be very toxic to young palms

initial establishment of the plantation, and thereafter at each occasion of replanting. There is need for educating the farmers who are taking up the oil palms right from planting. However, Yield Gaps 2 and 3 can be corrected in existing mature stands. The yield gap analysis is useful for the identification, selection and prioritization of best management practices (BMPs) to be implemented for intensifying yield.

BMPs chosen for implementation are those that have been proven over time to benefit palm growth and productivity, and to conserve soil, water and nutrients. They can be grouped into 3 broad categories according to purpose – (a) Crop recovery BMPs, (b) Canopy management BMPs, and (c) Nutrient management BMPs. Crop recovery BMPs included – (1) Harvest interval (HI) of 7 days, (2) Minimum ripeness standard (MRS) = 1 loose fruit (LF) before harvest, (3) Same day transport of harvested crop to palm oil mill, (4) Harvest audits to monitor completeness of crop recovery and quality (*i.e.* ripeness) of the harvested crop, (5) Good in-field accessibility (clear paths, bridges wherever needed) (6) Clean weeded circles, (7) Palm platforms constructed and maintained wherever needed, and (8) Minimum under-pruning in tall palms to ensure crop visibility. Canopy management BMPs included – (1) Maintenance of sufficient fronds to support high palm productivity, (2) Removing abnormal, unproductive palms, (3) In-filling unplanted areas, (4) Selective thinning in dense areas, and (5) Monitoring and management of pests (leaf eaters) and disease (Ganoderma), Nutrient management BMPs included – (1) Spreading pruned fronds widely in inter-row area and between palms within rows, (2) Eradication of woody perennial weeds, (3) Mulching with empty fruit bunches (EFB), (4) Management of applied fertilizers (*i.e.* type, dosage, timing and placement), and (5) Monitoring of plant nutrient status and growth.

Chapter 13

Oil Palm Based Cropping/Farming Systems

Oil palm being a plantation crop committed to land for 30 years with wider spacing at 9 m in triangular system has a long juvenile period of 2½ to 3 years, which provides ample opportunity to raise inter, mixed, multiple and multi-tier cropping and mixed farming system. Instead of waiting for some period to get income from oil palm the farmer can raise inter crops with annuals and biennials which will give some income to sustain the family particularly to small and marginal farmers. Again once the crop reaches maturity it will be feasible to go for mixed cropping and mixed farming which will help to increase the productivity of the plantation and also to help in getting sustainable income throughout the crop period besides providing employment opportunity continuously.

In view of continuous decrease in size of holding and per capita land availability, it is necessary to utilize the available space effectively both vertically and horizontally. Inter/mixed, multiple cropping, *etc.* can achieve the maximum utilization of available space. This will provide sustainable income throughout the period. Different Oil Palm Based Cropping System are discussion below:

13.1. Multiple Cropping

It is defined as the cropping system involving multiple species crop combination, both annual (ginger, turmeric, tuber crops, vegetables, legume, *etc.*) and perennial crops (banana, papaya, pineapple, cocoa, coffee, nutmeg, *etc.*) with an existing stand of perennial crop. In humid tropics, the basic sources of crop production *viz.* Solar radiation and water are abundantly available. Their high efficiency can be achieved by adopting multiple cropping system. Which generates more biomass, steady and higher income, additional employment opportunities for the farmer's family and meets the diversified needs of the farmer such as food, fruits, vegetables, fodder, fuel *etc.*

The multiple crop agro-ecosystem is designed in such a way that it improves production in a sustainable manner. The following principles need to be kept in mind in this system.

(*i*) It enhances recycling of biomass/crop residues and optimizes nutrient availability to ensure viable economic output of desirable commodities to satisfy market/consumer demand.

(*ii*) It secures favourable soil health/quality for optimum plant growth particularly by managing organic matter and enhancing soil biological activity.

(*iii*) It minimizes losses of nutrients and water through effective micro-climate, management, water harvesting and soil management through increased soil cover.

(*iv*) It promotes synergisms among agro-biodiversity components resulting in the speeding up of key ecological process and services.

These principles can be applied through various techniques and strategies.

a) Factors Affecting the Multi-Crop System

The multi-crop system is affected by many factors such as available solar radiation in the plantation, root distribution pattern, microclimate of plantation, choice of the crop, marketability and profitability as well as viability of the system. These are discussed as follows:

i) Availability of Solar Radiation

Solar radiation is an important factor affecting intercropping system. The solar radiation in oil palm plantation decreases rapidly in the initial years and by 6th year whole land of plantation is covered by the palms. Later on as the age of plantation increases the solar radiation available at ground level increases steadily and is given in Table 13.1.

Table 13.1: Solar Radiation Available at different Heights in 6 Year Old Oil Palm Plant

Height of Plant (m)	*Radiation Available Per cent*	
	10.30 AM	*5.00 PM*
9.0 Tree top:	100	100
7.5 height	90.9	86.5
6.0 height	72.3	61.4
4.5 height	49.3	36.3
3.0 height	29.0	17.6
1.5 height	20.6	11.0
Ground (0)	19.9	10.4

ii) Root Distribution

The roots of the oil palm plant are generally spread within the canopy of the palm. Most of the oil palm roots will remain within 3.6 m radius. A lot of space remains available for growing other crops in initial years. The space available in these three years for growing intercrops is given in Table 13.2.

Table 13.2: Space Available for Intercrops in Oil Palm Plantation at different Ages

Age (Years)	*Radius of Basin (m)*	*Area Available for Intercrops (Per cent)*
0-1	1	95
1-2	2	82
2-3	3	60

iii) Microclimate in Oil Palm Plantation

The microclimate of oil palm plantations favours luxuriant growth of shade-loving plants. Regular irrigation of oil palm provides enough water and sufficient humidity and decreases the temperature of oil palm canopy area. The high humidity, sufficient water and lower temperature create favourable environment for growth of weeds which needs lot of labour for their control. Thus, it is useful to have intercrops under such conditions. It helps in control of weeds as well as utilization of excess moisture and available land.

iv) Choice of Crops

The choice of crops depends on the region, expertise of the farmer, intercropping pattern, their intensity and space availability. The availability of irrigation facilities and amount of water available also affect the choice of crop to be grown. The choice of crop is also influenced by the availability of marketing and processing facilities. Soils of low fertility are not good for turmeric, ginger and cocoa but crops like tapioca and yams *etc.* can be grown successfully in such soils. The laterite soil of coastal region is suitable for growth of root crops, tobacco and vegetables especially water melon and cucumber. The crops like pineapple, banana, ginger, turmeric, clove, nutmeg, *etc.*, can also be grown successfully in oil palm plantations. There is also scope of growing flower crops, medicinal and aromatic crops, and tuberose, gladiolus, citronella grasses *etc.* in oil palm plantations.

The crop to be grown is greatly influenced by the grower's choice. A stable crop is always the first preference of the farmer, but a good rotation involving other crops rather than a single crop is always profitable. Basically, the crop selected should have shade tolerance and should not grow taller than Oil Palm. It should be complementary, exploit nutrients and water from different horizons of the soils.

13.2. Intercropping in Oil Palm

Although oil palm is a new crop for India, the experience of last 7-8 years shows that many annual and perennial crops perform well in oil palm plantation. These crops can be planted in two ways, namely, Inter row intercropping (between the

a) Lily in Oil palm

b) Marigold in Oil palm

c) Turmeric in Oil palm

d) Banana in Oil palm

c) Brinjal in Oil palm

Figure 13.1: Different Intercrops in Oil Palm.

rows of oil palm) and within row intercropping, *i.e.* in all the area except oil palm basins. While going for inter-cropping select a locally familiar crop which gives good yield and income.

Crops selected for intercropping should be compatible with the main crop and should not compete with the oil palm for light, water and nutrients. Suitable crops are vegetables, pulses, banana (Dwarf Cavendish), flowers, chillies, turmeric, ginger and pineapple. While raising inter crops, avoid tying of oil palm fronds which will reduce photosynthetic activity and ploughing close to the palm base which will cut the absorbing roots and thereby reduce intake of water and nutrients. Allow oil palm to grow freely. While growing pulses like green gram, black gram, cowpea and oilseeds like sesame, groundnut which do not require frequent irrigation, care should be taken to irrigate oil palm regularly so that oil palm does not suffer from lack of moisture.

In case if crops like banana, tobacco, chillies sugarcane *etc.* are to be grown, proper spacing should be given. Oil palm crop should not be affected due to shade and root competition of inter-crop. In one way growing sugarcane is considered as compatible in oil palm plantations because sugarcane crop will be irrigated throughout the year and thereby oil palm also will be irrigated. In other way it is incompatible as it shades the oil palm plants which are small in the initial years. So, propping and trashing the canes at least around oil palm basins should be done to avoid shading to oil palm.

The survey conducted in oil palm plantations in West Godavari district of Andhra Pradesh revealed that most of the farmers (93 per cent) are growing intercrops in the initial three years. The major intercrops grown were tobacco, groundnut, banana, maize, chillies, sunflower, vegetables, *etc.* These intercrops provided Rs. 5,000 to Rs.50,000/ha in the initial years. Wherever management was good and due care was taken for oil palm plants, there was no adverse effect of intercrops on growth and yield of oil palm.

Intercrops with aromatic plants like lemongrass, Palmrose, *Citrinella*, Pacholi, French Basila, Davana, *etc.* can be grown for extracting essential oils. Medicinal plants like *Piper longum*, Kacholar (*Kalmpferia galaga* L.), mango, ginger, periwinkle, *etc.* can also be grown (Figure 13.1).

The intercrop should always be treated as secondary crop and additional fertilizers and irrigation should be applied to them. Oil palm with 9x9 m spacing has ample space for inter/mixed crops. The intercrops should not create any type of competition to oil palm.

13.3. Mixed Cropping

Mixed cropping in oil palm is also possible after 3-4 years of planting. In mature plantation, shade loving and short statured crops can be grown. The crops like Cocoa (*Theobrama cocoa*), Coffee (*Coffea arabica*), Clove (*Engenia caryophyllus*), and Nutmeg (*Myristica fragran*) are suitable for this purpose (Figure 13.2).

a) Betelvine in Oil palm

b) Black pepper in Oil palm

c) Cocoa in Oil palm

d) Bush pepper in Oil palm

e) Vanilla in Oil palm

Figure 13.2: Various Mixed Crops in Oil Palm.

Since Oil Palm trees under irrigation will have a dense canopy system while selecting mixed cropping, one has to be more careful. Depending upon the sunlight falling on the ground the other crops should be selected. The population of mixed crop also should be rationalized. For example, planting cocoa or clove or nutmeg, only one plant should be planted in between the triangular plantation. While planting black pepper or live/dead standard the spacing should be carefully planned. Growing black pepper on the Oil Palm tree may sometimes be damaged by the falling of cut leaves or fruit bunches. When bush pepper is grown in rows adequate space of at least 2 m should be left between plants.

13.4. Mixed Farming

Mixed farming in oil palm involves establishment of pastures or grasses in oil palm plantation and maintenance of milch animals. This adds organic matter in soil and controls the weed growth. Fodder crops like *Stylosanthes* sp. and elephant grass (*Pennisetum purpureum*) can be grown in mature plantation. They provide sufficient fodder to rear 4-5 buffaloes per hectare (Figure 13.3). The basin of 3 metre radius should not be used to sow fodder to facilitate fertilizer application and easy harvesting. An additional income of Rs.6,000 to 10,000 can be earned this way in one hectare area.

The grazing area in each plantation may be divided into plots. The stocking density generally depends on the soil, climate, age of plantation, and management of plantation and grasses. In low stocking densities the live stock damage to plantations is no greater than a light pest or pathogen damage normally experienced

Figure 13.3: Mixed Farming in Oil Palm with Sheep, Goat and Cattle.

in plantations. If animal grazing is not done carefully, the land might be seriously puddled and drainage impeded. Rearing of sheep also can be done utilizing the grasses in the plantation.

Forage crops like *Paspalum conjugatum, Pennisetum purpureum, Cytococcum oxyphyllum, Setaria palmifloria, etc.,* Guinea grass, Hybrid Napier grass can be grown at 60x30 cm spacing leaving a basin of 3 m radius around the adult oil palm. This will help to increase water holding capacity and-porosity besides build up of organic C, N, P and K.

The mixed farming system greatly helps to increase income, provide employment opportunity, facilitate organic recycling, and improve soil health besides being eco-friendly.

13.5. Cover Crops

Cover crops are established and maintained in the plantation for conserving soil and for improving and maintaining the soil structure and fertility. Cover cropping is nothing but covering the entire space of oil palm plantations with a natural vegetation of non leguminous or leguminous cover with the objective of preventing weed growth, soil erosion, conserving soil moisture and humus. It also provides aeration to soil and recycles the nutrients through organic residues into the soil.

Leguminous creepers are often used as cover crops as these are easy to establish and are the most desirable species of plants that can be used as ground covers in plantations.

Usage of legume cover crops in the inter-row space between planted areas, particularly during the immature phase is advantageous. Leguminous covers have been proven to be effective and have good nutrient returns compared to natural covers. They have the ability to fix atmospheric nitrogen that can later be used by the immature plants, besides other characteristics common to all cover plants. Also, such legume covers keep down the soil temperature by moisture retention in the soil, suppression of weed proliferation, help in proper utilization of fertilizers applied to the plants and prevent soil erosion especially in hilly areas. A list of important cover crops is given in Table 13.3.

Cover crops should not compete with the oil palm to the extent that the yield of oil palm is affected. These crops should not be of fast growing nature, to the extent of shading young oil palm trees. They should need very little attention and form a substantial cover to the soil to yield good quality herbage which decompose very rapidly to add organic matter to soil which in tum becomes humus and supply nutrients. These crops should have root system which penetrates into soil to provide aeration.

The most common cover crops that can establish well in oil palm plantations are:

Pueraria phaseoloides, Calapagonium muconoides, Centrosema pubescens, Mimosa invisa, Mucuna sp. *etc.* These cover crops can be sown after taking intercrops during first two years. Cover crops are sown in the entire field leaving the basin.

Table 13.3: List of Cover Crops and their Characteristics

Sl.No.	*Name of the Cover Crop*	*Characteristic*
1	*Calapagonium caeruleum*	Excellent cover crop in humid-tropical tree plantations. Leguminous creeper, grows slowly at the beginning stage and spreads later. Shade tolerant, adapted to a wide range of soil textures. Grows best on well drained soils.
2	*Calapagonium muconoides*	Leguminous crop, grows rapidly and dies early can grow on a wide range of soil types, tolerant of inundation. Useful cover crops are legumes but have a short life-span.
3	*Centrosema pubescens*	Fast establishing legume but susceptible to pests, shade tolerant and slow starter. It can survive under stiff competition from *Calopogonium* and *Pueraria*
4	*Pueraria javanica*	A very popular cover crop. Vigorous, thick and dominant legume often suppressing the growth of undesirable plants. The plant can stand strong sun and smothers weeds. Wide adaptation to soil types, tolerance to soil acidity and high nutritive value.
5	*Pueraria phaseoloides*	Leguminous cover crop susceptible to pests
6	*Mucuna cochinensis*	A potential short term legume, fast growing in suitable areas prone to soil erosion, Leguminous creeper giving rapid coverage. Prefers well drained, medium to fertility soils but can be grown successfully on sandy soils. Prefers hot and humid climates. Requires high light intensity and improves soil fertility.
7	*Flemingia congesta*	Erect growth which improves the growth of palms
8	*A/nus nepalensis*	Non-leguminous nitrogen fixing tree
9	*Stylosanthus gracilis*	In the initial stages of oil palm growth may cause depressive effect
10	*Dolichos hosei*	Leguminous creeper, reported for persisting under trees
11	*Mucuna bracteata*	Deep rooted, shade tolerant and a long term legume, forms thick luxuriant cover and suppresses most weeds in plantations. Perennial creeper, vigourous in growth, tolerant to pests and diseases, Nitrogen fixing capacity of this legume was found to be high with good control against soil erosion.
12	*Psophocarpus palustris*	Leguminous creeper
13	*Desmodium ovalifolium*	Shrub, shows negative effect on oil palm growth
14	*Phaseolus calcaratus*	Commonly grown creeper, reported to die back early
15	*Arystasia intrusa*	Natural ground cover does not have deleterious effect on oil palm

In pure plantations these cover crops can be sown at the time of planting of oil palm seedlings. Cover crops help in soil and water conservation and check weed growth. When incorporated they also improve the organic matter content of soil and plant nutrient status.

Pueraria sp. is found to establish very well and spread fast in oil palm plantations.

Pueraria seeds before sowing require treatment with Conc. sulphuric acid (scarification) or soaking for 12-24 minutes and subsequent heat treatment for one hour at 39-40 °C.

Calapagonium sp. seeds are soaked in Conc. H_2SO_4 for 15-20 minutes and acid is repeatedly washed out with hot water. Inoculation of seeds with specific Rhizobium sp has to be carried out for better nodulation and rock phosphate has to be applied at the time of planting.

Maintenance of these cover crops consists of cutting back the adventitious growth to 30 cm as many as six times a year. Frequent clearing around the palm basin is needed so that the palms are not covered by the creepers especially the young palms. As the palms grow and shade the area, ground cutting can be reduced to one or two rounds depending upon the growth of the cover crops. Mixture of three cover crops, *viz., Calapagonium, Centrosema* and *Pueraria* can also be successfully raised in oil palm plantations instead of a single cover crop.

Chapter 14

Plant Protection

Plant protection in broad sense includes pests, diseases and weed control but in this chapter only pest and disease are covered.

14.1. Insect Management

Oil Palm like any other palm is attacked by several pests. The pests affecting coconut usually find a place in Oil Palm also. However, under good management conditions, these problems are very minimum and there is no need for using insecticides indiscriminately. Important pests of oil palm are discussed in this chapter:

a) Nursery Pests

Oil Palm like any other palm has the problems of pests. The pests of coconut always find a place in Oil Palm. However, under good management conditions, these problems are very minimum and there is no need for using the insecticides indiscriminately.

i) Red Spider Mite

It is most frequently encountered nursery pest, which occurs on the undersides of leaves, especially older ones. A few species of these mites, namely *Oligonychus* and *Tetranychus* have been found in most nurseries but in very low numbers which do not represent a serious economic hazard. The feeding wounds of a mite provide a way for entry of weak pathogens. Irregular orange red lesions are frequently seen in association with mite attack, hastening leaf decay. High volume sprays with Rogor (as Rogor 400) at 0.1 per cent, followed 10 days later by Tetradifon (as Tedion) in either a 0.2 per cent emulsion or 0.1 per cent wettable powder (0.2 per cent a.i.) have given effective control.

ii) Spindle Bug

Nymphs and adults of spindle bug (*Carvalhoia areca* Miller) suck sap from the spindle and tender leaves. A longitudinal, narrow, discoloured zone develops on the sides of the feeding marks resulting in necrotic lesions, which later, turn into dry brown patches. Its incidence is highest during June and lowest in February. The infestation can be brought down with 2 g of phorate 10 G (Thimet) taken in heat sealed polythene sachets, and placed on the top most two leaf axils (one sachet per axil). These sachets are transferred to the youngest leaf axils as and when new spindles emerge.

iii) Tussock Caterpillar

The larvae of *Dasyclira mendosa* Hubner feeds on young and mature leaves in oil palm nurseries. The caterpillars are polyphagous and can be identified by the presence of dense tufts of hair growing forward and backward. The caterpillars feed on the leaves causing defoliation. The percentage of seedling defoliation ranges from 3-20 in various places. Pest is present throughout the year. The larvae can be controlled by spraying with Carbaryl 50 per cent WP 0.0 I per cent after cutting and burning the badly affected leaves.

iv) Pink Shoot Borer

The shoot borer (*Sesamia inferens* Walker) occurs on primary and secondary nurseries. Larvae tunnels into the stem through the spindle leaf rachis and reaches the meristematic tissues arresting the growth, producing dead heart and little leaf symptoms.

Spraying of Carbaryl 50 per cent @ 0.01 per cent at bimonthly interval is recommended to bring down the pest incidence.

b) Pests of Adult Plantations

i) Rhinoceros Beetle (*Oryctes rhinoceros*)

One of the most serious pests of oil palm is the rhinoceros beetle. It commonly attacks Oil palm and sago palm. On oil palm this pest assumes importance right from the seedling stage.

- ☆ **Symptoms and Life Cycle:** The palm is affected by the adult beetle, which burrows into the cluster developing spears in the crown and bores its way throughout the petioles into the softer tissues of the younger unopened leaves. The most common and characteristic symptom is the wedge shaped gaps in the leaf silhouette (Figure 14.1). Permanently marked hole on the spear cluster with the chewed up fibre pushed outside the hole is also one of the characteristic symptoms of the damage. When the infestation is severe, damage symptoms are seen on fronds, crowns and spathes with reduced yields and sometimes lead to the death of the palms. The attack is most dangerous in young palms since the growing point may occasionally be damaged or a bud rot may develop which will kill the palm.

Figure 14.1: Rhinocerous Beetles (Adults) and Symptoms on Young Palm.

The pest is a prolific breeder and can multiply in any form of decaying organic matter.

The major breeding sites are:

- ✰ Cattle dung pits
- ✰ Decaying sawdust and woodbark heaps
- ✰ Decaying oil palm bunch refuse
- ✰ Rotting farm yard manure compost
- ✰ Tops of dead standing palms and logs of coconut and oil palm.

The female beetle lays 140-150 eggs in the decaying organic matter such as manure heaps, drying and dead palm trunks, decaying stumps and compost pits. The egg laying period is 8-18 days. A larva feeds on decaying matter and becomes full grown in 140-195 days. Pupation takes place in the soil. Adult is stout built, black beetle with a characteristic pronotal horn, which is more conspicuous in males than in females (Figure 14.2).

- ✰ **Management:** Integrated approach using all the possible methods of control is the only possible way to manage the pest effectively. These methods are discussed below:
- ✰ **Prophylactic:** Leaf axil filling with insecticide sand mixture is recommended as a prophylactic method to prevent the pest entry. Granular insecticides like Sevidol (8G) or Phorate (10G) mixed with fine sand in the ratio of 1:10 gives best results. This mixture should be applied in the leaf axils during April-May, Sep-Oct and Dec-Jan.

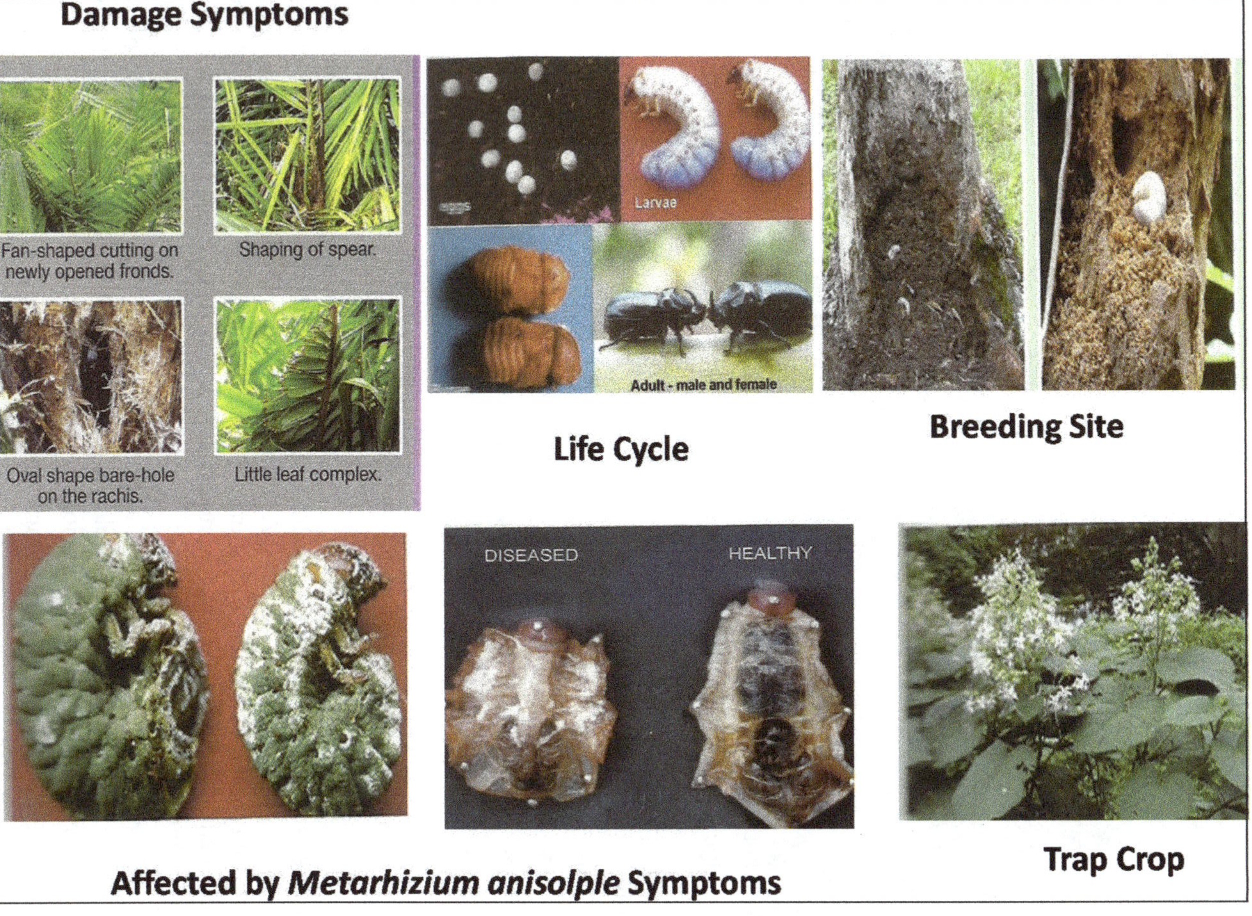

Figure 14.2: Symptoms, Life Cycle, Breeding Site and Bio-control.

- ☆ **Baiting:** Trapping the weevils using fermented castor cake is the best method to control this pest. For this purpose, 1 part of castor cake mixed with 2 parts of rice water and fermented for 3-4 days should be kept in wide mouthed pots at different places in the plantations. Adult beetles are attracted to this solution which should be removed everyday failing which these will give unbearable smell. The solution should be changed once in a week.
- ☆ **Sanitation:** Destruction of all possible breeding sites mentioned above and keeping the plantations and its surroundings in a clean condition are essential to keep the pest away from using it as breeding site. The decaying heaps of cattle dung, compost pits and other breeding sites should be treated with insecticide like Carbaryl 85 per cent WP (0.01 per cent) at regular interval which will control the immature stages of the pest. Fully decomposed farmyard manure (FYM) and compost should only be applied to the plants, otherwise the pest will migrate along with unrotten FYM and attack the palm.
- ☆ **Mechanical:** It involves extraction and killing of the beetle from the crown of infested palms with hooked pointed metal rod of 0.5 m length having a hook at one end and a handle at the base. After extraction the hole should be filled with sand.
- ☆ **Biological:** Larvae are attacked by fungal disease, caused by *Metarrhizium anisopliae* and nematodes. The virus *Baculovirus oryctae* has been found to be highly effective in suppressing the infestation. All the instars of the beetle including grub and adult are susceptible to the disease. The infected grubs become lethargic and cease feeding. Disease infected adults don't shows any external symptoms. However, these act as virus reservoirs spreading the infective virus in to the insect's natural habitats infecting the fresh ones.
- ☆ **Pheromone Trapping:** Adults beetles can be trapped by using pheromone material called Orycta pheromone (Figure 14.3).
- ☆ **Chemical:** Chemical control measures can be applied against grubs. Treatment of breeding sites with insecticides like BHC 50 per cent WP (0.01 per cent) or Carbaryl 85 per cent WP (0.01 per cent) at periodical interval can effectively control the immature stages.

ii) Red Palm Weevil (*Rhynchophorus ferrugineus* Oliver)

Red palm weevil is one of the important pests of oil palm. It also attacks other palms like coconut, palmyrah, *etc.* Larva (grub) is the damaging stage.

- ☆ **Symptoms and Life Cycle:** The damage is mainly due to the feeding activity of the grubs, which bore through and feed on the soft tissues of the stem meristem and mesocarp of fruits of ripe bunches. Infested palms show gradual wilting and drying of outer whorl of fronds. Presence of few holes on the trunk which chewed fibre protrudes, and oozes out the brown viscous liquid from the boreholes are the characteristic symptoms.

Figure 14.3: Pheramone Trapping.

Feeding sound of grubs can be heard by keeping the ear near the trunk portion. Death of young palms due to weevil attack have been recorded in many parts of the coconut growing countries. Female lays more than 200 eggs in 40-45 days period. Eggs are laid in holes, wounds or scars or crown or bole region. Grub takes 36-78 days to become fully grown. Pupal period is 12- 33 days. The total life cycle from egg to adult is 82 days (Figure 14.4).

- ☆ **Management:** Since the damaging stage of the pest is concealed in plant tissue, the pest can be detected only at a very last stage. Hence the prophylactic and sanitation measures are the best to control the pest effectively. Integrated approach using all the possible methods of control is ultimately the possible way to manage the pest effectively.
- ☆ **Prophylactic Measures:** Regular inspection of young palms; Avoiding of wounds on palms and treating, wounds and cut portions on palms with tar mixed with insecticides like Endosulfan 35 EC or Quinalphos 25 EC to prevent the entry of the pest in the palm trunk.
- ☆ **Sanitation:** Weevils find old felled palms good breeding sites, so clearing method will need to be considered at the time of replanting in areas where the weevil problem is severe.
- ☆ **Trapping:** Pheromones like Ferro lure or Rhynco lure can be used for trapping the adult beetles of red palm. Coconut toddy, macerated grapes, crushed sugarcane molasses can be used either single or in combination with yeast or acetic acid or both as attractants. This in combination with insecticides like Endosulfan (35 EC) or Quinalphos (25 EC) should be used which will attract and kill the adult beetles. Traps may also be made from recently cut potatoes or pieces of banana.
- ☆ **Chemical:** Root feeding with Monocrotophos (10 ml) mixed with equal quantities of water may be given only to affected palms. A method used for weevil control in coconuts in which a solution of Sevin (Carbaryl) is injected into the palm may find application in oil palm weevil control.

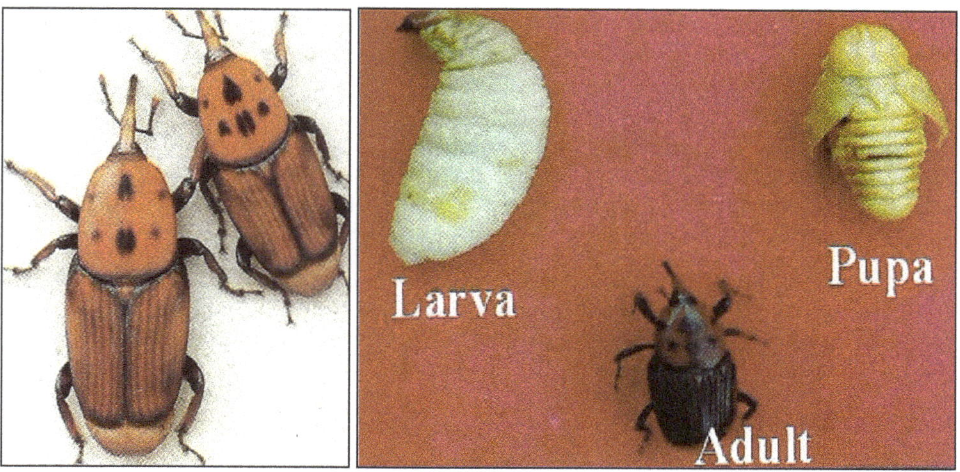

Figure 14.4: Red Palm Weevil Adult and Life Cycle.

c) Minor Pests

i) Nettle Caterpillar and Bagworm

The nettle caterpillars, *Thosea* sp. causes severe defoliation on oil palms, in Little Andaman Islands. Occasional infestation by the bagworm, *Manatha albipes* Moore and *Metisa plana* also cause defoliation. In Little Andaman, *Metisa* sp. and *Eumeta* sp. cause severe defoliation due to frequent outbreaks.

In severe case of damage, spraying with Carbaryl 50 per cent WP at 0.1 per cent is recommended after cutting and burning the badly affected and dried leaves. Root feeding with Monocrotophos (10 ml) mixed in 10 ml of water also will control these pests.

ii) Scales and Mealy Bugs

Scales, *Hemiberlesia lataniae, Chryhsomphalus aonodium* Linn. and *Pinnaspis aspidistrae* infest unripe and ripe oil palm fruits. *Dysmicoccus brevipes* also infests the pre-anthesising male and female inflorescences. Mealy bug species namely *Pseudococcus citriculus, Palmiculator* sp. and *Icerya aegyptiaca* infest the spear leaves of oil palm seedlings in the nursery and main field resulting in yellowing of young leaves and stunted growth of palms.

Scale insects can be controlled by spraying with Fenthion/Malathion 0.1 per cent. Damage caused by mealy bug can be controlled by spraying with Phosphamidon/Dimethoate at 0.05 per cent or Methyldemeton at 0.25 per cent.

iii) Aphids

Colonies of aphids like *Alysteroneura setariae* and *Schizaphis rottundiventris* occur on the under surface of the leaves and suck sap from them. In nursery and young palms they cause quite marked twisting and distortions of the spears.

Spraying with Dimethoate 0.04 per cent or Phosphamidon 0.02 per cent or Malathion 0.05 per cent on the under surface of leaves is recommended for the control of aphids.

iv) Hopper

Occurrence of this plant hopper is noticed in all the oil growing areas. They suck sap from the leaves. They attain importance, as they are vectors of mycoplasma like organisms (MLO's) in palms.

Since the plant hoppers breed on the decaying materials, field sanitation is essential for the control of plant hoppers. Spraying with Endosulfan 35 EC 0.1 per cent on the under surface of leaves (where the hoppers harbour) is recommended for the control.

v) Cockchafer Beetle

Damage caused by cockchafer falls into two categories:

- Damage to foliage by leaf eating adults. The two pests most frequently found are *Apogonia* and *Adoretus*, which make shot feeding holes in the leaves of nursery and field, planted young seedlings.
- Damage to roots by cockchafer larvae. The main species involved is *Psilopholis vestita*, previously recorded as a pest of rubber, which leads to reductions in both growth and yield.

Ploughing and digging during pre-monsoon and post-monsoon period and collection of adults during the period of emergence are effective control methods. In severe case of infestation insecticide application with Phorate 10 G (Thimet) @ 50 g/palm in May-June and Sep-Oct is recommended. The insecticide will have to be applied to the soil in the root zone and raked into the soil.

vi) Termites

Pericapritermes sp. and *Hypotermes* sp. feed on the roots of seedlings in polybags resulting in stunted growth. *Odontotermes* sp. infest the spear leaves, male inflorescence and fruit bunches in the field in Andhra Pradesh. Termites have caused the death of numerous palms where these have been planted in areas of peat soils, but elsewhere attacks are very infrequent.

The termites tunnel through the palm, preferring to attack the upper stem tissues. Ultimately, the trunk is so weakened that it collapses. Externally infestation is indicated by covered runs, which are made up of fine soil particles and particularly digested wood fragments, constructed from ground level to the point of entry into the trunk.

Control is by application of 0.5-1.0 per cent Dialdrin @ 0.5 lit/palm which has also been used successfully for termite control in coconut. Drenching with Chlorpyriphos 0.05 per cent is recommended for the control.

d) Vertebrate Pests

i) Birds

There are a number of birds, which eat mature oil palm fruits, occasionally, immature fruit is also attacked. Birds such as crows (*Corvus splendens profegatus; Corvus macrorhynchus culminatus*), mynah (*Acridotheres tristis*), babbler (*Turdoides affinis affinis*), parrots (*Psittacula krameri mannillensis*) feed on the mesocarp of fruits. Most are of minor nuisance only, although in restricted areas damage can be quite severe. In Malaysia, species include the long tailed parakeet (*Psittacula longicanda*), the blue ramped parrot (*Psittinus cyanurus*) and Malady lorikeet (*Loriculus galgulus*) which also cause damage to the fruits. In Colombia, losses of loose fruit left on roadside after harvesting through vultures has been quite considerable.

The ripe fruit bunches after 150 days of fruit set are to be covered with wire net of 1.25 cm mesh (60 x 90 cm size), red basket, plaited coconut leaf basket or oil palm leaves to avoid bird damage. Covering of bunches with oil palm leaves and tying with a piece of rope to keep them firm and impenetrable by the bird beak is found to be effective and cheap.

ii) Mammals

- ☆ **Rodents:** Rats have become pests of significance in parts of West Africa. Black rat, *Rattus rattus wroughtoni;* house rat, *Rattus rufescens;* lesser bandicoot, *Bandicota bengalensis;* larger bandicoot, *Bandicota indica;* Indian gerbil, *Tatera indica curvieri;* Western Ghat squirrel, *Funambulus tritriatus* and porcupines, *Hystrix indica* attack oil palm at various stages of its development. Among them the burrowing rat is more dangerous. It tunnels into the bole of the seedlings causing even their death. In mature palms rats eat the ripe bunches and gnaws the exposed pericarp of unripe and ripe fruits. Rats also destroy the male inflorescences while feeding on the larvae and pupae of pollinating weevils.

  Damage to young seedlings can be prevented by placing barriers consisting of 1.25 cm mesh (chicken wire mesh) collars around their base. They must be tightened to prevent the rats getting inside or underneath the guards. Baiting with Zinc phosphide, Bromodiolone and traps, bow trap *etc.* may be used as an integrated geared approach to minimise the rodents damage to oil palm.

- ☆ **Wild Boar:** Wild boar (*Sus scrofa*) digs up newly planted seedlings and chews them up. They also eat away the fruits from the bunches on the tree when they are accessible. A local wild boar scaring device has been developed to scar away wild boar from entering nurseries and young oil palm plantations. The plantation border is fenced with 18 gauge GI wire at 20 cm height on two lines parallel to the ground, supported on poles and kept in position with the help of guide hooks. The poles are positioned at 3-10 cm spacing depending upon the terrain of the field. At every five to six supporting poles, junction boxes are made. In the junction box, the plays, slabs and the crackers are arranged. These may be spaced at 5-15 meters apart depending on the landscape boundaries, roads *etc.* The two fencing lines arriving at the junction boxes from opposite sides are joined on to the oval plays and pulled closer and held in position with the help of a crushing slab hung from a third play kept on the first two plays. Underneath this crushing slab a cracker is kept. When the animal hits the fence, it will cause the first plays to pull apart resulting in the fall of the crushing slab kept directly underneath, making the cracker burst. The method has been found very effective in scaring away the animals.

14.2. Disease Management

a) Nursery Diseases

i) Blast Disease

Blast disease is initially caused by *Pythium* sp. most damage, however, is done by the secondary invader *Rhizoctonia*, Which is unable to penetrate undamaged root tips and first enters the root by parasitizing the Pythium thereafter rapidly destroying by cortical tissues.

Providing shade and adequate moisture to the bags are suggested to reduce the disease.

ii) Early Leaf Disease (Anthracnose)

Lesions caused by *Botryodiplodia theobroma* and *Glomeralla cingulata* leaf virus attack are frequently located at the leaf tip. Here, numerous spots, which are first pale in colour and then become brown, coalesce until the whole leaf tip is brownish grey and papery in texture. The border between healthy and diseased tissue is quite sharply defined, and becomes covered by large numbers of reproductive bodies of the fungus. Spores from there are washed down the leaf to initiate secondary infections.

Clipping of severely affected leaf portions and spraying of 0.2 per cent Dithiocarbamates (Indofil M-45) are said to be effective control measures.

iii) Leaf Rot

The fungus responsible for the disease, *Corticium solani*, is a very common soil inhabitant, in the asexual stage and is often referred to in this phase as *Rhizoctonia solani*. Infection of the youngest leaves occurs, causing a rot at the base of the unexpanded leaf, which, on opening, shows a transverse row of lesions. Lesion colour is a valuable diagnostic character. The lesions are first dark brown in colour and then they dry out to leave central tissues which becomes grey to almost white in colour, with a very prominent purplish brown margin. The central dead tissues are readily fragmented, leaving the affected area with a shot hole appearance. Badly affected seedlings should be removed and destroyed.

iv) Other Leaf Spots

Numerous dark brown leaf spots are caused, usually towards the leaf tip, surrounded by a chlorotic halo, by *Helminthosporium* all the tissue between lesions become a diffused yellow colour. As lesions enlarge they coalesce, and the leaf dies back from the tip at its margins. Secondary lesions are formed by spores washed down from the original infection points.

v) Curvularia Seedling Blight

Lesions associated with Curvularia (*Curvularia lauta* and *C. geniculata*) attack are usually easily recognizable. The dark brown spots have a sunken centre, and a narrow rim of slightly raised tissues develops between this and the very prominent, bright orange yellow halo around the lesion. When palms are growing in peat soil, the halo may be virtually absent. Lesions tend to remain discrete, up to 7-8 mm long, but in very heavy infection on older leaves a rapid die-back may occur. After death of the leaf, the original Curvularia lesions can still be seen quite clearly, allowing this to be used as a diagnostic feature.

Removal of badly diseased plants or leaves provides a partial control measure. Thiabendazol (80 per cent a.i. used at 0.1 per cent) and Thiram formulations at 0.2 per cent cone. (75-80 per cent a.i) have also been found to be effective in control of leaf spots.

b) Diseases in Adult Plantations

i) Bud Rot

Bud rot, the common fatal disease of oil palm, has been known since the earliest days of its cultivation and is reported from all the oil palm growing regions of the world. A number of micro-organisms isolated from the tissues affected by bud rot disease have been reported from various oil palm growing countries. Out of these only *Erwinia laythri* has been reported as pathogenic, but no circumstantial evidence of the disease development is proven. Hence, exact cause of the disease is still elusive.

The symptoms of the disease include yellowing of the spears leaves which subsequently turn to brown. Affected spear bends at the base and is seen hanging down among the healthy fronds. The rotting starts at the basal portion or collar region of the spear close to the meristem. The basal tissues of the spear completely rot, as a result it collapses and can easily be removed and comes out without much resistance when pulled out. The rotten tissues emit offensive odour. Rotting advances downward, the apical bud tissues get affected and a big crater is formed in the centre of the crown. Continuous and unchecked rotting leads to total destruction of meristem and ultimately death of the palm. However, outer leaves remain green for a long period even after the death of palm.

Low atmospheric temperature and high relative humidity coupled with rainfall aggravates the disease incidence. Often, the disease becomes rampant during the monsoon season when the inoculum build up becomes high due to non-eradication or non checking of the disease affected palms left over in the field. The young and juvenile palms are more susceptible. Rhinoceros beetle, is considered as the most important pest in promoting the disease and its incidence.

- ☆ **Management:** It is possible to cure the disease effectively, if it is detected in the early stages *i.e.*, when the spindle starts showing symptoms of withering, yellowing and dropping down. The affected spear should be pulled out along with the decayed tissues. The affected tissues and crown should be removed and drenched with fungicide solution, like Carbendazim or Thiram (0.1 per cent).

For advance stage disease affected palms, the leaves surrounding the spear should be first cut, so as to make sufficient space for free movement of operations. The affected tissues/parts of the meristem have to be removed layer by layer with a sharp knife till fresh unaffected tissues are seen.

While removing the affected tissues care should be taken that the central portion. *i.e.*, growing point of apical bud should not get damaged or split. Once the affected tissues are completely removed the crown should be cleared and drenched with 0.1 per cent Carbendazim solution. The exposed portion should be covered with perforated polythene sheet. In order to avoid invasion Monocrotophos @ 1 ml/lit of water can be added to the fungicide solution Early detection and timely curative measures can save the palms, which can otherwise be lethal.

Sprays: In the disease endemic areas as a prophylactic drenching with 0.1 per cent Carbendazim solution should be given at least once at the onset of monsoon and once during the end of monsoon. Where the beetle damage is predominantly high, it should be checked by filling of the inner whorl of leaf axils with BHC 5 per cent and sand mixture in 1:1 ratio.

ii) Basal Stem Rot (*Ganoderma*)

Basal stem rot is the major pathogenic disease reported from many countries causing significant losses to the oil palm industry. It is a serious disease of oil palm in Malaysia where the losses are to the extent of 46 per cent by the 15th year of planting. Generally the disease incidence is high in coastal estates where the old coconut plantations are already affected by the disease and soils are infected with the pathogen. Several species of *Ganoderma* fungus have been recorded from various parts of the world as its likely pathogen of basal stem rot. More recently, *Ganoderma* boninense Pat. is identified as the main causal agent of BSR.

Palms in the age group of 3-25 years are more susceptible. Species of *Ganoderma* pathogenic to oil palm have a wide host range such as coconut, rubber and several types of forest trees. In new plantations the primary infection arises from tissues of the former stand being colonised by *Ganoderma*, while the secondary spread occurs by root contact between diseased and healthy palms or through spores.

Symptoms of basal stem rot include- withering, yellowing and orange discolouration of the leaves followed by necrosis on one side of older fronds. Necrosis begins in the lower leaves and extends progressively to younger parts of the crown. The palm appears pale and produces excessive number of spears. Desiccated fronds drop or break at some point along the rachis to encircle the top of the trunk. Appearance of light brown lesions/rotting of the bole at the stem base is characteristic symptom at the advancement of disease.

Management: In order to prevent the spread of the disease and managing the disease affected palms in the initial stages itself, the following management practices are to be followed.

The stumps of the dead palms and disease affected palms giving uneconomic returns should be removed, shredded into pieces and burnt along with bole root system. The palms should be given normal recommended dosage of fertilizer. In addition each adult palm should be given 5 kg of neem cake/year. The organic matter content in the soil should be increased by addition of FYM or by growing leguminous cover crops or mulching of leaves *etc.* Sufficient soil moisture has to be maintained through irrigation coupled with mulches. While irrigating the gardens, care should be taken that, the water from the basins of the diseased palms should not go to the healthy palm basins. In ill-drained gardens, good drainage system has to be provided. The disease affected and apparently healthy palms should be treated with 10 ml Calixin (Tridemorph) or 10 G Aureofungin sol. (in 100 ml of water) per palm through root feeding. The selected root should be a young one having root tip. Before the root is dipped in the solution, the tip should be cut. The root feeding or soil drenching should be given thrice a year continuously for two years.

iii) Stem Wet Rot

Stem wet rot is a serious disease of oil palm reported from various oil palm growing countries *viz.* Malaysia, Indonesia, Papua New Guinea *etc.* A number of commonly occurring fungi have been isolated from diseased tissues but none of them proved as primary pathogen. The type of rot found in the stem suggests that a bacterium is involved but its pathogenicity is yet to be confirmed. Other reasons for the cause of the disease are suspected to be excessive soil moisture, possible physiological stress and susceptibility of palm variety during early years of fruit production.

Symptoms include, sudden and simultaneous death of all unexpanded spear leaves including the young expanded fronds surrounding the spear. Remaining fronds show yellowing discolouration and then rapidly wither and die. Sometimes, the older leaves die first and the symptoms progress to the younger fronds. When crowbar or a sharp rod is inserted into the affected stem just above the ground level, a lot of putrefied yellow brown thick fluid flows out which emits strong objectionable smell.

When the trunk of affected palm is cut and split open, it could be observed that, the internal parenchymatous tissues are completely destroyed except a narrow band around the stem periphery. The affected palms will usually have a cavity of variable size filled with fibrous mass at the centre of the stem. Rotting mass is generally bright yellowish in colour. However, at the base of the stem it is fibrous, usually black in colour and dry.

Management includes improvement in agronomic practices, providing drainage, avoiding flooding of the fields *etc.* In order to prevent the spread of disease, dead palms due to the disease should be excavated and burnt. Early detection of the disease and trunk surgery can save the palm.

Proper diagnosis of the disease affected palms can be made by hitting the palm with a wooden implement which gives a dull sound indicating underneath a soft area. For further confirmation, a sharp iron rod may be pierced into the stem base, which gives out some liquid. If the liquid is of putrefied smell, the palm should be treated immediately by trunk surgery.

Trunk surgery is done by excision of all affected fibrous tissues from inside the trunk. For excision of the diseased fibrous mass, a sharp harvesting chisel will be quite useful. First the outer stem tissues and frond, butts should be chiseled. The inner most diseased tissues including yellowish lesions which are generally seen along with the border of healthy and diseased tissues also should be removed. Since the rotten tissues are in irregular direction, the excision is time consuming method. It should be done with great care and patience and removal of healthy tissues should be avoided. When the surgery is completed a protective covering with Carbendazim (1 per cent) + Monocrotophos (1 ml) paste followed by hot coal tar should be given to prevent the invading micro-organisms and insects. Since it is not possible to remove all the diseased tissues from the very bottom of the palm in the first attempt, resurgence of infection is quite possible and treatment is essential after a gap of three months.

iv) Spear Rot

This is a highly infectious disease caused by Phytoplasma earlier called Mycoplasma Like Organisms (MLO). It is endemic to Kerala State. The first symptom of the disease is the chlorosis of leaves followed by necrosis, rotting of spear leaves and reduction in leaf size. In advanced stage the trunk of the pollen gradually tapers and arrests the inflorescence emergence leading to loss in productivity. Disease is infectious, vector borne and lethal.

The management of the disease involves rouging of affected palms and planting of barrier trees to isolate the infection source of other palms (RWD of coconut and YLD of arecanut).

v) Upper Stem Rot

This disease is more endemic in Andaman Islands. In early stages symptoms include gum exudation and bleeding and rotting of the stem, which occur at any point above 50 cm from the bole region. Decay of inner tissue is observed on chipping of bark as the disease advances. In due course rotting extends internally in an irregular pattern. The dark lesion also extends laterally causing dry rot of the stem. Externally presence of fibrous strands is noticed. As the rotting advances internally, the palm snaps at the site of rotting. The presence of healthy tissues above and below the lesion indicates that the infection is likely to be localised.

Management of the disease requires removal of rotten tissues followed by swabbing of 0.1 per cent Calixin; Painting with hot coal tar to the exposed cavity and Filling cavity with sand, cement and insecticide mixture.

vi) Bunch Failure

It is a condition in which bunches die at some stage during development. It is most frequent in 3-10 year old palms. The range of under development of fruits in a bunch may vary from zero to 100 per cent. Bunches produced during March-May are affected to the maximum extent. Failure in development of bunches can occur at any stage from anthesis to harvest. The tips of under developed fruits turn black and later become necrotic. Sometimes only a digital section of bunch may rot.

The bunch failure incidence has been found to be low in Little Andaman and at Palode in Kerala after the introduction of the pollinating weevil, *Elaeidobius kamerunicus*. Crown sanitation and Carbendazim spray (0.1 per cent) reduce the disease incidence.

vii) Bunch Rot

Bunch rot is an economically important disease in oil palm growing countries. It becomes severe during the monsoon or prolonged wet period.

Initial fruit attack by *Marasmius fungus* followed by secondary invasion by other microorganisms leads to this disease. Strands of mycelium seen spreading over the bunch surface and profuse at the back of the bunch. Mycelium grows over fruit surface and penetrates the mesocarp and leads to wet rot- fruit rot. Rotting results in increase of fatty acids content of fruit.

Management of diseases includes Palm sanitation besides Crown cleaning by removing aborted and dried bunches and inflorescences followed by spraying of 0.1 per cent Carbendazim all around the crown.

viii) Crown Disease

This juvenile disease occurring frequently on palms during the first 3 years after planting has been reported from many oil palm growing countries. In India the disease has been observed in the nursery stage and field palms in different states.

During the early phase of the disease brown lesions with water soaked margins are observed on the middle portions of the internal leaflets. When the affected spear unfurls, it develops a characteristic bending in the middle of the rachis. The leaflets on either side of the curved portion of the rachis are either completely rotten or a few stubs remain.

Diseased palms generally recover spontaneously within 1-2 years. To accelerate the recovery excision of rotten area and spraying of 0.1 per cent Carbendazim and 0.05 per cent Monocrotophos is recommended.

ix) Vascular Wilt Disease

Palms of all ages can be damaged, even in the nursery, but vascular wilt rarely becomes important until the palms are at least 10 years old. In palms less than six years old, the first symptom is the appearance of a single bright yellow leaf in the upper part of the crown. Other leaves in the vicinity of this leaf then become affected, resulting in desiccation and leaf death. Apparent recovery may occur, but affected palms usually die. Mature palms are affected to varying degrees and infection may be chronic rather than acute. Usually, the first obvious sign of the disease is the presence of a few old desiccated leaves. Symptoms may be more noticeable on one particular side of the canopy. Affected leaves normally fracture near their base and hang down. The symptoms spread through the older leaves resulting in a 'skirt' of dead leaves around the trunk. In some cases the palm dies within 12 months of the first dead leaf being visible. A section through the trunk shows marked browning of the vascular tissue, the vessels blocked by gums and other debris preventing the passage of water and explaining the desiccation of the leaves. Vascular discolouring is evident in both cross and longitudinal sections of the rachis/petiole.

Control measures include : a) do not plant on sites where the disease has been reported; b) purchase resistant planting material. Do not buy seeds produced in areas where the disease is present unless accompanied of a recognized phytosanitary certificate; c) ensure that the seedlings are healthy and that field nutrition is optimized. Adopt hygienic planting methods; d) clean pruning/felling tools that have been used on a diseased palm with undiluted alcohol or bleach diluted 1:1 with water before they are used on another palm. Progress of the disease cannot be arrested once the palm is infected. Felling and removal of the palm is only beneficial if it prevents an infestation of rhinoceros beetle or other pests. There is some evidence that good agronomic practice retards disease spread and, certainly, maintenance of a high potassium status is beneficial. Replanting on land where palms have died from 'vascular wilt', without some form of crop rotation, is wasteful. In regions where vascular wilt disease is present, resistant varieties of palms should be used for all new plantings

Chapter 15

Harvesting, Yield and Handling

Oil palm starts bearing bunches from 2½ to 3 years after planting. Proper and timely harvesting of fruit bunches is an important operation which determines the quality and quantity of oil to a great extent. When the bunch is mature and ready for harvesting (a) fruits in the bunch turn yellowish orange (b) 5-10 fruits from each bunch drop on their own (c) when pressed hard with the fingers orange coloured oil exudes from the fruits. While harvesting, a stalk length of 5 cm alone should be left.

15.1. Degree of Ripeness

Fruit ripeness is related to the oil content in the mesocarp and is the most important requirement for harvesting to be carried out. The fruits ripen progressively from the outer and top parts of the bunch. In the early stages, the mesocarp has high water and carbohydrate content, and very little oil, but as the fruit ripens, the oil content and the formation of FFA increases. The highest concentration of oil occurs during the final weeks of development of the fruit. A week before harvest, the oil content can increase up to 80 per cent.

The term 'ripening' as on today refers to the change of surface colour from black to yellowish orange or the number of loose fruits of a bunch fall on the ground observed before harvest, and have to be used as an indicator of oil content which is the important variable. The colour change is not sufficiently sensitive, and the present criterion used in most plantations is the loose fruits per bunch or per kilogram of bunch. In fact, the oil formation and fruit abscission processes are quite the exact standards by which 'ripeness' is judged. The period over which fruit loosening occurs varies from 11 to 20 days, depending on bunch size. The commercial objective is to determine the best compromise between obtaining the maximum amount of oil in the bunch, and having only a few loose (fully ripe) fruits. This avoids large numbers of loose fruits that have to be picked up and are easily lost.

Ripe fruits that are shed from the bunch or remain on its outer surface may be physically damaged because the exocarp becomes softer, so that bruising occurs more easily, liberating an enzyme that de-esterifies the oil and releases FFA. The lower the amount of acid, the higher will be the quality of the oil. The FFA within a ripening fruit is about 0.5 per cent, and the oil from a bunch that was harvested and milled promptly, at a correct level of ripeness, should have an FFA of around 2 per cent. Currently, 5 per cent FFA is the official limit for premium quality in Malaysia. This also means that the harvesting rounds must be reasonably frequent, otherwise there will be many overripe bunches with large numbers of loose fruit, whatever standard is applied. The only practicable measure is to observe the number of loose fruits, and with palms of any height, that means loose fruit on the ground because the bunch cannot be examined closely. There will be many more fruits that are technically 'loose', but are still attached to or resting on the bunch. There is extensive evidence that the oil content continues to rise after fruit abscission commences and this rise is an artefact caused by the method of measurement, in that the mesocarp loses moisture so that its oil content rises, when it is expressed as a percentage of the total weight. The total weight of the mesocarp decreases as moisture is lost, but the weight of the oil increases after the loss of the first fruit. However, the oil/ bunch ratio remained fairly constant for bunches between one and nine loose fruits and those between ten and 50 detached loose fruits, and for practical purposes a bunch might be regarded as ripe if it had one detached fruit. It must be borne in mind that, once it starts, the abscission process continues throughout the harvesting interval, so that even if the least ripe bunch harvested on a particular round has only one loose fruit on the ground, the average bunch harvested at that time must have many more. On are average, complete detachment of fruit takes less time on small bunches than on large bunches, so harvesting small bunches on time is more critical. However, this lower MRS undoubtedly means that some oil is lost. Detachment of fruit increases rapidly after the one loose- fruit stage, and up to 70 per cent of a carrier's time can be spent on collecting loose fruits if the harvesting interval is extended from 10 days to 16 days The number of loose fruits that are found after the bunch has been harvested and the loose fruits are all detached may be up to ten times as many as those initially seen on the ground, and this delays the cutter greatly. The time to collect loose fruits with even this one-loose-fruit standard was five times longer than the harvester took to cut the bunch and associated fronds, suggesting that this may be the limiting step in harvesting. The number of loose fruits that were missed and therefore not collected during harvest increased sharply with palm age, presumably because the bunches were larger and fell from a greater height, so scattering larger numbers of fruits over a larger area. With labour shortages as in Malaysia, it may be more cost-effective to set a low loose fruit number as the MRS, even if this means accepting a lower oil/bunch ratio, because the higher standard implies more time spent on picking up loose fruits, and greater losses of these concluded that the Malaysian industry was losing the equivalent of US$250 million because of the low oil extraction ratio (OER) that may result from current harvesting standards, and suggested that work on mechanising loose fruit collection in a practical way, to allow a higher standard to be used, should have high priority. A fruit collection machine has been designed, working on the suction principle, but it does not appear to have been adopted in practice. There is no doubt that mills in Indonesia processing fruit from young plantations are obtaining OERs of up to 24 per cent, whereas most Malaysian mills now have OERs around 20 per cent, with substantially the same type of planting material.

15.2. Bunch Ripeness

Optimum ripeness measured as a function of fruit detachment varies according to the plant age, genetic material and environmental conditions. A faster rate of ripeness has been observed in the tenera types, in the smaller bunches and in bunches of younger palms.

The criteria used in determining the degree of ripeness based on the fruit detachment are as follows:

i. Fallen fruits: 10 detached or easily removable fruits for young palms and 5 for adult palms.
ii. Number of fruits detached after the bunch is cut: 5 or more fruits/kg of bunch weight.
iii. Quantity of detachment per bunch: Fruit detachment of 25 per cent on the visible surface of the bunch. These criteria should be applied with flexibility.

15.3. Pre-harvest and Harvest

15.3.1. Pre-harvest

Since the first bunches produced by the palms are small and have low oil content, they are not harvested until they reach a larger size and have a higher extraction rate, or they are castrated when the environmental conditions are adverse. In plantations where castration has not been practiced, it is necessary to carry out a sort of preliminary pruning (sanitary) to eliminate dry leaves and overripe or rotten bunches and residues must be removed from the weeding circle.

The oil extraction rate of bunches is low during the first two or three years, but it reaches maximum values when bunches weigh more than 4 kg on are average.

One of the main objectives of general pruning is to make ripening bunches visible to the harvester when he does his rounds. Once a bunch that is ripe by the current criteria has been missed, it will be very overripe by the time the next harvesting round arrives. There will then be a large number, perhaps 200 or more, of loose fruit scattered on the ground, and the FFA in these fruits will already be rising. For the same reason, epiphytes on the trunk and old male inflorescences will usually be removed, or collected from the ground if they have fallen off, as the weeded circle is cleared to make the observation and collection of loose fruit easier. There are, however, reasons for leaving the epiphytes, as it has been suggested that they support a beneficial insect population and generally increase biodiversity. When pruning or harvesting is carried out, a considerable number of fronds may be cut off. These should be stacked together in a way that covers the spiny petioles, which otherwise are a danger to the workers, and in a way that minimises erosion. The use of the knife and pole has been normal in plantations in South-east Asia for many years. In west Africa the harvesting pole has been brought into use, but some wild palms may still be harvested by the harvester climbing up to the ripe bunch and cutting it loose with a machete. Because wild palms only yield reasonably well when their crowns are above any surrounding trees, and are rarely cut down, harvesting may mean climbing up to 20 m or more, using a rope slung around the palm for support.

15.3.2. Harvest of Bunches and Frequency

Harvesting rounds should be carried out as frequently as possible in order to reduce the number of over-ripe bunches to be harvested, which have a high degree of fruit detachment and oil acidity. Harvesting must take place in such a way that a bunch, which is almost ready for one cycle, will not be over-ripe for the following cycle. In lean period of production harvesting can be made less frequent and it should be more frequent in peak periods. Harvesting rounds of 10-12 days are generally practiced but during rainy season, it should be done at closer interval of 6-7 days as ripening is hastened after rains.

Factors determining harvest intervals basically are age, labour availability, worker's skill and knowledge, transportation capability and extraction capacity of the mill.

15.3.3. Harvesting Tools

The tools used for harvesting of FFB are shown in Figures 15.1–15.3.

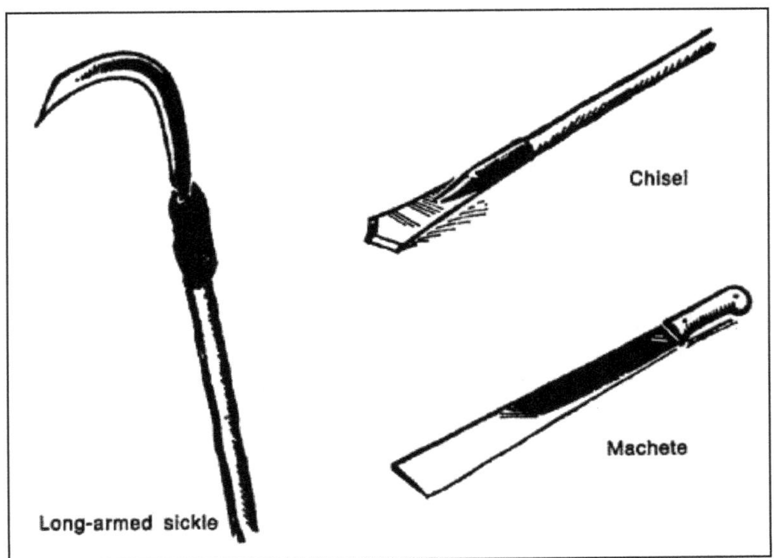

Figure 15.1: Harvesting Tools.

Figure 15.2: Various Harvesting Tools and Accessories for FFB Harvest (*Source*: Sawit and Karet, Malaysia).

15.3.4. Harvesting Young Palms

The method of harvesting young palms is normally by a chisel on a short wooden pole or light hallow aluminum pipe (Figure 15.4), which is convenient for gaining access to the peduncle (bunch stalk), even when the subtending leaf is still attached. The frond bases that make harvesting difficult are not those truly subtending the bunch, but to one side, as the bunch rests on these. The general rule now is to avoid cutting down any green frond unless it is unavoidable. This is particularly important for the young palms, because they have few leaves and a small leaf area index (L) once bearing starts, and it is usually recommended that the

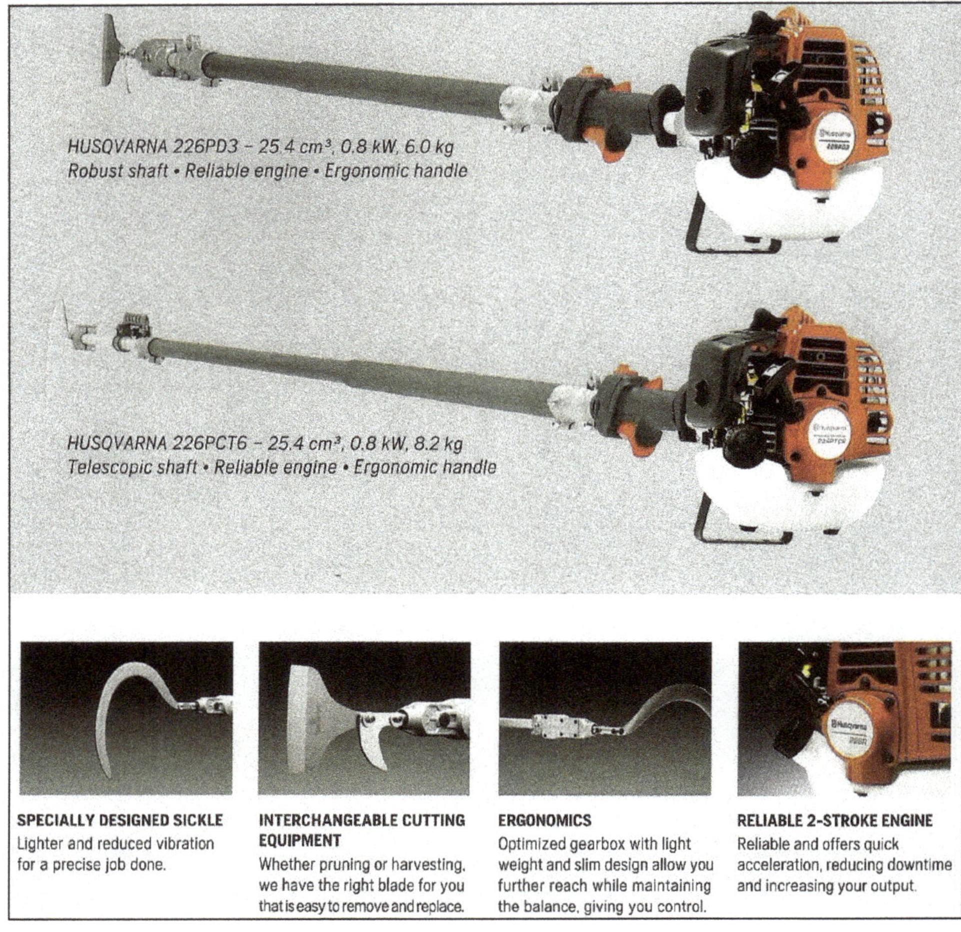

Figure 15.3: Power Operated Harvesting Tools (*Source*: Husqvarna, Malaysia).

next two leaves below the harvested bunch should be allowed to remain if possible. The stalk length of harvested bunches should not be more than 5 cm. Narrow chisel (6-9 cm wide) is generally used till the palm reaches two meters above the ground. For palms up to 4 m height a wider chisel of 14 cm is used.

15.3.5. Harvesting of Mature Palms

The major difference between young and mature palms is that the latter are taller, which increases the effort of cutting, and makes it more difficult to see ripening bunches and estimate when they are ready for harvesting. Against this, the less dense growth of cover plants makes access easier. From about 5 years onwards it becomes more difficult to cut out the bunch with a chisel, and an alternative tool is used. In tall trees, bunches are cut by using Malaysian knives. This curved knife is attached

Figure 15.4a: Harvesting Three Year Old Palm with Chisel.

Figure 15.4b: Harvesting in Young Oil Palm.

to the end of a long, thin and light pole, which could be made of bamboo, aluminum or any other light and strong material. The knife can be fixed at the end of the pole with steel wire or with screws. While cutting with Malaysian knives, a good cut depends not so much on strength but on the position angle of the knife blade. In uneven stands, an adjustable, telescopic type of pole is used (Figures 15.5 to 15.8).

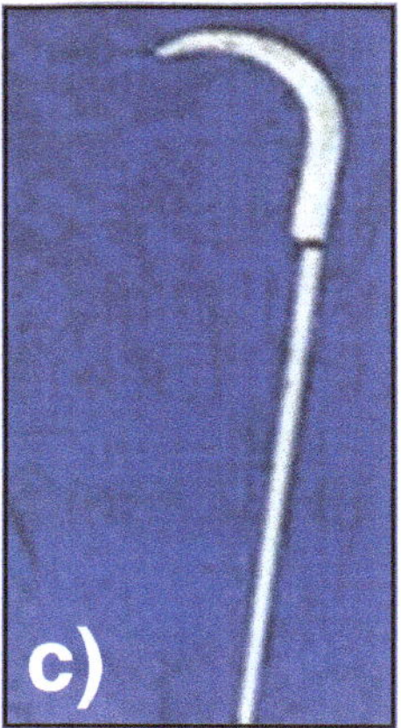

Figure 15.5: (a-b) Harvesting of Matured Palms and (c) Harvest Pole.

The sickles about 60 cm long, are mounted on the end of a bamboo or aluminum pole.

Possibility of using carbon fibre poles has been considered, but so far none is in routine use. With these, palms of over 12 m tall can be harvested although the latest forms of extensible poles can be extended upto 16 m. These are normally very effective, and they can increase bunch recovery greatly, with considerable cash savings for old palms. In tall palms it is essential to cut the fronds that subtend the harvested bunch before cutting through the bunch peduncle, to give vision and

Figure 15.6: Harvesting with Power Operated Harvester.

Figure 15.7: Lightest Harvesting Pole in Operation.

access. This is acceptable, because the leaf area in old palms is still adequate after a few fronds have been cut off. The handling of the pole is a skilled and heavy job.

15.3.6. Harvesting, Gathering and Transport of Bunches

Once the bunches are cut, it is very important that they should be gathered and taken to the extraction mill quickly without damaging them, so that they are sterilised on the same day of the harvest and thus, halt the decomposition process of oil into glycerin and free fatty acids (FFA). The fresh fruit bunches (FFB) and loose fruits (LF) are normally carried from the fields to the collection sites along the roads using small carts drawn by animals and small tractors which can easily travel between the rows of palms and cause only little compaction to the soil.

a) Harvest Using Ladder

b) Using cycle tyre for climbing for harvest

Figure 15.8: Harvesting FFB Using Local Knowledge.

From the collection sites, the fruits are loaded and taken to the extraction mill on transportation units of various types having diverse capacity like tractors, trucks *etc.* (Figure 15.9). The fruit receiving system in the mill should facilitate unloading in accordance with the type of transportation used.

Figure 15.9: Gathering and Transport of FFB.

15.3.7. Gathering and Transport of Loose Fruits

The loose fruit deteriorates much more rapidly than the fruits in bunches, so its gathering and transportation to the factory should be given due priority on the same day of the harvest. At every harvest lot inside the field, in addition to the bunches, all the loose fruits should be gathered. At the collection centres, it is advisable that the fruits be unloaded and placed on canvas sheets, bags or some similar material so as to prevent losses and facilitate gathering of loose fruits (Figure 15.10).

Figure 15.10: Collection and Gathering of Loose Fruits.

15.3.8. Harvest Organization

Once the harvest interval has been selected, the area to be harvested every day must be determined in case of large-scale plantations. For this purpose, the total area is divided among the number of days of the planned interval (not including holidays). Then the area to be harvested is divided by the worker's harvesting capacity (which must be known). The worker's harvesting capacity depends on factors like the height of the palm, number of bunches in an area, average bunch weight *etc.* The taller the height, greater is the skill required to handle the cutting tools. When there are a few bunches per area unit, the harvester must walk longer distances (more rows) to complete his task. If bunches are more per unit area, the harvester need not walk more. If bunch weight is more, carrying and transportation needs more effort. Field conditions having variations in topography, poor drainage and weed growth also affect the workers efficiency. However, in small holders' plantations the harvest interval depends on the maturity of bunches.

15.3.9. Factors Affecting the Harvest

a) Pruning and Epiphyte Growth

Palms having too many dry leaves due to lack of pruning, too many ferns and climbing plants on their stems and weeds in the circles, hamper visibility making it difficult to find the ripe bunches to be cut.

b) Plants Located Next to Drains or Natural Ditches

Sometimes palms are planted in areas with old drainage systems or next to natural ditches and when bunches are cut, some of them fall in these ditches. In this case, it is a good idea to have a worker in charge of recovering these bunches.

When palms are planted in areas whose drainage system was designed for other purposes, it is preferable to stake the planting sites in such a way that the rows are centered in the areas between the drains, even if this means not following the planting patterns of neighbouring plots.

15.3.10. Distribution of the Area to be Harvested

Cutting can be distributed either by rows or plots. Work assigned by rows makes the harvest proceed evenly, but there are no incentives for work quality. When a whole plot is assigned to a worker, he will be responsible for his work and poor performance can be detected easily. Besides, workers tend to take better care of the plot as if it were their own. The disadvantage of assigning plots is that if the area is large, supervision is more difficult; therefore there is less control and less efficiency in other related activities such as loading and transporting of the fruit. This problem can be reduced if each worker is assigned several small plots, so as to achieve better distribution of labour in heterogeneous plantations.

15.4. The Ripening Process

15.4.1. Criteria of Ripeness

The most important requirement in harvesting is to carry it out at the time when the bunch is ripe. The fruits ripen progressively from the outer and top parts of the bunch. The term 'ripening' can be ambiguous, as applying to oil content, surface colour change or the number of loose fruit of a bunch.

The important variable is the oil content. Only the colour change from black to orange and number of loose fruit can be observed before harvest, and have to be used as an indicator of oil content. The colour change is not sufficiently sensitive, and the last criterion is used in most plantations, as loose fruit per bunch or per kilogram of bunch. In fact, the oil formation and fruit abscission processes are quite the exact standard by which 'ripeness' is judged and should be clearly defined in all research. The period over which fruit loosening occurs varies from 11 to 20 days, depending on bunch size. The commercial objective is to determine the best compromise between obtaining the maximum amount of oil in the bunch, and having only a few loose (fully ripe) fruits.

This avoids large numbers of loose fruits that have to be picked up and are easily lost (Gan *et al.*, 1994, 1995).

Ripe fruits that are shed from the bunch or remain on its outer surface may be physically damaged because the exocarp becomes softer, so that bruising occurs more easily, liberating an enzyme that de-esterifies the oil and releases FFA. The lower the amount of acid, the higher will be the quality of the oil. The FFA within a ripening fruit is about 0.5 per cent, and the oil from a bunch that was harvested and milled promptly, at a correct level of ripeness, should have an FFA of around 2 per cent.

Currently, 5 per cent FFA is the official limit for premium quality in Malaysia, although a stricter standard will shortly be introduced. This also means that the harvesting rounds must be reasonably frequent, otherwise there will be many overripe bunches with large numbers of loose fruit, whatever standard is applied. However, there is a pressing shortage of labour in the Malaysian plantation sector (Malek and Mohammed Nasir, 1995}, so that the frequency of harvesting rounds has risen from 10 to 16 or more days in many estates.

15.4.2. Minimum Ripeness Standards

There has been a lengthy debate about the minimum ripeness standard (MRS} for fruit to be harvested in plantation practice. The only practicable measure is to observe the number of loose fruit, and with palms of any height, that means loose fruit on the ground, because the bunch cannot be examined closely. There will be many more fruit that are technically 'loose', but are still attached to or resting on the bunch. There is extensive evidence that the oil content continues to rise after fruit abscission commences (Corley and Law, 2001). It has been suggested by Rajanaidu *et al.* (1988} that this rise is an artefact caused by the method of measurement, in that the mesocarp loses moisture so that its oil content rises, when it is expressed as a percentage of the total weight. The total weight of the mesocarp decreases as moisture is lost, but the weight of the oil increases after the loss of the first fruit (Corley and Law, 2001).Gan *et al.* (1995) concluded that the oil/bunch did increase with bunch ripeness, as bunches with 50-200 loose fruits after cutting had oil/ bunch 1.9 per cent higher than a bunch with one loose fruit. However, the oil/ bunch remained fairly constant for bunches between one and nine loose fruits and those between ten and 50 detached loose fruits, and for practical purposes a bunch might be regarded as ripe if it had one detached fruit.

Corley and Law (2001) and Rao *et al.* (2001) examined all the evidence, and both concluded that there was increase in the bunch oil content after the one-loose-fruit stage.

15.4.3 Bunch Counting Systems

In those places where animal hauled carts or light motor vehicles are used for carrying the fruit within the plantations, the bunches are counted as they are carried to the road where they are counted again as they are being loaded onto trunks, wagons or carts. At some of the transportation units, the bunches must be counted to compare the figures reported by both the person counting and the loaders.

15.5. Grading Procedures

15.5.1. Sampling Procedures

i) Select about 5-100 bunches at random as sample from each consignment to be graded. The sample taken should represent the top, middle and bottom portions of the consignment.

ii) The minimum sample size of each consignment to be graded should be determined based on the following criteria:

★ If the net weight of the consignment is less than 5 tons, the minimum sample size should be SO bunches.

★ If the net weight of the consignment is five tons or more, the minimum sample size should be 100 bunches.

iii) The sample size should be economical, practical and able to detect any change in the bunch quality, especially the degree of ripeness at 95 per cent level of confidence. Separate the bunches that have been sampled for grading from the rest of the bunches.

15.5.2. Grading Frequency

i) The minimum grading frequency for each supplier of fresh fruit bunches with long term contract should not be less than 10 per cent of the total consignment or at a ratio of 1:10 lorries. If there is variation in the quality of fresh fruit bunches supplied or doubts regarding the bunch quality, the grading frequency should be increased to fifty per cent of the total consignment or at a ratio of 1:2 lorries.

ii) For suppliers without long-term contracts, grading should be done on all consignments.

15.6. Bunch Classifications

Fresh fruit bunch can be classified and graded according to the following criteria (Figure 15.11).

a) Ripe Bunch

A bunch which has reddish orange colour, and the outer layer of the fruitlets mesocarp is orange in colour. This bunch has at least 10 fresh sockets of detached fruitlets and more than fifty per cent of the fruits still attached to the bunch at the time of inspection at the mill. The bunch and the loose fruits are to be sent to the mill within 24 hours after harvest.

b) Under Ripe Bunch

A bunch which has reddish orange or purplish red colour and the outer layer fruitlets mesocarp is yellowish orange in colour. This bunch has less than 10 fresh sockets of detached fruitlets at the time of inspection at the mill. The bunch and the loose fruits are to be sent to the mill within 24 hours after harvesting.

c) Unripe Bunch

A bunch which has black or purplish black fruits and the outer layer fruitlets mesocarp is yellowish in colour. This bunch does not have any fresh sockets of detached fruitlets at the time of inspection at the mill. The sockets, if any, on the bunch are not due to normal ripening process.

d) Over Ripe Bunch

A bunch which has darkish red coloured fruits and has more than fifty per cent of detached fruitlets but with at least ten per cent of the fruits still attached to the bunch at the time of inspection at the mill. The bunch and the loose fruits are to be sent to the mills within 24 hours after harvesting.

e) Empty Bunch

A bunch which has more than ninety per cent of detached fruitlets at the time of inspection at the mill.

f) Rotten Bunch

A bunch which is partly or wholly ripen and together with its loose fruits has turned blackish in colour, rotten and mouldy.

g) Long Stalk Bunch

A bunch which has a stalk of more than 5 cm in length, is considered in this category.

h) Unfresh Bunch

A bunch which has been harvested and left at the field for more than 48 hours before being sent to the mill. The whole fruit or part of it together with its stalk has dried out. Normally, this type of bunch is dry and blackish in colour.

i) Old Bunch

A bunch which has been sent to the mill, the fruitlets of which still remaining on the bunch are dry and brownish black in colour. The stalk is also dry, soft, fibrous and blackish in colour.

j) Dirty Bunch

A bunch with more than half of its surface covered with mud, other dirt particles and mixed with stone or other foreign matters.

k) Small Bunch

A bunch which has small fruits and weighs less than 2.3 kg (5 lbs).

i) Pest Damaged Bunch

A bunch in which more than thirty per cent is damaged by pest attack such as rats, birds *etc.*

m) Diseased Bunch

A bunch which has more than fifty per cent parthenocarpic fruits and is not normal in terms of its size or its density.

n) Dura Bunch

Dura bunch fruits have the following characteristics:

- ★ Shell thickness: 2-8 mm
- ★ Ratio of shell to fruit: 25-50 per cent

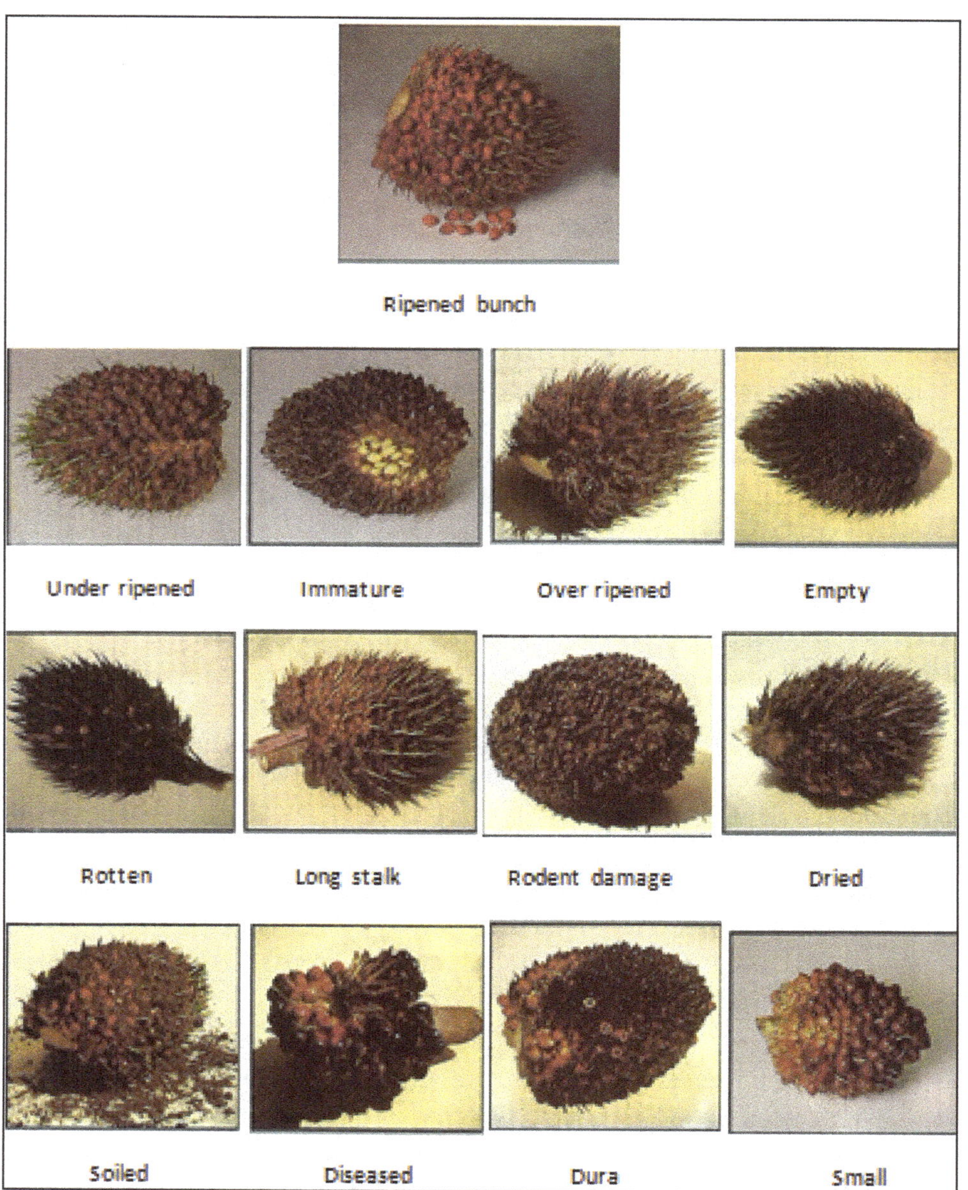

Figure 15.11: Types of Branches.

- ★ Ratio of mesocarp to fruit: 20-60 per cent
- ★ Ratio of kernel to fruit: 4-20 per cent
- ★ No fibre ring around the shell

o) Loose Fruit

A fruit detached from a fresh fruit bunch because of ripeness and is reddish orange in colour. All loose fruits have to be in the mill within 24 hours after harvesting.

p) Wet Bunch

Wet bunch refers to a consignment of fresh fruit bunch, which has excessive free water.

15.7. Grading Methods

The sample that has been selected will be graded to determine the quality of bunches and the extraction rate that can be given to the supplier. During grading the following practices should be carried out.

- ★ Inspection and assessment of the bunch quality
- ★ Calculation of penalty for poor quality bunch
- ★ Determination of the basic extraction rate
- ★ Calculation of the graded extraction rate

15.7.1. Inspection and Assessment of the Bunch Quality

- ★ The grading of the consignment of fresh fruit bunches should be done in the presence of lorry driver or his attendant.
- ★ The lorry with the consignment of fresh fruit bunches which has been selected to be graded is directed to unload on the platform near the loading ramp. Ensure that the bunches are evenly laid out and no overlapping or layering should occur.
- ★ Count the number of bunches in the consignment and calculate the average bunch weight with the following formula.

$$\text{Average bunch weight} = \frac{\text{Net weight (kg) as per weighbridge}}{\text{Total number of bunches}}$$

Record the information in the grading sheet form as shown in Appendix.1.

i) From these bunches, select at random 50-100 sample bunches and separate them from rest of the bunches.

ii) Grade, classify and count the sample bunches into 5 groups based on the criteria of bunch classifications as follows.

- ★ Ripe bunch
- ★ Under ripe bunch
- ★ Unripe bunch
- ★ Empty bunch
- ★ Rotten bunch.

Inspection and assessment of the bunch quality should be done quantitatively.

i) Record the number and the percentage of each group in the grading form as shown in Appendix 1. The total percentage of the 5 groups must be one hundred per cent.

ii) Grade, classify and count again all the sample bunches into 5 groups as follows:

   ★ Long stalk bunch
   ★ Dirty bunch
   ★ Dura bunch
   ★ Old bunch
   ★ Wet bunch

Record the number and percentage of each group in the grading form as shown in the Appendix l:

15.7.2. Determination of the Basic Extraction Rate

The basic extraction rate is the theoretical extraction rate which is also the maximum extraction rate of the oil and kernel. This extraction rate can be determined in two ways; that is by the age of the palm and the bunch weight.

a) Determination of the Basic Extraction Rate Based on the Age of the Palm

The basic extraction rate can be determined by the age of the palm, provided that the information regarding the year when oil palm was planted is known. This method is suitable for mills that receive fruits from their own estates.

b) Determination of the Basic Extraction Rate Based on the Bunch Weight

This is suitable for mills that receive their supplies from outside estates and dealers who do not have information regarding the age of the oil palm. The average bunch weight can be calculated by dividing the net weight with the total number of bunches.

15.7.3. Calculation of the Grade Extraction Rate

The graded oil and kernel extraction rate that can be given to supplier after the grading of the fresh fruit bunches can be calculated based on the discount system as follows.

Graded extraction rate = Basic extraction - Penalty rate

The graded extraction rate should be recorded in the monthly grading form as stated for the purpose of payment to the suppliers.

15.7.4. Grading Report

a) Grading Form

i) All observations and calculations during grading must be recorded in the grading form particulars that have to be recorded are as follows:

- ★ Net weight
- ★ Number of bunches
- ★ Average bunch weight
- ★ Number and percentage of unripe bunch
- ★ Number and percentage of under ripe bunch
- ★ Number and percentage of ripe bunch
- ★ Number and percentage of empty bunch
- ★ Number and percentage of rotten bunch
- ★ Number and percentage of long stalk bunch
- ★ Number and percentage of dirty bunch
- ★ Number and percentage of dura bunch
- ★ Number and percentage of old bunch
- ★ Number and percentage of wet bunch
- ★ Observations on bunch quality
- ★ Name and signature of grading officer

ii) Use a separate grading form for each grading consignment.

iii) This is to be filled in duplicate and the original copy is to be given to the supplier or his representative.

b). Monthly Grading Summary Form

i) All observations and calculations recorded in the grading form have summary form. The particulars that have to be recorded are as follows.

- ★ Amount of fresh fruit bunches received
- ★ Amount and percentage of fresh fruit bunches graded
- ★ Crude palm oil produced
- ★ Oil and kernel extraction rates achieved
- ★ Average bunch weight and age of palm
- ★ Percentage and penalty (if any) for unripe bunch
- ★ Percentage and penalty (if any) for under ripe bunch
- ★ Percentage and penalty (if any) for ripe bunch
- ★ Percentage and penalty (if any) for empty bunch
- ★ Percentage and penalty (if any) for rotten bunch
- ★ Percentage and penalty (if any) for long stalk bunch
- ★ Percentage and penalty (if any) for dirty bunch
- ★ Percentage and penalty (if any) for dura bunch
- ★ Percentage and penalty (if any) for old bunch

★ Percentage and penalty (if any) for wet bunch

★ Name and signature of the mill manager

ii) Only one copy of this form is to be filled for record purpose and retained by the mill.

15.8. Yield

Oil palm is the most productive oil bearing plant species known in the world. One hectare of oil palm under good management conditions produces on an average 4.5 tons of oil/ha/year, 0.45 ton palm kernel oil and 0.45 ton palm kernel cake. The yield of oil palm depends on climatic conditions, variety, age and management practices followed. Oil palm FFB yields reported from different countries are given in Table 17.1. From fifth year onwards the average yield may be 20-25 tons FFB/ha/ year. It can also yield up to 30-40 tons FFB/ha/year under very good management conditions using high yielding tenera planting material. The average weight of a harvested fresh fruit bunch is 20-25 kg and average number of bunches would be 10-12/tree/year. Bunch weight up to 60-100 kg is also possible. The commercial Fresh Fruit Bunches according to the age of palm are given Table 17.1.

Table 17.1: Commercial Fresh Fruit Bunches (FFB) According to Age in Oil Palm

Age (years)	*Fresh Fruit Bunches (t/ha/year)*	*Oil (t/ha/year)*
3	7.9	1.6
4	16.5	3.6
5	23.6	5.4
6	25.8	6.0
7-18	28.0	6.7
19	28.0	6.6
20	28.0	6.6.
21	27.0	6.6
22	26.2	6.1
23	25.9	5.9
24	24.7	5.5
25	23.5	5.2

The actual yields achieved in the 16 largest oil-palm producing countries in the world in 2013 are shown in Table 17.2.

15.9. Cost of Cultivation

The cost of cultivation of Oil Palm as per NABARD worked out during 1990 is given below: NABARD regularly workouts the cost of cultivation and makes it available to the financial institutions (Banks) for adoption to issue crop loans (Table 17.3).

Table 17.2: Fresh Fruit Bunch (FFB) and Crude Palm Oil (CPO) Production and Yield per Harvested Hectare in the Main Palm-Oil Producing Countries in 2013

Country	*Area Harvested*[1] *(Mha)*	*Annual Production (MT)*		*Yield ($t\ ha^{-1}\ yr^{-1}$)*		*OER*[2] *(per cent)*	*Data source*
		FFB	*CPO*	*FFB*	*CPO*[3]		
Indonesia	7.1	120	26.9	17	3.8	22.4	FAO, unofficial figure
	8.1		30.5		3.8		USDA
Malaysia	4.6	95.7	19.2	21	4.2	20.0	FAO, unofficial figure
	4.5		20.2		4.5		USDA
Nigeria	3.0	8.0	1.0	2.7	0.32	12.0	FAO, estimate
	2.5		1.0		0.39		USDA
Thailand	0.63	12.8	2.0	20.5	3.1	15.1	FAO, official data
	0.66		2.0		3.0		USDA
Colombia	0.45	5	1.0	20	3.5	17.5	FAO, official data
	0.34		1.0		3.1		USDA
Ghana	0.36	2.1	0.12	5.8	0.30	5.2	FAO, estimate
	0.37		0.49		1.3		USDA
Guinea	0.31	0.8	0.05	2.7	0.20	7.4	FAO, estimate
	0.31		0.05		0.16		USDA
DRC (Congo)	0.28	1.8	0.30	6.6	1.1	16.7	FAO estimate
	0.18		0.22		1.2		USDA
Côte d'Ivoire	0.27	1.7	0.42	6.5	1.5	23.1	FAO, unofficial figure
	0.27		0.42		1.5		USDA
Ecuador	0.22	2.3	0.33	10.6	1.5	14.2	FAO, official data
	0.22		0.57		2.6		USDA
Papua New Guinea	0.15	2.1	0.50	14	3.3	23.6	FAO, unofficial figure
	0.15		0.50		3.4		USDA
Cameroon	0.14	2.5	0.23	18.2	1.7	9.3	FAO, unofficial figure
	0.13		0.29		2.2		USDA
Honduras	0.13	2	0.43	16	3.4	21.3	FAO, unofficial figure
	0.13		0.46		3.7		USDA
Brazil	0.11	1.3	0.34	11.5	3.1	27.0	FAO, official data
	0.12		0.34		2.8		USDA
Guatemala	0.07	1.5	0.40	22.8	6.2	27.2	FAO, unofficial figure
	0.10		0.43		4.3		USDA
Costa Rica	0.07	1.3	0.30	17.5	4.0	22.9	FAO, estimate
	0.06		0.21		3.5		USDA
World	18.1	266.5	54.4	14.8	3.0	20.3	FAO, aggregate
	18.6		59.4		3.2		USDA

1. Area harvested excludes immature area. 2. Oil extraction rate (OER) was calculated from the yield data (ton CPO/tonFFB×100). 3. CPO yield was calculated by dividing production over harvested area (MTonCPO/mHaharvestedarea).

Sources: FAO (2013); USDA-FAS (2016). Numbers must be viewed with some caution, as good-quality data on harvested area and yiest of Cultivationld is difficult to obtain, especially for smallholder plantations.

Table 17.3: Cost of Cultivation NABARD (Rs. Per ha)

Particulars	*1st Year*	*2nd Year*	*3rd Year*	*4th Year*	*Total*
Land preparation	550				550
Pit preparation	660				660
Plant material/seeds	7900				7900
Plantain and stalking	330				330
Manuals	1100	1540	1980	2200	6820
Fertilizers	1320	2590	3850	3850	11610
Pesticides/applying cost	420	690	830	830	2770
Labor	330	440	660	750	2180
Irrigation cost	1320	1320	1320	1320	5280
Inter cultures	440	440	440	530	1850
Inter cropping	1200				1200
Harvesting				500	500
Miscellaneous	1200				1200
Total	16770	7020	9080	9980	42850

There is lot of variations in cost of cultivation worked out between NABARD's and actual. The actual cost of cultivation and the expected returns as per the feed back obtained from the farmers fields have been compiled and presented by Kochu Babu and Prabhakara Rao and published in National Seminal on Research and Development of Oil Palm in India – 2005 and is reproduced below:

Table 17.4 also gives the expenditure details up to Fifth year, during which period the farmer might have got a subsidy of Rs.21832/ha for planting material and plantation management. The authors have stressed the need to support small holders with subsidies without which it may be difficult to manage the juvenile period due to financial constraints. They also have indicated that that the minimum size holding should be 1.0 ha with which it is possible to get Rs. 3000 per month. It was also indicted that with proper inter crops and proper maintenance thereafter, Rs. 25000/ha as net profit can be obtained.

Table 17.4: Actual Cost of Cultivation of Oil Palm and Expected Returns (as per the feed back from the farmers of A.P.)

Sl. No.	*Item of Expenditure*	*Year (Figures in Rs.)*						
		I	*II*	*III*	*IV*	*Total*	*V*	*VIII*
1	Cost of Seedlings @ Rs.55x143	7865	0	0	0	7865	0	0
2	Gap Fillings @ Rs.55x7	0	365	0	0	385	0	0
3	Saplings transport @ Rs.7	1001	49	0	0	385	0	0
4	Manures (FYM, Organic Compost)	750	1200	1500	1750	5200	2250	2750
5	Fertilizers	2200	3700	5240	6098	17332	6098	6098
6	Plant Protection	250	400	550	550	1750	550	550
7	Intercrop	1200	1200	0	0	2400	0	0
8	Land Preparation	1500	0	0	0	1500	0	0
9	Lining and Pitting	750	0	0	0	750	0	0
10	Inter-culture	750	750	1500	1500	4500	2500	1500
11	Basin Making/twice/year	715	858	1001	1287	3861	1430	1430
12	Pruning	250	250	500	500	1500	750	750
13	Labour for Irrigation	1000	1250	1500	1750	5500	2000	2000
14	Electricity (including depreciation)	2000	2000	2000	2000	8000	2000	2000
15	Harvesting	0	0	375	750	1125	1250	1500
16	Collection and Transportation	0	0	350	750	1100	1250	1500
17	Misc. watch and ward *etc.*	9000	9000	9000	9000	36000	9000	9000
	TOTAL (A)	29231	21042	23516	25935	99724	28078	29078
	Subsidy on seedlings	6750						
	Fertilizer	1545	3097	5220	5220	21832		
	(B)	8295						
	(A-B)	20936/ ha	18945/ ha	21796/ ha	24125/ ha	77892/ ha		
	Income from intercrop (C)	4500	4500	3000				
	Income from FFB (D)				20500		28700	73800
	Net Realization	-16436	-14445	-18796	-3625		622	44722

Source: Kochu Babu and Prabhakara Rao, 2005.

Based on the available information in the West Godavari District, where the oil palm was taken up as small holder crop with lot of intercrops, the income from oil palm crop including intercrops has been worked out and reproduced in Table 17.4. Summary of the out come is produced in Table 17.5.

Table 17.5: Income from Oil Palm Cultivation, (Andhra Pradesh) (Rs./ha)

Year	*Actual Net Income from Inter-crops*	*Expenditure in Oil Palm*	*Net Income Rs./ha*
1st	50000	20346	29654
2nd	50000	15346	34654
3rd	--	14396	-5696
4th	--	16714	12286
5th	--	22588	23812
6th	--	23576	40224
7th	--	25171	61829
8th	--	29799	74601
9th to 17th yrs	--	30679	73721
18th	--	31366	64334
19th to 25th yrs	--	30954	56046

Source: FFF (personal communication).

From the above table it is very clear that investing for first four years and maintaining the crop for another 20-25 years, not only gives a sustainable income but also provide continuous employment. The detailed cost of cultivation with profit for 25 years is given in Table 17.6.

15.10. Labour Productivity and Employment Opportunity

Oil Palm cultivation and production is highly labour-intensive and can be used for job creation on a large scale in developing rural economies with adequate supplies of otherwise unused land and manpower. Exact comparative input data are not yet available and should be the matter of national investigations, as should be the comparative production cost studies. The country whose oil palm sector has been most thoroughly analyzed and is best known, Malaysia, has more than 320,000 persons employed in the oil palm sector. In total 8,60,000 people were employed both directly and indirectly. As this caused a relative labour shortage in Malaysian agriculture, the government imported manpower from neighboring countries.

Manpower shortages also led to capital-intensive changes in the nature of work; while planting, tending, and direct harvesting are clearly best performed by manual labour and will remain so, land preparation, removal of old palm or rubber trees, transportation of bunches, manuring, weeding and other functions have been mechanized, not uniformly throughout the entire sector but progressively as individual estates intensify their cultivation activities, and the high profit margin allowed it.

15.10.1. Labour Requirement for Oil Palm Cultivation

Oil palm being a perennial crop requires one person for 5 ha of land on a plantation scale. But in India oil palm being the small holders 'crop, the small areas of 1-2 ha can be managed by the family labour. The labour productivity is higher in

Table 17.6: Cost of Cultivation of Oil Palm for 25 Years

Sl.No.	Item of Expenditure	Irrigated (per hectare in Rupees)								
		1st Year	2nd Year	3rd Year	4th Year	5th Year	6th Year	7th Year	8th Year	9th Year Onwards
1	Land preparation	2500	0	0	0	0	0	0	0	0
2	Pitting and planting	1430	0	0	0	0	0	0	0	0
3	Seedling cost (After Subsidy)	1430	0	0	0	0	0	0	0	0
4	Manures	3000	3000	3000	3000	3000	3000	3000	3000	3000
5	Fertilizers cost (After Subsidy 4 Years)	1186	2001	563	1495	5783	5783	5783	8940	8940
6	Pest, disease and weed control chemicals	600	600	800	1000	1200	1400	1600	1800	2500
7	Labour for application (Manures, Fertilisers and Weedicides)	400	400	400	500	600	700	800	900	1000
8	Basin maintenance (Labour)	1430	1430	1430	1430	1430	1430	1430	1430	1430
9	Irrigation	4000	4000	4000	4000	4000	4000	4000	4000	4000
10	Electricity charges	1520	1520	1520	1520	1520	1520	1520	1520	1520
11	Motor repairs	1000	1000	1000	1000	1000	1000	1000	1000	1000
12	Harvest and transport (FFBs)	0	0	375	1250	2000	2750	3750	4500	4500
13	Interest on working capital	1850	1395	1308	1519	2055	1993	2288	2709	2789
	Total Cost	20346	15346	14396	16714	22588	23576	25171	29799	30679
	Returns:									
	Yield from oil palm (MT)	0	0	0	1.5	5	8	11	15	18
	Gross income of oil plam (Rs)	0	0	0	7500	25000	40000	55000	75000	90000
	Expenditure on oil palm (Rs)	20346	15346	14396	16714	22588	23576	25171	29799	30679
	Net income on oil palm (Rs)	-20346	-15346	-14396	-9214	2412	16424	29829	45201	59321
	Average net income from inter crop	50000	50000							
	Net income (Rs.)	**29654**	**34654**	**-14396**	**-9214**	**2412**	**16424**	**29829**	**45201**	**59321**

* FFB rate taken at Rs.5000/mt. * Price Fixing Committee of State Government fixes price every time.

case of oil palm compared to any other agricultural crop. Harvesting, collection of fruit and transport to the collection centre can be handled on contract basis, which otherwise means the increase in labour productivity. This can be compared with that of the mechanical harvesters for rice, wheat and other crops where labour is less involved. Approximate labour requirements for field operations in major Oil Palm Growing countries is given in Table 17.7.

Table 17.7: Approximate Labour Requirements for Field Operations in Major Oil Palm Growing Countries

Operation	*Labour Requirements (man-days/ha)*						
	Malaysia		*Indo-nesia*	*Thailand*	*Ghana*	*Ivory Coast*	*Colom-bia*
Sources	*1*	*2*	*3*	*4*	*5*	*6*	*7*
Weeding: immature areas							
Circle weeding	2.4	-	2	1.7-2.8	2	6-9	-
Interline weeding	2	-	11	2.8-4.0	3.6	2.5	-
Imperata control	0.5	-	-	0.2-0.9	-	7.5	-
Cleaning palm crowns	-	-	-	-	-	1.5	-
Weeding: mature areas							
Circle weeding	0.17-0.4	-	0.7	0.2-0.7	1.5	3-6	-
Interline weeding							
Chemical	0.25	-	1	0.6-1.4	1.1	2.4	-
Manual	-	-	3.8	-	3.1	-	-
Pest and disease control (if needed)							
Pest and disease censum	0.1	0.03	1	-	0.03	0.04	0.03
Rat baiting, 5 rounds/year	0.4	0.22	0.7-1.3	0.4-0.6	-	-	-
Trunk injection of insecticide/round	0.3	0.15	1.1	0.6	-	-	1.33
Insecticide fogging/round	-	-	-	-	0.2	-	0.03
Other operations							
Pruning	-	0.3-1.3	4.9	1.3-5.4	1.6-3.1	-	0.7
Fertilizer application							
Manual	-	1.5-3	1.3-2.1	1.5-2.3	0.6	-	0.6
Tractor	0.2	-	-	0.35	-	-	0.2
EFB mulching							
Manual spreading	-	1-1.5	-	2.8	-	-	-
Mechanics	-	0.1-0.3	-	0.5	-	-	-
Harvesting and fruit transport							
Total	8-11	-	13.7	Dec-24	-	-	-
Separate operations							
Cutting	4-8.5	-	-	4.3-7.2	-	-	-
Loose fruit collection	-	-	-	-	-	-	-

Contd.

Operation	*Labour Requirements (man-days/ha)*						
	Malaysia		*Indo-nesia*	*Thailand*	*Ghana*	*Ivory Coast*	*Colom-bia*
Sources	*1*	*2*	*3*	*4*	*5*	*6*	*7*
In-field transport							
Minitractor	0.9-1.7	-	-	-	-	-	-
Iron buffalo'	-	-	-	1.3-5.7	-	-	-
Buffalo	3.8	-	-	-	-	-	-
Harvesting, fruit transport processing (tFFB per man-day)							
Total	1-1.9	1.2-2	1.7	-	-	-	-
Separate operations							
Cutting	3-4.8	-	-	3.5-5.5	2.3	-	-
Loose fruit collection (tLF/ man-day)	-	-	-	-	0.25	-	-
In-field transport	-	-					
Minitractor	15-18	-	-	7.4	-	-	-
Iron buffalo'	-	-	-	-	-	-	-
Buffalo	3.8	-	-	-	-	-	-
Wheelbarrow	-	-	-	-	2.1	-	-
Processing	8.5	-	7	-	-	-	-

Sources: 1: Pamol Plantations Sdn Bhd; 2: Rankine and Fairhurst (1998c); 3: London Sumatra; 4: Univanich Group; 5: Benso Oil Palm Plantations; 6: Jacquemard (1998); 7: Unipalma. LF: Loose fruit.

15.10.2. Employment Opportunity by Developing Oil Palm Industry in India

In today's world where rising unemployment is a serious concern in both urban and rural areas, the rise of oil palm industry can be looked as a welcome opportunity for many. Mentioned below are some points which will support the fact that oil palm plantations provide much needed employment opportunities:

- It gives assured employment to the agricultural workers round the year as it is not a seasonal crop.
- Skilled labourers (harvesters) get a good remuneration during the peak harvesting period.
- Processing mills will support a major workforce.
- It is a guaranteed income (after 3 years gestation period) for the absentee landlords, for whom it will act as an extra income/alternative source of employment with zero headache.
- Crop is a boon for three subgroups of casual labours *i.e;* cultivators, share croppers and lease holders.

- ★ Timely fertigation or fertilizer application being an important prerequisite of oil palm, villagers can make clusters and take care of fertilizer supply to the farmers on time by opening fertilizer depot (thereby doing a business).
- ★ Collection centres serve as a hub for labourers, as huge quantity of produce needs to be unloaded, cleaned, weighed and reloaded, thus serving as a good source of income.
- ★ Oil palm cultivation having an extended full phased process starting from nursery raising to refining and marketing, gives a golden scope for the Infrastructure development with it.
- ★ Building up of axillary industries generates further employment.
- ★ Oil palm cultivation also makes local people more dependent on agricultural firms since they no longer grow their own food.

Note: According to findings for every 1000 hectare -

At the company level-

Direct labour requirement = 100

At the farmer's level -

Labour requirement = 200 (1 labour required to maintain 5 ha)

Indirect labour requirement = 100

During procurement of FFB, labourers, drivers, cleaners, small shop establishers at the developed area (Tea/grocery/beetle nut) etc. are the key players in the indirect group of labourers counted for the movement of the above mentioned.

For every 1000 hectares, around 100 labourers get an employment indirectly.

Total = 400

In today's context, availability of labour for agriculture work becoming difficult and need for mechanization is very important. However, one has to find out skilled labourers for harvest by setting up of Village or Mandal level Labour Banks or Harvest Banks, providing them the harvesting tools, training to operate efficiently, and ensuring them fairly good salary and giving them health Insurance.

Chapter 16

Processing of Oil Palm Fruits

Palm oil is derived from highly perishable fleshy mesocarp of its fruits of right maturity. The fresh fruit bunches (FFB's) and the raw material after harvest, are required to be processed preferably within 24 hours for the production of good quality palm oil. Seasonality of the fruiting *i.e.,* 60 per cent of the fruits are harvested during the peak season spanning about 4 months leaving only 40 per cent for the remaining 8 months, is another problem. Further, oil palm starts yielding only 3 years after planting and progressively increases up to eighth year when it stabilizes. The factors mentioned above put severe constraints on the capacity utilization of palm oil mill *viz.,* only during the peak four months when the mill runs at installed capacity while in the remaining months the capacity utilization will be at much lower level. A captive plantation area, therefore, is an essential pre-requisite for the viability of the mill. Capacity of the plant and various equipments should be designed for a given area of plantation carefully during the early stages of plantation development itself. Presence of an extremely active enzyme lipase in the mesoecarp renders the oil inedible unless proper care is taken from harvesting through transport to subsequent processing of the fruits. Laboratory studies have shown that this enzyme is bound to the membrane of fat globules and any damage to the membrane triggers the activity and the free fatty acid (FFA) released thus, could be as high as 60 per cent depending on the severity of damage. High FFA not only makes the oil inedible but also makes rate of rancidity faster, fixes the colour and increases refining loss. Palm oil mills, therefore, should be located in the palm plantation in order to process the fruits immediately after harvest. Modem mills have been designed to match the size of plantation and to cater to the fruit yield of the captive area. Unlike in Malaysia and elsewhere smaller plantations with mills of matching capacities are suitable in India.

Harvesting of fruits at the right maturity is important with respect to yield and quality of palm oil since immature fruits yield less oil and over mature fruits have high FFA. Excessive bruising of the fruit bunches should be avoided during harvest and transport. Processing of the FFB within 24 hours after harvest is essential to obtain edible quality raw palm oil. Various devices and methods are available now to avoid damage to the fruits during harvest and transport.

16.1. Milling Technology

Oil palm processing technology has undergone tremendous changes over a period of time as industry started growing on a large scale. The traditional processing of oil palm is very crude country method of extraction of palm oil as seen in the Figure 16.1.

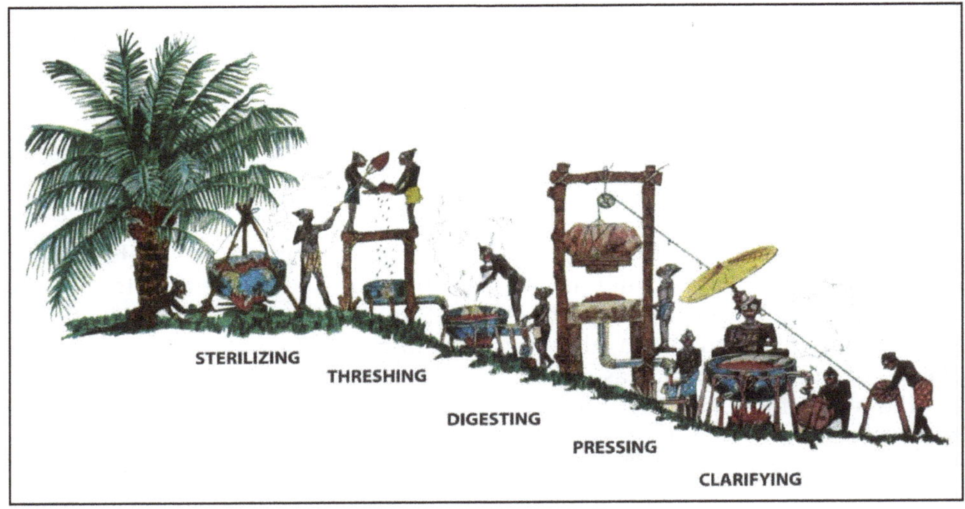

Figure 16.1: First Palm Oil Factory in Africa.

The above schematic drawing is a typical illustration of the primitive processing unit. A careful study of above will draw a parallel to current method of production of palm oil, which evolved through many years of trial and practice, resulting in the current method involving the use of modern equipment.

While the steps involved in the processing remained the same, the refinement in developing machineries took place with increased capacity to increase the mill efficiency and cogeneration of energy in boilers, sterilizers, strippers, shredders, hydraulic to screw press, *etc.*

Various phases of development s in processing facilities have taken place over a period of time as indicated below:

- ★ **Early 1950's:** Mini mill of capacity 1.0 to 1.5 t/ha, mini sterilizers, hand operated presses and locomotive boilers,

- ★ **Early 1960's:** Mini mills of 5.0 to 10.0 t/ha, horizontal sterilizers, automatic press and tire tube boilers.
- ★ **Early 1970's:** Mills of medium capacity of 15 to 20 t/ha, horizontal sterilizers, screw press, water tube boilers and steam engines/turbines.
- ★ **Early 1980's:** High milling capacity of 20 to 40 t/ha with double processing lines.
- ★ **Early 1990's:** High mill capacity of 60 t/ha, enabling better mill layout,easier, higher efficiency, improvement in fruit handling, increased size of fruit cages to 3.5 tons.
- ★ **Early 2000:** Mill capacity expanded beyond 60 t/ha, introduction of bigger size cages (5 to 20 tons) and tipple system replacing hoisting cranes.

The origin of today's processing method is an adoption of age old concept, involving primitive methods of processing and extraction of oil. This concept has been brilliantly reflected in a pictorial form, which appears in any publication tracing the origin of the palm oil processing industry.

Introduction of new equipments and systems has enabled smoother operation and higher efficiency. Larger capacity mills with automation reduced the work force, achieved plant efficiency, improved working conditions with high level safety, addressed environmental issues and paved way for by product utilization and value additions (Figure 16.2).

Figure 16.2: Modern Palm Oil Processing Mill.

16.2. Steps Involved in Processing

Once the FFB is transported to the factory the processing is started immediately. The steps involved in the process are narrated below:

16.2.1. Sterilization

This serves the dual objectives of inactivation of the enzyme lipase and loosening of the fruits from the bunch. It also softens the cell wall and coagulates the proteins that facilitate oil extraction in the subsequent stages. A horizontal type steriliser is adopted to minimise bruising of fruits. The FFB are loaded in a perforated cage and placed in the steriliser. The steriliser is a long cylindrical vessel provided with a steam inlet and outlet, pressure gauge *etc.* Sterilization of FFB for 60 minutes of 45 psi has been found to be optimum to achieve the objectives mentioned above.

16.2.2. Stripping

The process of sterilization only loosens the fruits but does not separate them from the bunch. The loose fruits are stripped off from the bunches with the aid of a mechanical bunch stripper. This device is a rotary drum with baffles and perforations. When the sterilized bunches pass through the rotating drum (20 rpm), the fruits are separated from the bunches and loose fruits fall through the perforations and the empty bunches pass to the other end of drum. The loose fruits and empty bunches are collected separately.

16.2.3. Digestion

Purpose of digestion is to convert the loose fruits into pulp and in the process the cell walls are broken facilitating release of oil with the help of thermal and mechanical energy. The digestor is a vertical jacketed cylindrical vessel fitted with a centrally mounted agitator having specially designed blades rotating at slow speed (25 rpm). The loose fruits are charged into the digester from the top with the aid of an elevator. Live steam is injected into the jacket and into the vessel to maintain the temperature at 95°C. The digested mash of semi-solid consistency is discharged through a discharge chute at the bottom.

16.2.4. Pressing

The digested mash consisting of oil, water, seed, fibre and other suspended matter is charged into a perforated cage and subjected to pressing. The hydraulic press is fitted with a plunger matching with the cage diameter. The pressing is done while the mash is hot (80-90°C) at 750 psi. The hot oil water mixture with suspended solids is expelled through the perforations leaving the solid press cake in the cage.

16.2.5. Clarification

The oil water mixture is filtered to remove the fibrous debris and is collected in the clarification tank. Clarifier is a vertical cylindrical vessel fitted with steam coil. The oil-water mixture is diluted with hot water (1:2) and heated to 95°C. The oil being lighter rises to the top and is decanted continuously into a collection tank. The water sludge at the bottom is discharged as waste.

16.2.6. Purification

The crude palm oil from the clarifier still contains 0.4 to 0.6 per cent of water, 0.1 to 0.2 per cent sludge and other impurities which affect the quality of oil. So it is passed through a high speed centrifuge (8,000 rpm) at 80°C to remove the traces of solid impurities and water. The pure raw palm oil thus obtained is stored in tanks.

16.2.7. Vacuum Drying

Excess moisture in the oil leads to FFA release and quality deterioration on storage. Vacuum drying is adopted to remove the excess moisture and for attaining the optimum moisture of 0.1 to 0.15 per cent.

16.3. Oil Storage

The raw palm oil is stored in tanks provided with steam coils.

16.4. Nut Recovery

Palm kernel oil is derived from the palm nut recovered from the press cake. After extraction of palm oil, the press fibre has a moisture content of 45-50 per cent. Mechanical recovery of the seeds involves breaking and drying the press cake, followed by separation of the nuts. The unit consists of a steam jacketed paddle, conveying and cake breaking section. During the passage, the press cake is broken with the help of the rotating paddles and is also dried as it is heated in the steam jacket. The dried press cake is separated into fibre and nut in a vertical separating column using suction. The lighter fibre is drawn into a cyclone and the heavier nuts are allowed to fall into a horizontal rotating nut polishing drum. The fibre thus separated is used as boiler fuel. Final removal of residual fibre from the nuts is achieved in the polishing drum along with grading of the nuts based on size. A small palm oil mill envisages only up to the nut recovery stage and the nut as such is a saleable product.

16.5. Kernel Oil Extraction

Palm kernel oil is derived from the palm nut recovered from the press cake. The cleaned, polished nuts are fed to the cracking section which usually consists of revolving screens to grade the nuts, the crackers, and screens or columns to separate un cracked nuts and/or dust and small shell particles from the mixture. Palm kernel and shell mixture is separated into kernel and shell in a clay bath maintained at a particular specific gravity. The shell being heavier, sinks and the kernel being lighter floats and is skimmed off. The separated kernel is dried to a final moisture of 6-8 per cent. The kernel is powdered and steam conditioned followed by extraction of oil in expeller. The palm kernel oil is similar to coconut oil in fatty acid composition and it fetches a premium price.

16.6. Refining

Oil extracted from the fleshy orange red mesocarp is known as crude palm oil (CPO) which on refining becomes RBD palm oil and on fractionation is called as RBD Palmolein. CPO has thick consistency, red colour and high smoke point. Hence, refining aims to convert the crude oil to quality oil. The purified product had a wide acceptance compared to CPO since it has a fairly good colour, flavour and consistency. Nutritive value of palm oil mainly depends on the presence of minor components such as carotenoids, vitamin E sterols, phosphatides, triterpenic and aliphatic alcohols. However, in the refined palm oil all the components except the carotenoids are present. Carotenoids are generally removed or destroyed while refining.

16.7. Recent Advances in Milling Technology

Having looked at the historical background of the development of the state of milling technology, some of the new technologies, design and operational concepts are going on since 1980 in Malaysia. A few technologies that are now considered as best practices in milling technology are summarized below:

16.7.1. Fruit Reception and Handling

- ★ Major changes were introduced to the front end operation of the conventional mills involving reception, conveyance and sterilization of the FFB.
- ★ Unloading of FFB on the ramp, discharging on too cages and transporting them through network or rail tracks to sterilizers posed monumental! task and a real set back to expansion of milling capacity, apart from the effect on the quality due to bad handling with the introduction of "transfer carriage system" and the "tippler" a major breakthrough was made in the front end operation of the mill.
- ★ Elimination of the messy rail track system, Easier and faster means of conveying FFB to the sterilizers by using conveyors,Increase of size of cages from 2.5MT to 20MT,use of double door sterilizers, elimination of the use of hoist or gantry cranes, and most importantly the expansion of milling capacity to reach 120 MT per hour and Cleanliness and safety are important features.

16.7.2. Sterilization

- ★ Installation of larger size sterilizers, of capacity of say 4 cages 20 MT each
- ★ Stainles steel sterilizers to reduce wear and corrosion
- ★ Easier to operation maintenance and less manpower
- ★ Cleanliness and safety
- ★ Automation of control sterilizing cycle- triple peak cycle and for charging and discharging of cages.
- ★ Use of vertical/spherical or use of continuous sterilization which eliminates the use of sterilizer gauges,rail tracks, overhead cranes,tippers, transfer carriages and tractors.

Three types of sterilizers are now in use. Their salient features of the above -systems are:

i) Vertical

- ★ Simplified front end operation of the mill
- ★ No rail tracks and cages are used
- ★ Steel vessel in vertical position

- ★ FFB feeding by conveyors and loaded directly into the sterilizer
- ★ Low maintenance cost
- ★ Smaller factory space (claim)
- ★ A few mills are in operation using this system, and the same is further being improved.

ii) Spherical

- ★ Simplified front end operation of the mill
- ★ No cages and rail track used
- ★ Steel ball-like vessel installed at an elevated position
- ★ FFB feeding by conveys and loaded directly into the vessel thru central opening.
- ★ Sterilizer fruits are emptied by rotating vessel 180 degree.
- ★ Superior aeration system (claim)
- ★ The system is in operation in two mills in Philippines. Currently, a new mill is being built in Malaysia using this system. The plant is designed by KONPRO and will be operational by April/May 2008.

16.7.3. Continuous Sterilisation

- ★ Simplified front end operation of the mill
- ★ No rail track cages used
- ★ Sterilization process uses steam at atmospheric pressure or at low pressure.
- ★ This system uses moving chains as a medium to hold FFB for sterilizing
- ★ Volume of condensate is lower
- ★ Lower maintenance cost
- ★ The system was actually developed by Malaysian Palm Oil Board (MPOB) and the commercialization of the system is being undertaken by a local engineering concern Modipalm Engineering Sdn.Bhd. The system is in operation in a number of new palm oil mills in Malaysia and Indonesia.

16.7.4. Thresher and Threshing Technique

Basic design of the thresher has not changed, but improvements have been made to facilitate higher operating capacity and longer retention time.

- ★ Due to Weevil effect, bunches become more compact and more difficult to strip leading to use of bunch crusher, for second stage threshing
- ★ Double and triple peak sterilization
- ★ Better material of construction
- ★ As mentioned earlier the technique of squeezing oil from the mesocrap of the sterilized palm changed drastically in the 1970's, with the advent of screw presses replacing the hydraulic presses

- ★ Since then, the screw presses have gained wide acceptance and further improvements have been made to increase the processing capacity from 9 MT per hour to the presently 20 MT per hour.
- ★ The bigger range unit at 20 MT per hour was introduced in early 2000, and is gaining acceptance amidst earlier reservation about its performance and efficiency.

Now the capacity has been increased considerably to even 90 MT FFB/hour

16.7.5. Clarification

- ★ In the area of oil clarification, much change have been made, involving installation of new types of oil separators and purifiers, contributing to more efficient oil recovery.
- ★ These are less compact and easier to operate
- ★ Emphasis on cleaner oil room operation
- ★ Use of decanters as standard palm oil processing equipment. Decanters, if properly operated can contribute to higher oil recovery, about 0.05 per cent to FFB
- ★ Latest Decanter System D3-PRO (Alfa Laval) reduce dilution (an effluent problem) and ECO-D2 (Westfalia Separator)
- ★ Above improvements and practices are ongoing and promoted by major equipment manufacturers like Alfa Laval and Westfalia

16.7.6. Kernel Recovery

The typical arrangements of a kernel recovery stations in a modern palm oil mill will have the following:

- ★ Ripple mills, replacing the nut cracker have enabled automation of the kernel plant, apart from the need to drying of nuts.
- ★ Some mills have introduced "continuous kernel tray dryer" equipment which can reduce moisture content consistently making it easy to operate and maintain.
- ★ The time tested dry separation using winnowing system, alone or in combination with hydro cyclone or hydro clay bath (wet system) is also widely used.
- ★ Good sterilization and optimum standard of operation and maintenance of kernel recovery plant will ensure over 95 per cent efficiency of recovery.

16.7.7. Automation System for Palm Oil Mills

Plant wide automation systems can improve process efficiency and increase productivity as in all other industries.

Palm oil industry is slow due to various problems encountered.

- ★ High cost of installation, operation and maintenance

- ★ Skill required to operate
- ★ Expensive cost of replacement
- ★ Availability of cheap labour

However, automation is sufficiently applied m the following areas of mill operation, contributing to efficiency and savings:

- ★ Weighing operations
- ★ Cage handling
- ★ Sterilization operation
- ★ Press operation
- ★ Kernel separation and transport (using pneumatic system)
- ★ Boiler operation
- ★ Office automation (use of ICT)
- ★ Preparation of accounts
- ★ Store operations
- ★ Continuous improvements and higher level of automation and computerization are taking place in this area.

16.7.8. Steam Management

- ★ Steam management is essentially the management of steam generation and utilization by installation of well-designed and suitably sized boiler to provide steady supply of steam during normal operation and to meet any sudden high demand.

The following aspects need consideration while selection of boiler and capacity

- ★ Establish steam balance carefully
- ★ Type and construction
- ★ Fuel requirements
- ★ Level of automation
- ★ Safety features
- ★ Useful life
- ★ Air pollution control equipment
- ★ Ash disposal system

16.7.9. Hygiene and Food Quality

- ★ Palm oil is edible and looked upon as food, thus demanding a high degree of hygiene. Accordingly palm oil manufacturing must comply with international guidelines and requirements such as Hazard Analysis Critical Control Points (HACCP), EurepGAP, *etc.*
- ★ To achieve this future mill design must ensure it is well laid and ventilated.

- ★ Machines installed must meet safety standards and be easy to operate and maintain cleanliness.
- ★ Health and Safety procedures must be drawn up and enforced strictly. Good working environments to ensure safety and welfare of workers.
- ★ Mills must meet food quality standards in order to ensure marketability of its product in the face of increasing consumer awareness for safe products.
- ★ Clean FFB from the plantations and clean mill operation will no doubt equate to high efficiency, minimum wastage and good marketability of its products.

All the above concepts are currently being adopted and in operation in India with varying degree of success.

Chapter 17

Food and Non-Food Uses of Palm Oil for Nutrition and Health

Palm oil is one of the 17 edible oils, which have been declared as wholesome and nutritious edible oil is suitable for human consumption (FAO/WHO food standards). Palm oil and its other derivatives are today widely used in both edible foods including Oleo foods and non-edible applications all over the world. Palm oil and palm kernel oil have many of the required characteristics suitable for food applications like cooking oil, frying fats, Vanaspati, margarines specialty fats for cooking, cookies, cake mixes, instant noodles, toffee fats, biscuits *etc.* It greatly contributes to the health and nutritional security of millions around the world. Nonfood uses include soap, detergents, personal care products, surfactants, lubricants, polymers, paints and surface coating, pharmaceutical products *etc.* (Figure 17.1).

Global consumption of palm oil is given in the Figure 17.2. While 71 per cent of palm oil is used for food, 24 per cent is used for cosmetics and 5 per cent for energy.

17.1. Food Uses

17.1.1. Household Cooking Oil

RBD palmolein (the liquid fraction of palm oil) is a suitable general-purpose cooking oil in the household sector. Due to its low rate of degradation, palmolein is indeed an economical cooking oil since it can be used over and over again without turning rancid. An additional advantage for housewives is that with palmolein no sticky varnish-like film forms on the surface of fryers, as it happens with many other oils. Furthermore, palmolein has fewer tendencies to smoke and foam. It can

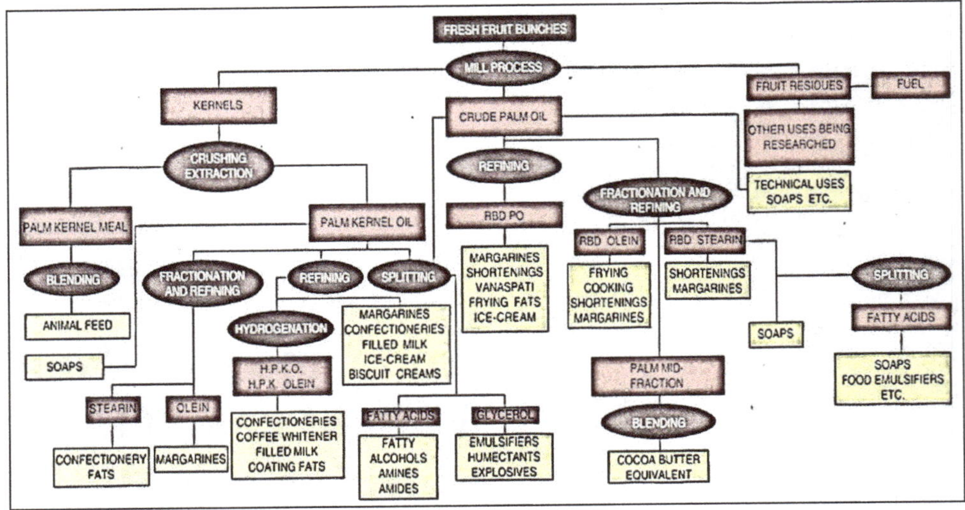

Figure 17.1: Palm Oil Utilisation Chart.

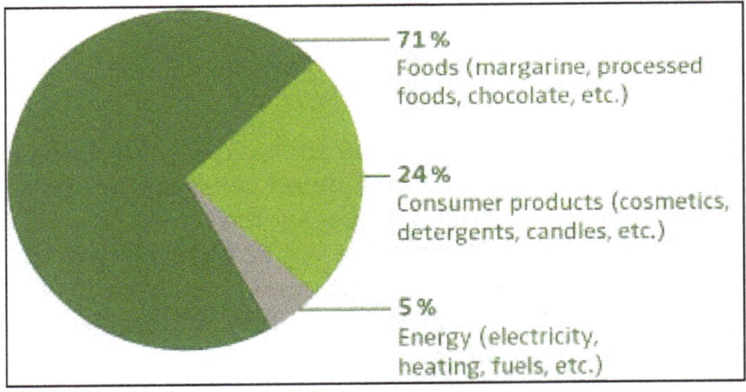

Figure 17.2: Global Consumption of Palm Oil.

also act as the perfect partner for blending with any other vegetable oils to maintain traditional tastes

17.1.2. Industrial Frying

RBD palmolein is most suitable for frying and it has been used the world over for the processing of potato chips, French fries, instant noodles, doughnuts, fried chicken and snack foods. The relative stability of palmolein at frying temperature and its lower rate of oxidation makes it an excellent medium for frying. These distinctive properties also extend the frying life of the oil, making it very economical.

a) Margarine

Palm oil is an ideal ingredient for manufacturing margarine which is an alternative to butter. Palm oil and palm kernel oil are available in various processed grades, which have a desirable tendency to crystallize into an appropriate form.

Palm based margarine can be easily spread at room temperature and is resistant to rancidity. These margarines are preferred in making cakes.

b) Soap Noodles

Soap noodles are made commercially by blending various ratios of the palm oil: Palm kernel oil, for example 80:20, 70:30, and 60:40. The advantages of Palm based Soap noodles are that these produce a high quality white soap, give a consistent composition, and use a well-established and cleaner technology for production, convenient for soap manufacture and reliable supply of raw material (Palm oil and palm kernel Oil).

c) Shortening

Palm oil can be used to produce shortenings of excellent quality with wide applications. Palm-based shortening is suitable for making bread, cakes, pastries, cream and sweets as well as for frying. Pastries and bread with palm-based shortening possess extra softness, better texture and good storage stability. Likewise, palm-based shortening imparts easy to break and crumbly structure to biscuits which melt readily in the mouth.

d) Vegetable Ghee

Palm oil can be used alone or blended with other vegetable oils to produce good quality vegetable ghee. Palm oil in its natural state needs minimal or no hydrogenation and is close to many of the specifications for vegetable ghee without any treatment other than refining.

e) Ice Cream

Another special use of palm oil is in ice-cream. Although milk fat is an ideal ingredient for ice-cream, carefully blended vegetable oils can act as an excellent alternative. Palm oil and palm kernel oil not only give a well-structured and smooth-textured ice-cream but they are also resistant to rancidity.

f) Palm Based Chocolate Products

Palm oil and palm kernel oil are ideal raw materials for the production of specialty fats due to their excellent physico-chemical properties. They can further be modified to extend the range of utilization. This is used in preparation of chocolates. The advantages of using Palm based fats in chocolate products are, they don't require tampering, form stable crystals and are less subject to bloom, have ecological properties suitable for modern and highly flexible production lines are stable in hot climates, have consistent quality, reliable supply, besides being versatile and economical.

g) Confectionery

Cocoa-butter replacers (CBR) are fats which replace cocoa-butter partly or wholly in chocolates and confectionery. By fractionation processes, palm oil and palm kernel oil can be used to produce cocoa butter replacers.

h) Palm Oil Based Oleo Foods

Indonesian Oil Palm Research Institute,Medan had developed technologies to prepare Oleo foods like Palm ghee, sweet condensed milk enriched with Omega 3, Red Margarine, red palm oil, Pal mega (Palm oil enriched with Omega Palm frying shortening Rispa *etc.*

i) Non-dairy Creamers

A blend comprising palm or palm kernel oil and other oils has been used widely to replace milk fats in non-dairy coffee whiteners or creamers. They replace cream in coffee and offer great advantages of shelf-life and convenience. Palm based non-dairy creamers give a smoother, tastier and a more satisfying cup of coffee.

17.2. Non Food Uses

Nonfood uses of Palm oil and Palm Kernel Oil are given in Figure 17.3.

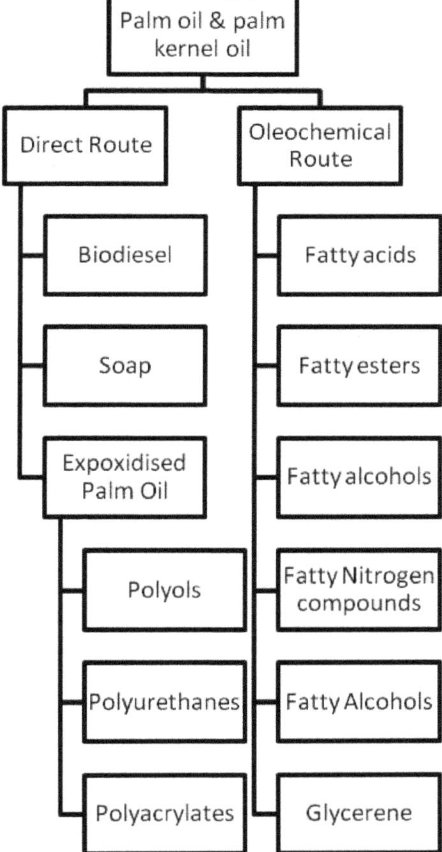

Figure 17.3: Non-Food Applications of Palm Oil and Palm Kernel Oil.

17.2.1. Palm Based Polyurethane Products

Currently, almost all of the chemicals used for the production of polyurethane are derived from petroleum. Therefore, any increase in the price of the crude oil will strongly affect the price of the polyurethane chemicals. The world consumption of polyurethane in 2005 was around 1.8 MT. There is huge opportunity for partial replacement of petroleum based polyols with natural oil polyols in polyurethane applications. Thus, for the past 10 years, MPOP has been actively involved in R&D activities on the production of palm based polyols concurrent with the development of polyurethane foam.

The advantage of palm based polyurethane products is that palm oil based polyols are formulated into polyurethane products using only water as the blowing agent; whereas most commercial foams are produced using CFC chemicals. Polyurethane products produced are VOC free or are able to achieve zero emission when tested using the GC – MS with head space. The price of palm based polyurethane is competitive to the price of petroleum based polyurethane foam.

17.2.2. Soap and Detergents

The single largest non-edible use for palm oil is in the manufacture of soaps and detergents. Palm stearin (the solid fraction of palm oil) has been widely accepted as a raw material by the industry, because of its price competitiveness and its excellent soap-making properties, which impart stability and consistency to the structure of the final product. Palm stearin is available in fully refined, bleached and deodorized form and can be used directly in soap manufacturing without any pre-treatment. Palm oil based soaps possess good detergency, solubility and foaming properties and good stability of colour, glossiness and texture.

17.2.3. Oleo-chemicals

Oleo chemical products derived from downstream processing of palm oil and palm kernel oil have a large variety of end-uses. The basic oleo-chemicals that can be produced are fatty acids, methyl esters, fatty alcohols, fatty amines and glycerol. These products find applications in various fields, including the manufacture of soaps, detergents, cosmetics, pharmaceuticals, *etc.*

Stearic acid is used for making hot melt adhesives and in candle manufacture to control flame brilliance, eliminate smoke, obnoxious odours and promote burning up action to prevent overflow of melted wax.

An important characteristic of palm oil based oleo-chemicals is that it is biodegradable and does not contribute to pollution.

17.2.4. Personal Care Products

Personal care products like cream, lotion and milk are commonly formulated to contain 15–25 per cent of oil or fat soluble components, with the rest as aqua soluble component. Palm oil is used in manufacturing of the personal care products.

17.3. Palm Oil for Nutrition

Palm oil is one of the seventeen edible oils possessing FAO/WHO food standards. and declared as wholesome and nutritious edible oil suitable for human consumption. Palm oil is not a fully saturated oil, as it contains equal amount of saturated and unsaturated fatty acids. It has been a safe and nutritious source of edible oil for healthy humans for thousands of years. Palm oil like all other oils and fats provides 9 kilocalories of energy per gram compared to 4 kilo calories each from proteins and carbohydrates. It helps to maintain healthy skin and hair, ensures proper growth and enables body to absorb vitamins. It is known as tropical oil. The other tropical oils are palm kernel oil and coconut oil.

17.3.1. An Excellent Dietary Energy Source

Palm oil is easily digested, absorbed and utilized in normal metabolic process. It plays a useful role in meeting energy and essential fatty acid needs in many regions of the world.

17.3.2. Rich in Carotenoids

Crude (unrefined) and red or golden (specially refined) palm oils are the richest natural source of carotenoids (Table 17.1). Beta carotene in crude palm oil is a potent antioxidant and pre-cursor of Vitamin-A. Carotene content in crude palm oil is 500-800 ppm which is about 15 times more than that present in carrot and is recommended to overcome Vitamin-A deficiency. It promotes good night vision.

Table 17.1: Carotene Composition of Palm Oil

Carotenoid	*Crude (ppm)*	*Red Palm Oil (ppm)*
Phytoene	10.4	6.5
Cis β-carotene	5.6	3.5
β-carotene	448.0	280.0
α-carotene	281.6	176.0
Cis α-carotene	20.0	12.5
Lycopene	10.4	6.5
Others	29.6	18.5
Total	800	500

Source: Regional Research Laboratory, CSIR, Trivandrum.

17.3.3. Rich in Vitamin E and Antioxidants

Palm oil is a natural source of antioxidants. Vitamin E constituents tocopherols and tocotrienols (Table 17.2). These natural antioxidants play a protective role in cellular ageing, atherosclerosis and cancer.

Table 17.2: Vitamin E Content of Edible Oils

Name of the Oil	Tocopherol (ppm)				Tocotrienol (ppm)				Total
	α	β		δ	α	β		δ	
Palm oil	164	–	–	–	174	–	313	80	731
Palm Olein	196	–	–	–	201	–	372	96	865
Palm stearin	79	–	–	–	81	–	168	44	372
Soybean oil	101	–	593	264	–	–	–	–	958
Corn oil	112	50	602	18	–	–	–	–	782
Sunflower oil	487	–	51	8	–	–	–	–	546
Cocoa butter	11	–	170	7	–	–	–	–	188

Source: Selected Readings on Palm oil and its Uses, PORIM, Malaysia, 1994.

17.3.4. Balanced Fatty Acid Composition

Palm oil has almost perfect 50:50 balance of saturated and unsaturated fatty acid composition compared to other vegetable oils. The principal saturated fatty acid present in palm oil is the C16:0, palmitic acid, which is regarded as less hypercholesterolemic than saturated fatty acids in the range of C6 to C14 of which palm oil has only trace amounts. Fatty acid composition of some common edible oils and fats are given in Table 17.3.

Table 17.3: Fatty Acid Composition of some Common Edible Oils and Fats

Type of Fat	Saturated (Per cent)	Mono-unsaturated (Per cent)	Poly-unsaturated (Per cent)
Coconut	94	4.5	1.5
Palm oil	50	39	10
Palmolein	46	43	11
Cotton seed	83	10	1.5
Peanut	19	39	41
Olive	13	79	8
Corn	16	30	53
Soybean	14	25	60
Butter	66	30	4
Beef tallow	52	44	4
Lard	41	47	12

Source: Selected Readings on Palm oil and its Uses, PORIM, Malaysia, 1994.

17.3.5. Palm Oil and Coronary Heart Diseases

Palm oil and RBD olein contain 10-13 per cent of linoleic acid, an essential fatty acid. A possible action of linoleic acid in the diet is its ability to lower the LDL (low density lipoprotein) cholesterol and raise the levels of protective HDL (high density lipoprotein) cholesterol in the blood stream and thus, lower the risk of coronary heart diseases. The cholesterol levels of edible fats and oils used commonly is given in Table 17.4. It is clear from the table that the cholesterol level of palm oil varies between 13-19 ppm which is either less or almost equal to many other edible oils. According to FAO information, oils having 20 ppm cholesterol or less are considered safe for cooking and the choleterol level in palm oil is insignificant. The countries all over the world especially African nations have been using crude palm oil since ages and there are no complaints of increased heart related ailments.

Table 17.4: Cholesterol Levels in Crude Oils and Fats

Oil Type	*Range (ppm)*	*Oil Type*	*Range (ppm)*
Coconut	05-24	Cotton seed oil	28.108
Cocoa butter	Upto 59	Rapeseed oil	28.108
Palm kernel oil	09-40	Maize oil	18-95
Palm oil	13-19	Lard	3000-4000
Sunflower oil	08-44	Butter	2200-4100
Soybean oil	20-35	Bee fat	800-1400

Source: Selected Readings on Palm oil and its Uses, PORIM, Malaysia, 1994.

17.3.6. Palm Oil and Cancer

Palm oil does not cause an increase in tumour cells which may be attributed to the presence of naturally occuring tocotrienols and thus, reducing the possibility of cancer.

17.3.7. Palm Oil and Thrombosis

Arterial thrombosis is the formation of blood clot in the artery. Palm oil diet lowers the tendency towards inducing arterial thrombosis.

17.3.8. Palm Oil and Atherosclerosis

Atherosclerosis is the process or condition in which fat or cholesterol contributes to the formation of fatty acid deposits in the artery which may result in heart-attack. Palm oil inhibits the formation of such deposits or the relative surface area is smaller compared to other groups of oils.

17.3.9. Absence of Trans Fatty Acid

Palm oil and its fractions can be used directly in a variety of food uses without the need to undergo hydrogenation and does not contain any transfatty acid isomers compared to other hydrogenated oils. Hydrogenated oils have been associated with increase in blood cholesterol and can also result in metabolic and nutritional disturbances.

Chapter 18

Waste Utilization and Environment

Besides palm oil and palm kernel oil, palm oil industry generates enormous quantities of waste from two sources, the field and the palm oil mills. Biomass produced in the field and by products obtained during processing of Palm oil were once considered as wastes. These are now being utilized for making many products of economic uses which are ecofriendly and environmentally sustainable. Besides making wealth from waste also helps to solve the environmental issues. The chart given in Figure 18.1 shows the utilization of oil palm waste for making wealth.

18.1. Types of Wastes

Oil palm tree is the greatest source for vegetable oil as well as bio mass as could be seen in the following pictorial diagram. Produce Oil Palm plantation only 10 per cent oil and 90 per cent bio mass as shown in Figure 18.2. Waste from the field comprises palm trunks and fronds. The fronds are obtained when pruning is done. Waste fronds are also produced along with trunks, when palms are felled for replanting. In the standing palm trees annual pruning of leaves will amount to about 10.4 t/ha.

The principal wastes produced during the operation of an oil palm mill are empty bunches (21 to 22 per cent), mesocarp fibers (12 to 16 per cent), shells (5 to 7 per cent) of the total percentage of fresh fruit bunches. Other biomasses from the mill are decanter solids and effluent treatment sludge which are comparatively in small quantities. Palm kernel meal or palm kernel cake and mill effluent (POME) are other very important waste products.

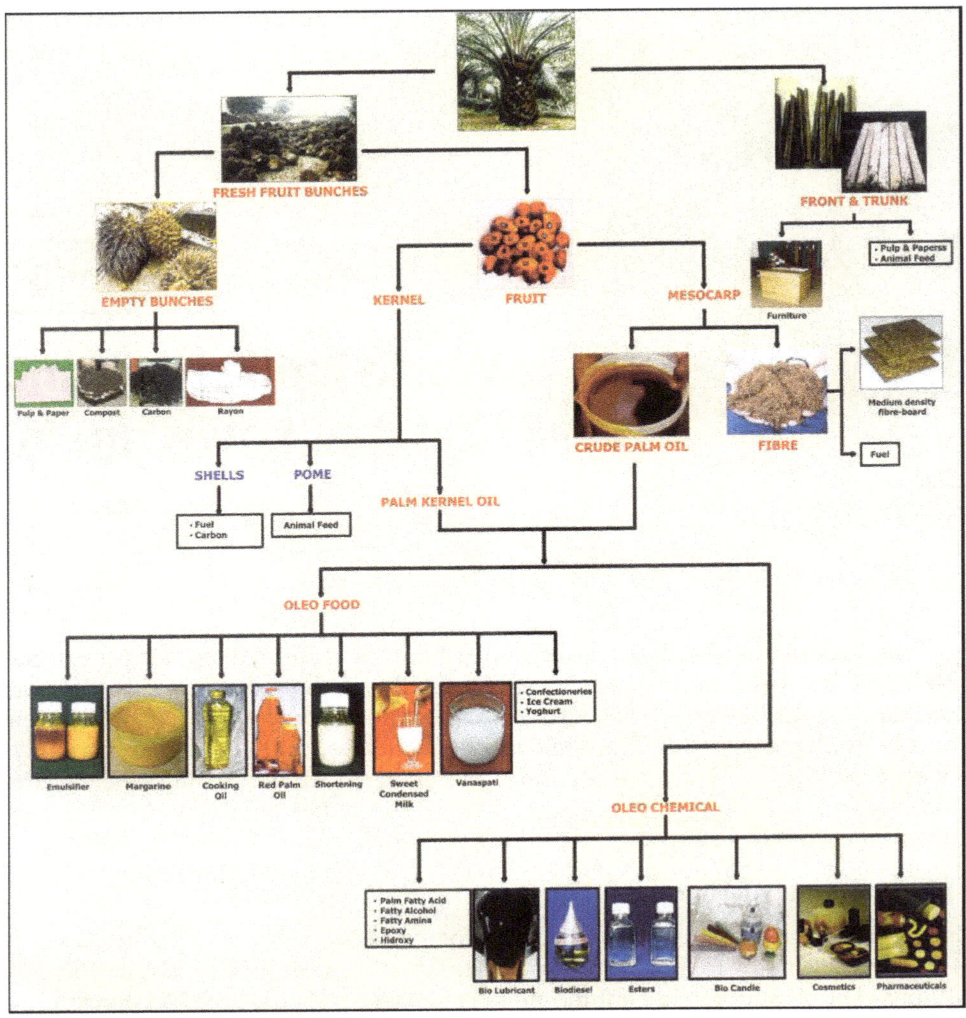

Figure 18.1: Oil Palm by-product Utilization.

18.1.1. Trunks and Fronds

The trunks and fronds are the bulk biomass source with large reserve of plant nutrients.When these are recycled to the plantation, the soil physico-chemical and biological properties increased, thereby increasing water holding capacity, and also yield. Trunks are shredded and composted also. The nutrient composition and the fertilizer equivalent as well as monetary values from trunks and fronds are given in Table 18.1.

Trunks, frond, petiole and Empty Fruit Bunches are rich sources of lignocellulose and have vast scope for using the same in wood based industries for making particle boards, medium density fibre boards, gypsum boards, roof tiles, pulp and paper

Figure 18.2: Distribution of Various Components in Oil Palm.

etc. EFB fibre strands for the manufacture of mattresses, rubberized foam for car seat *etc.* It can be blended with coconut fibre and lot of value added products can be prepared.

Oil palm trunks and fronds can be used for the industrial production of building materials by processing into boards, panels for walls and ceilings and other similar products. These can also be used directly as fuel and for the production of chemicals. Oil palm trunks contain carbohydrates which when allowed to degrade in the field, supply nutrients to the soil, increasing its fertility as well as conserving moisture and reducing soil erosion. Research aimed at production of vitamin-E from oil palm leaves is also in progress. Technologies are available for preparing Ply wood and oil palm lumber from Oil Palm Trunks and modulated particle board from oil palm fronds in Malaysia.

The oil palm trunks can be used to make block wood which is comparable in strength to other commercial chip boards. Other technological uses include roof materials made from a mixture of oil palm by-products and condensed fuel pellets of low sulphur value and thus of low pollution rate.

18.1.2. Empty Fruit Bunches, Fibres and Shells

Empty bunches are used in the field for mulching, where they also act as a good source of organic material. Mushroom cultivation can also be taken up on empty bunches. The fibres and most of the shells produced by the mills are used as fuel for the boiler. In the nursery, fibres and shells are used as mulching materials. High quality medium density fibre boards (MDF) are being prepared from briquettes.

Table 18.1: Nutrient Composition, Fertilizer Equivalent and Monetary Value of Trunks and Fronds

Residue	*Reference*	*Nutreients on Dry Matter (kg.ha)*				*Fertiliser Equivalent* (Kg/ha.)*	*Monetary Value (US$)*
		N	*P*	*K*	*Mg*		
Trunks	Chan *et al.* (1981)	368.2	35.5	527.4	88.3	Urea 800	163
		219.6	21.2	314.5	52.6	CIRP 235	17
	Mohd. Hashim *et al.* (1993)					MOP 1059	153
						Kies. 542	64
							397
Fronds at felling	Chan *et al.* (1981)	150.1	13.9	193.9	24.0	Urea 326	66
		119.8	11.0	109.7	23.3	CIRP 92	7
	Mohd. Hashim *et al.* (1993)					MOP 389	56
						Kies. 147	17
							146
Fronds at annual pruning	Chan *et al.* (1981)	107.9	10.0	139.4	17.2	Urea 235	48
						CIRP 65	5
						MOP 280	40
						Kies. 106	12
							105

The Empty Fruit Bunches constitute about 22 per cent of Fresh Fruit Bunches (FFB) and are generally wasted in many mill areas. Sometimes used as mulch in the palm basins, these are a good source of organic manure. This can be used as supplementary energy source in the mill or composted or incinerated for ash.

As a mulch,the EFB can be arranged as a circle in palm basins about 50 cm away from base of the trunk. Table 18.2 indicates that one ton of EFB would give a fertilizer equivalent of 7.0 kg urea, 2.8 kg super phosphate,19.3 kg Muriate of potash and 4.4. kg kieserite. The rate of application varies with the age of plantation and soil type. Generally 15 to 20 tons/ha for young plantation and 25 to 30 tons for mature plantations is suggested.

Table 18.2: Composition of Empty Fruit Bunches

Parameter		*Dry Matter Basis*		*Fresh Weigh Basis* (Mean)*
		Range	*Mean*	
Ash	(per cent)	4.8-8.7	6.3	2.52
Oil	(per cent)	8.1-9.4	8.9	3.56
C	(per cent)	42.0-43.0	42.8	17.12
N	(per cent)	0.65-0.94	0.80	0.32
P_2O_5	(per cent)	0.18-0.27	0.22	0.09
K_2O	(per cent)	2.0-3.9	2.90	1.16
MgO	(per cent)	0.25-0.40	0.30	0.12
CaO	(per cent)	0.15-0.48	0.25	0.10
B	(ppm)	9-11	10	4
Cu	(ppm)	22-25	23	9
Zn	(ppm)	49-55	51	20
Fe	(ppm)	310-595	473	189
Mn	(ppm)	26-71	48	19
C/N ratio			54	54

* Moisture content 60-65 per cent.

Source: Gurmit Singh *et al.* (1990).

18.1.3. Products from Fibres of Empty Fruit Bunches

A number of products can be prepared from fibres of empty fruit bunches. These are shown in Figure 18.3. These are all bio-degradable.

a) Bunch Ash

Bunch and its produced during the incineration of EFB is a valuable fertilizer and its composition is given in Table 18.3. Based on its NPK contents, one ton of bunch ash, has a monetary value, considering the high cost of fertilizers. However, in recent years, with greater awareness of the need for clean pollution free environment, bunch ash production has steadily declined in countries like Malaysia and Indonesia.

Fiber products

EFB FIBRES

COMPOSITE MATERIAL

MOLDED WARES

PULP MOLDED WARES

PULPS & PAPERS

COMPOSTING

BRIQUETS & PELLETS

EFB FIBRE BOARDS

Figure 18.3: Empty Bunch Fibre Utilization.

Figure 18.4: Fibre Briquetes.

Figure 18.5: Compressed Laminated Products.

Many of the companies have progressively decommissioned their incinerators and embarked on bunch mulching.

Table 18.3: Nutrient Composition and Fertiliser Equivalent of Bunch Ash

Nutrient	Per cent on Dry Ash	Fertiliser Equivalent (Kg/tonne)	
K_2O	41.4	Muriate of Potash	690
P_2O_5	3.7	Rock Phosphate	123
MgO	5.8	Kieserite	215
CaO	4.9	-	

Source: Toh *et al.* (1981).

b) Mesocarp Fibre, Shell and Boiler Ash

The mesocarp fibre and shell are being used in the mill as a boiler fuel to produce steam and generate power for the various mill processes. The calorific values of the above materials are given in the Table 18.4. The energy generated during the burning of both of these residues is sufficient to meet the energy demand of the mill, with some excess electricity available for distribution to the domestic use in the estate.

Surplus shells are used for mulching oil palm polybag seedlings and for surfacing the farm/mill roads. It is possible to take activated carbon also.

Table 18.4: Calorific Value of Oil Palm Solid Fuels

Fuel Source	Dry Weight Basis	Fresh Weight Basis
	Calories/g	
Mesocarp fibre	4,461	2,676
Shell	4,893	4,404
EFB	4,282	2,568

Source: Adapted from Lim and Ratnalingam (1980).

c) Palm Kernel Cake or Meal

Palm kernel cake or meal is a byproduct of kernel after extracting the oil Table 18.5 & 18.6. The cake is dry, greety and is a valuable source of protein and energy. Though protein content is low the quality is high. It is mainly used in dairy cow for increasing fat and milk.

18.1.4. Palm Oil Mill Effluent (POME)

The major source of waste water generation from palm oil mill is sterilizer condensate, hydro cyclone waste, and separator sludge. On an average 0.9–1.5 m^3 of POME is generated for each MT of CPO produced. The POME is rich in organic carbon with a biochemical oxygen demand (BOD) value higher than 20 g/L and nitrogen content around 0.2 and 0.5 g/L as ammonia nitrogen and total nitrogen (Gurmit *et al.*, 1999 and Ma *et al.*, 2001).

Table 18.5: Composition of Palm Oil Meal

Parameter	*Per cent*
Moisture	<7
Ash	20
Silica	8
Ether Extract	12
Crude fibre	12
Crude protein	12
Crude cellulose	20
NFE	28
P	0.3
K	2.5
Mg	0.7
Ca	0.8
Gross energy	4,400 Kcal/kg

Source: Adapted from Gurmit Singh *et al.* (1990).

Table 18.6: Composition of Palm Kernel Cake

Parameter	*Per cent on Dry Matter*
Dry Matter	90.6
Crude Protein	19.0
Crude fibre	16.0
Ether Extract	2.0
Ash	4.2
N-free extract	58.8
Ca	0.34
P	0.69
Mg	0.16
Gross energy	17.3 MJ/kg

Source: Devendra (1977).

i) POME is characterized by acidic, thick brownish colloidal slurry of water, oil and suspended water.
ii) POME after biological treatment is increasingly applied to oil palm land.
iii) Decanter drier solids from the clarification station are processed into organic fertilizer or animal feed.
iv) Application of POME to land as fertilizer supplement was found beneficial to crop productivity, physical and chemical soil properties.

v) Application of POME in palm plantations is widely accepted as an economically viable and environmentally acceptable waste management technique.

vi) The POME treatment can help in alleviating soil moisture especially on areas with regular water deficit in South Kalimantan.

vii) Application of POME can increase leaf nutrient content like N,P,K and Mg. It also stimulated root proliferation and increased soil humidity, C-organic, available P, exchangeable Mg and Ca and Soil pH. It can contribute in promoting low soil porosity and increased bulk density.

viii) POME contains fair amount of nutrients which can be profitably utilized for making fertilizers (Table 18.7).

Table 18.7: Characteristics of Palm Oil Mill Effluent (Ma, 2000)

Parameter	*Concentration (mg/L)*	*Element*	*Concentration (mg/L)*
Oil and grease	4000-6000	Potassium	2,270
Biochemical oxygen demand	25,000	Magnesium	615
Chemical oxygen demand	50,000	Calcium	439
Total solid	40,500	Phosphorus	180
Suspended solids	] 8,000	Iron	46.5
Total volatile solids	34,000	Boron	7.6
Total nitrogen	750	Zinc	2.3
Ammonicals nitrogen	35	Manganese	2.0
		Copper	0.89

Table 18.8: Chemical Analysis of Various Types of POME

Type of POME	*Composition*				
	BOD	*Total N*	*P*	*k*	*Mg*
Raw*	25,000	948	154	1,958	345
Digested					
Stir red tank	1,300	900	120	1,800	300
Supernatant	450	450	70	1,200	280
Supernatant**	215	201	29	1,076	90
bottom sludge (anaerobic)	1,000-3,000	3,552	1,180	2,387	1,509
bottom sludge (aerobic)	1,500-3,000	1,495	461	2,378	1,004

* Not permitted for land application. ** Pond supernatant from the Decanter-drier system.

Source: Kanagaratnam *et al.* (1981).

18.1.5. POME for Biogas Production

Bio gas produced during the anaerobic digestion of POME is a clear and renewable source of energy. It consists of a mixture of 60-65 per cent methane, 30-40 per cent carbon dioxide and traces of hydrogen sulphide. For every cubic meter of POME digested, about 28.3 cum of biogas is produced. This is one area, the palm oil mills in the country can exploit. Chemical composition of pome is given in Table 18.8.

Table 18.9: Systems of Land Application of Digested POME

System	*Terrain*	*Reference*
Sprinkler	Flat to gently undulating	Koh and P'ng, 1981
		Lim *et al.*,1983
Straight furrows	Flat Undulating	Gurmit Singh, 1990
		Kanagaratnam *et al.*, 1982
Long beds	Flat	Wood, 1981
		Quah *et al.*, 1982
Cascading flat beds	Undulating	Gurmit Singh, 1990
		Mohd. Tayeb *et al.*, 1987
		Lim, 1987
Tractor-tanker-pump	Flat to gently undulating	Tam *et al.*, 1981

Table 18.10: Composition of Dried Decanter POME

Nutrient	*Range (per cent)*	*Mean (per cent)*	*Nutrient*	*Range (μg/ml)*	*Range (μg/ml)*
C	35-40	36.5	B	14-22	20
N	1.8-2.3	2.00	Cu	20-50	45
P	0.3-0.4	0.33	Fe	3,000-6,000	4,800
K	2.0-3.2	2.50	Mn	50-80	60
Mg	0.5-0.8	0.70	Zn	20-100	60
Ca	0.6-1.0	0.80			

Table 18.11: Composition of Raw and Composed Decanter POME

Component	*Per cent on Dry Matter*	
	Raw	*Composted*
C	36.8	30.8
N	2.08	2.48
P	0.33	0.41
K	3.02	3.46
Mg	0.72	0.82
Ca	0.75	0.86
C:N Ratio	18.1:1	12.4:1

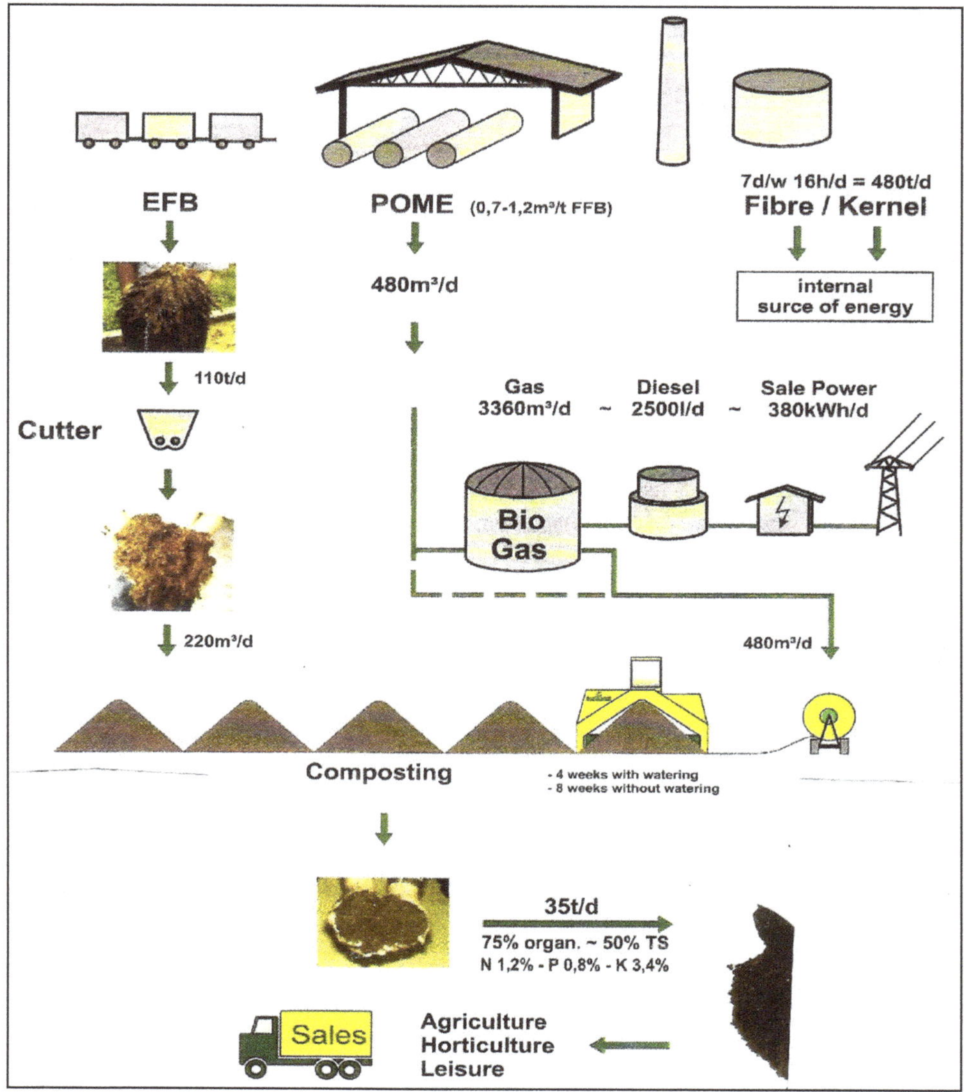

Figure 18.6: Palm Oil Mill Waste+Sewage Management (30t FFB/h).

The Indian Palm oil industry is very young and hence most of the residues are being wasted. However, palm oil industry in Malaysia has developed lot of technologies in the recent past. Lot of machineries and equipment manufacturers have been developed and are available which are ecofriendly and environmentally sustainable. The palm itself is an efficient scrubber of atmospheric carbon dioxide compared to many other oil seed crops.

18.2. Systems Approach for Zero Waste and Zero Discharge in Palm Oil Mill

A diagrammatic representation and description of zero waste/zero discharge is given in Figures 18.6, 18.7 & 18.8.

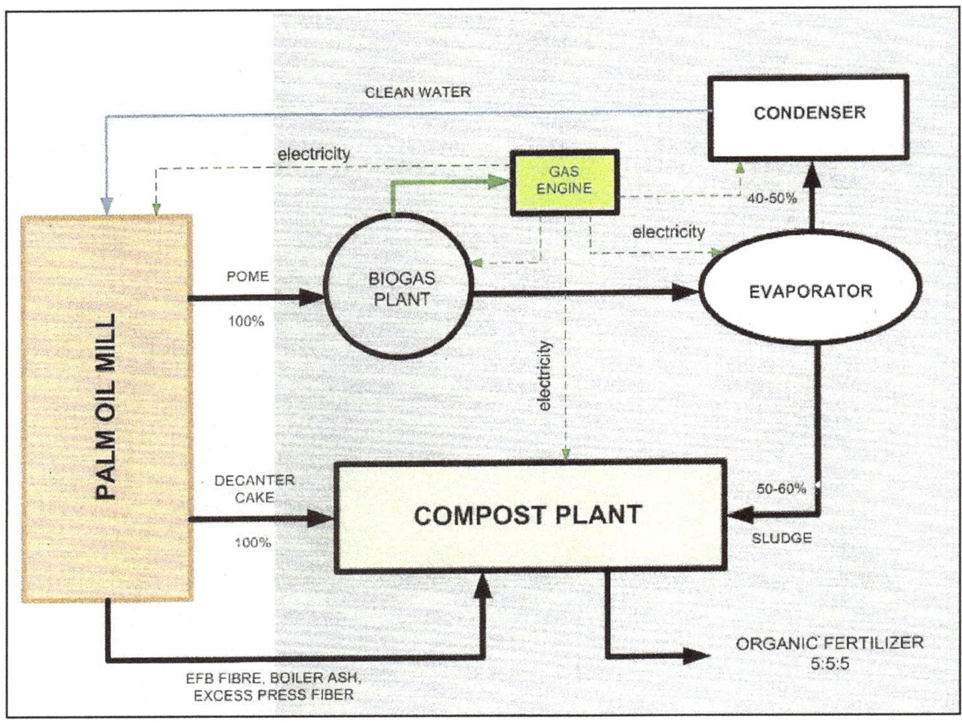

Figure 18.7: Zero Waste/Zero Discharge.

- POME

  Palm oil mill effluent is treated biologically using advance anaerobic digestion process to produce biogas for electricity generation. The digested sludge is then evaporated in a mechanical vapor recompression evaporator leaving behind 50% of sludge as the bottom fraction. The water vapor is condensed and the resulted clean water is returned to the mill

- EMPTY FRUIT BUNCH

  Empty fruit bunches are shredded, pressed and cut into a size suitable for composting. A modern and environmental friendly composting process is used to convert a mixture of decanter cake, EFB fiber, excess press fiber, digested sludge and evaporator sludge to organic fertilizer. Additional nutrient may be added to adjust the final product to contain 5:5:5 NPK

- UTILITY

  Electricity requirement for the complex will be generated from biogas containing more than 60% methane using the latest technology of gas engine. Excess electricity will be connected to the mill to replace diesel for electric generation.

Figure 18.8: System Approach–Zero Waste/Zero Discharge.

18.3. Management of Environmental Issues

Today most of the palm oil mills are moving towards zero energy concept to save the environment. Accordingly, the machineries required are available in the leading palm oil producing countries like Malaysia and Indonesia. This zero energy concept gives us the additional benefits of Bio-gas and compost which can be enriched to have a balanced nutrient content. The two diagrammatic process are given in Figure 18.9.

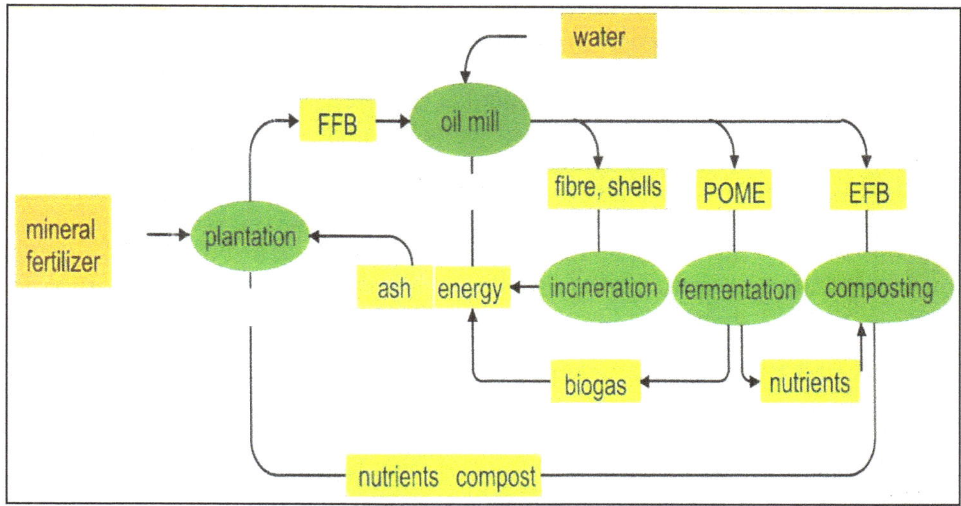

Figure 18.9: Zero Energy Concept.

Any sustainable development should have four essentials like being socially acceptable, economically viable, environmentally compatible and technically adoptive. Oil palm is one of the world's most efficient bearing crops in terms of land utilization, efficiency and productivity. A single ha produces upto 10 times more oil than other oilseeds. Oil palm yields an average of 3.68 MT of oil/ha/year (projected to rise to 6 MT within the next decade) compared to soybean, sunflower seed, and rapeseed.

Oil palm, with an area of 14.538 million ha and production of 59.6 MTones in the world during 2014 is the best suited crop for increasing the vegetable oil economy and Malaysia and Indonesia are major partners in this. As a vegetable oil yielding crop, the oil palm is unsurpassed in its ability to convert solar energy into oil yield. Compared to any other oil yielding crop oil palm has highest oil with 4-8 t of crude palm oil/ha and 0.4 – 0.8 palm kernel oil/ha/year. Thereby, the crop is very viable both for farmers as well the countries through its export earnings. The whole Malaysian economy over the period of time has changed drastically due to palm oil export. It also contributes to the nutritional security besides the food uses since it contains many vitamins like A (Carotene), 600 ppm and Vitamin E, 900 ppm.

Oil palm is an efficient scrubber of atmospheric CO_2 (100t/ha/year), as efficient as tropical rain forest. The thick canopy like umbrella shape provides shade and protects the soil from degeneration (Gurmit Singh, 2003). The growing of green manure crops right from planting, inter crops like pulses, oil seeds, vegetables, flowers *etc.*, during the pre-bearing period and placement of pruned leaves along the palm rows, proper water management, discriminatory use of fertilizers applied in several doses coupled with organic manures, recycling of organic wastes available in the oil palm plantation, integrated weed, pest and disease management in the plantations maintains soil fertility and crop productivity. The other very important issue pointed out by Environmentalists is that by growing oil palm the rivers will dry and ground water level will go down since oil palm requires 200 litres of water per palm per year. The water requirement of two crops of paddy in a year, banana and sugar cane require 1800 -2000 mm water which is almost the same. With the organic residue management available in the oil palm plantations the organic matter content will definitely reduce the water requirements further. In the processing sector the management and utilization of palm oil residues and wastes like palm oil mill effluent (POME) use of empty fruit bunches (EFB), Palm Kernel Cake (PKC), mesocarp fibre and shell, trunks at replanting, leaf fronds *etc.* contribute much to replenishment of the soil.

18.4. Oil Palm and Carbon Sequestration

Oil palm is a C_3 plant, which possesses high photosynthetic capacity. The carbon assimilated by oil palm fronds, through the process of photosynthesis will be assimilated to fronds, stem and roots and the remaining assimilates are then directed towards bunches. The photosynthesis in oil palm reveals considerable variations ranging below 5 $\mu mol^{-2}.s^{-1}$ (Hirsch, 1975), between 4-9 $\mu mol^{-2}.s^{-1}$ (Suresh and Nagamani, 2006), between 6-9 $\mu mol^{-2}.s^{-1}$ (Potulski, 1990), between 14-20 $\mu mol^{-2}.s^{-1}$ (Corley, 1983). This variation could be attributed due to differences in planting material, environmental conditions, age of the palm and position of the frond or leaflet measured. Photosynthesis also varies depending on where the frond lies in the crown. A crown contains 35 to 42 fronds and is regularly pruned while harvesting bunches. Variations in the photosynthetic rate in the different fronds have been observed (Corley, 1983; Suresh and Nagamani, 2006). The first fronds (1-7) recorded higher photosynthetic rates which decreased with frond age. The SIMIPALM model has simulated the carbon fixation in palms grown under two different ecological conditions namely Africa and South East Asia. An average mature palm in Ivory Coast having a crown of 35 fronds can potentially fix 300 Kg $C.yr^{-1}$ from a captive area of 315 sq.m of laminae. Similarly in South East Asia, the palms fixed 350 kg C. yr^{-1}. for a capture area of 340 sq.m. Henson (1994) gave an estimated root respiration of 32 kg $C.yr^{-1}.palm^{-1}$ for a 10-year-old plantation with a less developed root system in Malaysia. Hence it can be estimated that under optimum conditions, a mature palm would release 750 kg $CO_2.yr^{-1}$, which will be equivalent to 200 kg $C.yr^{-1}$ through the process of respiration.

Various studies have reported that the total annual carbon release from the soil into the atmosphere varied from 1170 g $C.m^{-2}.yr^{-1}$ in North Sumatra to 1610 g C. m^{-2}.

yr^{-1} in Benin, indicating a greater respiratory loss from roots in Benin. Similarly the total carbon allocation to the roots ranged from 1025 g C. $m^{-2}.yr^{-1}$ in North Sumatra to 1438 g C. $m^{-2}.yr^{-1}$ in Benin.

Perusal of the photosynthetic rate in the different leaves of oil palm indicated that it was maximum in the 9th leaf (9.18 $\mu mol^{-2}.s^{-1}$), which differed significantly from that of other leaves (Suresh and Nagamani, 2007). Diurnal variations in the photosynthetic rate indicated it ranged from 0.64 to 10.07 $\mu mol^{-2}.s^{-1}$. The highest photosynthetic rate was observed between 9:00 to 10:00 AM and decreased thereafter with increased leaf temperature and low stomatal conductance. (Suresh, 2006). In another study, Nagamani and Suresh (2006) have reported that oil palm accumulated as high as 20-22 $t_{dm}.ha^{-1}.Yr^{-1}$ and a comparison among the different hybrids indicated that it was highest in Papua New Guinea hybrids followed by Costa Rican and Palode hybrids. Promising yields, as high as 35-40 MT FFB per year, are being realized in the best plantations of Andhra Pradesh. Taking a cue from the global estimates of the carbon storage potential of oil palm as done in other countries, the estimated storage of carbon by oil palm under Indian conditions could be to the tune of 4.86 MT $C.yr^{-1}$ from the potential 0.8 m ha identified. In real terms, if we take the current area of oil palm in India into consideration, the estimated carbon storage by the oil palm would be to the tune of 0.43 MT $C.yr^{-1}$ from 0.07 m ha.

Oil palm is a prime candidate for storing carbon in the tropical countries like Malaysia and Indonesia, where it is grown and is also eligible for the CDM. The annual biomass productivity of oil palm amounts to 50 $t_{dm}.yr^{-1}$ during its 25-year lifespan. Studies on the carbon balance in oil palm under different ecosystems have indicated that the annual storage of a mature oil palm plantation (excluding bunch harvesting) amounts to 1340 g $C.m^{-2}.yr^{-1}$ (13.4 t $C.ha^{-1}$). Harvesting and continuous export of FFB causes this carbon storage to fall (250 g $C.m^{-2}.yr^{-1}$), but still it remains higher than that of a tropical forest (43 g $C.m^{-2}.yr^{-1}$). A comparison of the carbon storage potential in the different plantations indicated that the carbon storage was highest in eucalyptus (390 - 470 g $C.m^{-2}.yr^{-1}$) followed by oil palm and coconut (125 - 530 g $C.m^{-2}.yr^{-1}$). When the trunks of oil palm are utilized at the end of the cycle, the carbon balance would be in favour of oil palm than eucalyptus. An oil palm plantation can "sequester" up to 15 t CO_2 from the atmosphere for each hectare planted, thus contributing to mitigate the greenhouse effect. Thus oil palm plantations can sequester more carbon (C) per unit area and could be an effective carbon sink in the coming years.

The oil palm plantations contribute towards carbon sequestration and to provide a moderating effect on the rising global temperature, which most annual agricultural crops do not. Carbon sequestration in Oil Palm is eligible for Clean Development Mechanism. Global carbon storage by oil palm is estimated to be 74 MT C yr^{-1} for 12 million hectares. If 1 m ha in India is brought under oil palm, it is possible to sequester about 6 MT C yr^{-1} (Table 18.12).

Carbon Trading is another area coming up fast in India. There is good scope for getting good additional amount of money from oil palm plantations which has to be looked into.

Table 18.12: Estimated Carbon Fixed by Oil Palm

Age (Years)	*Total Stock (106 MT)*
1-3	2.522
4-8	17.113
9-13	16.455
14-18	21.327
19-24	21.104
25>	10.753
Total	**89.274**

In the past, the conventional biological system, using anaerobic and aerobic or facultative processes to treat palm oil mill effluent (POME) to approved statutory limits for discharge, was successful. However, with the mills becoming bigger and having to deal with large volumes of POME, the system has become less efficient and problematic. There is now greater emphasis on reduced environmental pollution through emission of particulates, black smoke and GHG into the atmosphere.

Over the last 10 years, the thinking on palm oil mill environmental issues, has evolved from 'treatment of waste for disposal' to 'beneficial utilization of by-products" POME contains substantial qualities of valuable plant nutrient which can be used as fertilizer for the crop.

Traditional treatment of these wastes contributes to much environmental as well as economic concern. Mulching of EFB creates solid (slow degradation of EFB head) and air pollution (GHG) emissions from anaerobic degradation in the field while treatment of POME using pondin g system releases much waste water (final discharge) as well as air pollution (GHG emissions). As these wastes are currently under utilized, and a good source of organic matter and plant nutrients, maximizing nutrient recovery from such waste is desirable in terms of cost saving and yield.

In Malaysia and Indonesia, various projects are being undertaken by large plantation companies using composting technology to convert EFB, POME, decanter cake and other solid palm oil waste into value added nutrient rich natural fertilizers. The resultant waste water is subjected to further treatment through an "Ozonation System" to enable discharge via water course or land application (at acceptable BOD levels) or recycle for mill processes.

The advent of Clean Development Mechanism (CDM) under the Kyoto Protocol has also created additional opportunities for waste management. By re-claiming GHG emissions from POME ponding system and using them as a source of energy, Certified Emission Reductions (CERs) can be obtained in the process and traded for carbon credit.

With these systems in place, palm oil mills are geared effectively to eliminate solid, liquid and gases waste pollution and in the process produce nutrient rich natural fertilizer for the plantation, reclaim and recycle bio-gases as source of additional energy for the mill, and achieve eco-friendly discharge of waste water. This is sure step towards sustainable waste management practice and as an extension towards the production of SUSTAINABLE PALM OIL. Much of the development work and applications in this field are taking place within the above frame-work. The technology and process systems are many and varied. KONPRO is involved in the development and application of some of these systems in collaboration with technology providers and experts.

References

Anonymous, 1988. Potentialities of oil palm Cultivation in India. Report of the working group constituted by Department of agriculture and Co-operation, Ministry of Agriculture, Government of India, Published by the Indian Council of Agricultural Research, New Delhi. p. 72.

Anonymous, 2002. Impact Evaluation of Oil Palm Development Programme (OPDP) (Andhra Pradesh, Karnataka and Tamil Nadu)-Final Report. Sponsored by Technology Mission on Oilseeds, Pulses and Maize) (TMOP and M), Department of Agriculture and Cooperation, Ministry of Agriculture, Government of India. Agricultural Finance Corporation, New Delhi. p.128.

Arya, Krishna, 1998. Oil palm development in India: S mission mode approach. In National Seminar on Opportunities and Challenges for Oil Palm Development in 21st Century, Souvenir. Published by SOPOPRAD, Vijayawada, A.P., India. pp. 20-25.

Asmono, Dwi, 2007. Indonesia's Global Scenario of Oil Palm Planting Materials and Future Strategies to Meet Demand.;74-85. In *Proc. National Conference on Oil Palm*. Edit. P. Rethinam *et al.* p. 221.

Aye Pe, Nampoorhiri, K.U.K., Achuthan Nair, M., Mya Mya Win, and May San., 2008. Oil Palm research, development and production in Myanmar – an overview. *J. of Plantation Crops* 36 **(3)** 310-314.

Barlow, J., Gardner, T.A., Araujo, I.S., Avila-Pires, T.C., Bonaldo, A.B., Costa, J.E., Esposito, M.C., Ferreira, L.V., Hawes, J., Hernandez, M.I.M., Hoogmoed, M.S., Leite, R.N., Lo-Man- Hung, N.F., Malcolm, J.R., Martins, M.B., Mestre, L.A.M., Miranda Santos, R., Nunes-Gutjahr, A.L., Overal, W.L., Parry, L., Peters, S.L., Ribeiro-Junior, M.A., da Silva, M.N.F., da Silva Motta, C., Peres, C.A., 2007. Quantifying the biodiversity value of tropical primary, secondary, and plantation forests. *Proceedings of the National Academy of Sciences* 104, 18555–18560.

Carter, C., Finley, W., Fry, J., Jackson, D., Willis, L., 2007. Palm oil markets and future supply. *European Journal of Lipid Science and Technology* 109, 307–314.

Chadha, K. L. 2006. Progress and Potential of Oil Palm in India. Report of the committee to reassess fresh/potential areas of oil palm in India. Department of Agriculture and Co-operation, Government of India, New Delhi. Xxiv, p. 216.

Choudhary *et. al*, 1995. Comparison of palm olein and olive oil: effect on plasma lipids and vitamin E in young adults. *Am, J. Clin. Nutr.*, **61**:1043-51.

CODEX Alimentarius, 1983. vol. XI. FAO/WHO, Rome:115-130.

Corley, R. H. V. and Tinker, P. B., 2005. The Oil Palm. Blackwell Publishing. p. 562.

Corley, R. H. V., 2009 How much palm oil do we need ? Environmental science and policy 12.134-139.

Corley, R. H. V. How much palm oil do we need? Environmental Science and policy 12 (2009) 134-139.

da Costa, R. C., 2004. Potential for producing bio-fuel in the Amazon deforested areas. *Biomass and Bioenergy* 26, 405–415.

Desmarest (1967), 2: de Taffin and Daniel (1976); 3: Ochs and Daniel (1976); 4: Chan (1979); 5: Corley and Hong (1982); 6: Chan *et al.* (1965), 7: Chuah and Lim (1989); 8: Prioux *et al.* (1992); 9: Corley (1992); 10: Unipalma (unpubl.); 11: Lim *et al.* (1994); 12: Palat *et al.* (2000); 13: Kee and Chew (1993); 14; Mite *et al.* (2000).

Fact Sheet Malaysian Palm Oil MPOB, 2007.Pub.Malaysian Palm Oil Council, Malaysia p. 63.

FAO, 1994. Fats and oils in human nutrition: report of a expert consultation. FAO Food and Nutrition Paper No. 57, 1994.

FAO, 2006, Global forest resources assessment http://www.fao.org/forestry/32039/en/, accessed June 4, 2008.

FAO, 2007. Resource STAT-Land. http://faostat.fao.org, accessed September 8, 2008.

Fearnside, P. M., 2001. Soybean cultivation as a threat to the environment in Brazil. *Environmental Conservation* 28, 23– 38.

Garrity, D. P., Soekardi, M., Van Noordwijk, M., de la Cruz, R., Pathak, P. S., Gunasena, H.P.M., Van So, N., Huijun, G., Majid, N.M., 1996. The Imperata grasslands of tropical Asia: area, distribution, and typology. *Agroforestry Systems* 36, 3–29.

Germer, J., Sauerborn, J., 2008. Estimation of the impact of oil palm plantation establishment on greenhouse gas balance. *Environment, Development and Sustainability* 10, 697-716.

Gurmit, S., Lim Kim H, Teo L, David Lee K, Oil palm and the environment a Malaysian perspective. vol. 1. Malaysian Oil Palm Growers Council, 1999. p. 253.

HHS/USDA, 2005. Dietary Guidelines for Americans 2005. U.S. Department of Health and Human Services and U.S. Department of Agriculture.

http://usda.mannlib.cornell.edu/MannUsda/ view Document Info.do? document ID=1289, accessed March 27, 2008.

Kennedy, G., 2001. Global trends in dietary energy supply from 1961 to 1999. Paper prepared for FAO/WHO/UNU Expert Consultation on Energy in Human Nutrition, 62 p.

Kochu Babu, M. and Prabhakara Rao, K. J., 2005. Economics of Oil Palm. *Proceedings of National Seminar on Research and Development of Oil Palm in India*. pp. 142-147.

Koh, L. P., Wilcove, D.S., 2008. Is oil palm agriculture really destroying tropical biodiversity? *Conservation Letters* 1, 60–64.

Koh, L. P., 2007. Potential habitat and biodiversity losses from intensified biodiesel feedstock production. *Conservation Biology* 21, 1373–1375.

Koh, L. P., 2007. Potential habitat and biodiversity losses from intensified biodiesel feedstock production. *Conservation Biology* 21, 1373–1375.

Koh, L. P., Wilcove, D.S., 2008. Is oil palm agriculture really destroying tropical biodiversity? *Conservation Letters* 1, 60–64.

Ma AN. Environmental management for the palm oil industry. *Palm Oil Dev* 2000; 30:1–9.

Ma AN, Chow CS, John CK, Ibrahim A, Isa Z. Palm oil mill effluent—a survey. In: Proceeding PORIM regional workshop on palm oil mill technology and effluent treatment, palm oil research institute of Malaysia (PORIM) Serdang. Malaysia, 2001. pp. 233–69.

Malaysian Palm Oil Council (MPOC). Oil palm tree of life (free edition book). MPOC Official report 3, 2006.

Mistry, Dorab E., 2009. Global Vegetable Oil Price Outlook 2009-2010. Paper presented at *Fourth China International Oil Conference* held at Shangri-La Hotel, Guangzhou, China.

Mistry, Dorab E., 2011. Palm and Lauric Oils Price Outlook. 2011 with Special Reference to India. POC 2011, held at The Shangri La Hotel, Kuala Lumpur, Malaysia.

Morton, D.C., DeFries, R. S., Shimabukuro, Y. E., Anderson, L. O., Arai, E., de Bon Espirito-Santo, F., Freitas, R., Morisette, J., 2006. Cropland expansion changes deforestation dynamics in the southern Brazilian Amazon. *Proceedings of the National Academy of Sciences* 103, 14637–14641.

Naidu, Raja N. 2007. Status of Oil Palm planting material in the world and future challenges. :38-57. In *Proc. National Conference on Oil Palm*. Edited P. Rethinam *et al.* p.221.

Nesaretnam, K. *et al.*, 2005.Tocotrienol-richfraction from palm oil affects gene expression in tumors resulting fromMCF-7 cell inoculation in athymic mice. *Lipids*. 39 **(5)**: 459-467.

Ng, T. K. W., Heyes, K. C., *et al.*, 1992. Dietary palmitic and oleic acidsexerts similar effects on serum cholesterol and lipoprotein profiles in normocholesterolemic men and women. *J. Am. Coll. Nutr.*11 **(4)**:383-90.

Oil World, 2007. Oil World Annual 2007. ISTA Mielke GmbH.

Ong, A. S. H., 1993. Natural sources of Tocotrienol. In Lester Packer and Jurgen Fuchs (eds).Vitamin E in health and disease. Mereel ekker, Inc.New York.

Ong, A,S.H and Goh,,S.H., 2002. Palm Oil a healthful and cost effective dietary component. *Food Nutr. Bull.* 23; 11-22.

Pagiola, S., Agostini, P., Gobbi, J., de Haan, C., Ibrahim, M., Murgueitio, E., Ramirez, E., Rosales, M., Ruiz, J.P., 2004. Paying for biodiversity conservation services in agricultural landscapes. FAO Environment Dept Paper 96, Rome, 32.

Palanisami, K., Chandrasekaran, B., Devasenapathy, P., Santhi, R. 2003. Water Management Technologies. Published by Water Technology Centre, Tamil Nadu Agricultural University, Coimbatore.

Qureshi, A. A. *et al.*,1995. Response of hypercholesterolemic subjects to administration of tocotrienols. *Lipids.* 30(12):1171-7.

Rethinam, P. , 1998. Historical perspective of Oil Palm research and development in India. In National Seminar on Opportunities and Challenges for Oil Palm Development in 21st Century, Souvenir. Published by SOPOPRAD, Vijayawada, A.P., India. pp. 16-19.

Rethinam, P., 1998. National Seminar on Opportunities and Challenges for Oil Palm Development in 21st Century, Souvenir. Published by SOPOPRAD, Vijayawada, A.P., India.

Rethinam, P., Reddy, V. M., Mandal, P. K., Suresh, K., Prasad, M. V., 2009. Oil Palm for Farmers' Prosperity and Edible Oil Security. Proceeding of National Conference on Oil Palm. Published by the Society for Promotion of Oil Palm Research and Development, Pedavegi, India. p. 221.

Rethinam, P., Reddy, V.M., Prasad, M.V., Mandal, P.K., Suresh, K., 2008. Summary Proceedings – The Vijayawada declaration on oil palm. *National Conference on Oil Palm.* Oil Palm for Farmers' Prosperity and Edible Oil Security. p. 27.

Rethinam. P and Suresh, K., 1998. Oil Palm Research and Development. *Proc. of the National Seminar – Opportunities and Challenges for the Oil Palm Development in the Twenty First Centaury.* SOPOPRAD, Pedavegi, AP. p.114.

Salleh, Mamat, 2007. Key note Address in the Planting Materials of oil palm organized at Kulalumpur, Malaysia in 2007.

Schmidt, J. H., Weidema, B.P., 2008. Shift in the marginal supply of vegetable oil. *International Journal of Life Cycle Assessment* 13, 235–239.

Singh, Gurmeet, 2003. Oil palm and the environment. A Malaysian perspective – Personal communication.

Sumathi, S., Chai, S. P., and Mohamed, A. R., 2007. Utilization of oil palm as a source of renewable energy in Malaysia Renewable and Sustainable Energy Reviews 12 (2008) 2404–2421.

Sundaram, K. M., French, M.A. *et. al.*, 2003. Exchanging partially-hydrogenated fat for palmitic acid in the diet increases LDL-cholerterol and endogenous cholesterol synthesis in normocholesterolemic women. *Eur. J. Nutr.* 42 (4):188-94.

Tomich, T.P., Kuusipalo, J., Menz, K., Byron, N., 1997. Imperata economics and policy. *Agroforestry Systems* 36, 233–261.

UNPD, 2006. World Population Prospects: The 2006 Revision. http://esa.un.org/unpp/, accessed March 27, 2008.

USDA, 2005. Oilseeds: World Markets and Trade. Circular Series FOP 9-05.

USDA, 2007. Economic Research Service: Oil Crops Yearbook.

Yusof, B., 2006. Sustainable palm oil production in Malaysia. London: Symposium on sustainable resources development; 18, May, 2006.

www.coconutboard.gov.in. Coconut Development Board, Kochi, India.

Index

A

Ablation 158, 159
Achievements Research 40
- Crop Improvement 40
- Crop Physiology 42
- Cropping System 42
- Disease Management 43
- Germplasm 40
- Hybrid Seed Gardens 41
- Nursery Management 41
- Nutrient and Water Management 41
- Pest Management 42
- Post-Harvest Technology 43
- Recent Advances 32
  - *Breakthroughs in Biotechnology 34*
  - *Crop Improvement 32*
  - *Planting Materials 33*
  - *Tissue Culture 34*

African oil palm Development Association 30

B

Bag Handling 142
Basic Extraction Rate 257
Basin 196
- Application 196
- Management 158

Bio-efficacy 165
- Crop Injury 165
- Phytotoxicity 165

Biofertilizers 195
Biogas Production 295
Botany 95
- Flowering 99
- Fruit 102
- Leaf 98
- Root System 98
- Seed 95
- Stem 96
- Taxonomy 95

Breeding 105
- Fruit Types 108
- Genetics 106
- Methods 111
  - *Hybridization 111*
  - *Inter Specific Hybridization 112*
  - *Molecular Breeding 112*
  - *Objectives 110*
  - *Selection 111*
- Production Components 108
- Reproductive Biology 106

Bunch 253
- Composition 108
- Counting Systems 252
- Dirty 254
- Diseased 254
- Empty 254

- Long Stalk 254
- Loose 256
- Old 254
- Over Ripe 253
- Pest Damaged 254
- Quality 256
- Ripe 253
- Rotten 254
- Small 254
- Under Ripe 253
- Unfresh 254
- Unripe 253
- Wet 256

C

Callus Induction 124
Carbon Sequestration 300
Climate 73
- Humidity 74
- Mitigating Effects 84
- Rainfall and Distribution 73
- Sunshine and Solar Radiation 74
- Temperature 73
- Wind 74
- Zones 74

CPCRI Committee 48
Criteria of Ripeness 251
Cropping 215
- Cover Cropping 185
- Intercropping 217
- Mixed Cropping 219
- Multiple Cropping 215

Crop Production 35
- Agronomy 35
- Crop Physiology 38

Crop Protection 38
- Diseases Management 38
- Pest Management 38

D

Disorders 209
- Bunch Failure 211
- Chimeras 212
- Leaf Base Wilt 210
- Leaf Breaking 210
- Peat Yellows 210
- Plant Failure 209
- White Stripe 209

Drainage Management 185
DRDA 54
DRIS 204

E

Embryogenesis 124
Explants 124
Exports 9
- Animal Oils 10
- Vegetable Oils 10

F

Factors Affecting 250
- Epiphyte Growth 250
- Natural Ditches 250
- Pruning 250

Fertigation 197
Fertilizers 141, 194
- Inorganic Fertilizers 194

Field Nursery 125
Food Quality 277
Fruit Bunches
- Fibres 291
  - *Boiler Ash 293*
  - *Bunch Ash 291*
  - *Kernel Cake 293*
  - *Meal 293*
  - *Mesocarp Fibre 293*
  - *Shell 293*

G

Grading 252
- Form 257
- Frequency 253
- Report 257
- Sampling 252

H

Harvesting 241
- Mature Palms 246
- Tools 244
- Young Palms 245

I

Imports 9
- Animal Oils 9
- Vegetable Oils 9

Inter Row Application 196
Irrigation 184
- Methods 184
    - *Basin 184*
    - *Drip Irrigation 184*
    - *Flood Irrigation 184*
    - *Micro-Sprinkler Irrigation 184*
- Sources 175

K

Kernel Recovery 276

L

Leaf Analysis 199, 200, 202
- Choice of Leaf Material 200
- Frequency 201
- Leaf Sampling 200
- Preparation of Leaf Samples 200
- Sampling Density 201
- Sources of Variation 201

M

Major Producers 18
- Africa 26
- Indonesia 22
- Latin America 26
- Malaysia 22
- Myanmar 24
- Thailand 24

Methods of Weed Management 162
- Biological Control 165
- Chemical Weeding 164
- Integrated 165
- Legume Covers 162
- Manual Weeding 164
- Mulching 186
- Natural Ground Covers 163
- Plastic Mulching 165

Mill Effluent 293
Milling Technology 274
- Continuous Sterilisation 275
- Fruit Reception 274
- Handling 274
- Sterilization 274
- Threshing Technique 275

Mixed Farming 221
Moisture Conservation 185
Mulching 139, 159
- Empty Bunches 160
- Oil Palm Leaves 160
- Organic Wastes 159
- Plastic Mulches 160

N

Nava Bharat Plantations 54
Non Food Uses 282
- Detergents 283
- Oleo-chemicals 283
- Personal Care Products 283
- Polyurethane Products 283
- Soap 283

Nursery Establish 133
- Choice of Soil 134
- Diseases 234
    - *Blast Disease 234*
    - *Curvularia Seedling Blight 235*
    - *Early Leaf Disease 234*
    - *Leaf Rot 234*
    - *Leaf Spots 234*
- Fencing 134
- Field Nursery 125
- Filling of Bags 134
- New Seedlings 167
- Pests 225
    - *Pink Shoot Borer 226*
    - *Red Spider Mite 225*
    - *Spindle Bug 226*
    - *Tussock Caterpillar 226*
- Polybag Nurseries 135
    - *Double Stage Nursery 135*
    - *Single Stage Nursery 135*
- Pre-nursery 125
- Rooting 124
- Site Selection 133

Nutrient 189
- Boron 208
- Calcium 191
- Magnesium 207

- Micronutrients 192
- Nitrogen 190
- Phosphorus 190
- Potassium 190
- Sulphur 192

Nutrition 284
- Antioxidants 284
- Carotenoids 284
- Diseases 285
  Atherosclerosis 286
  Cancer 286
  Coronary Heart Diseases 285
  Thrombosis 286
- Energy Source 284
- Fatty Acid 285
- Trans Fatty Acid 286
- Vitamin E 284

O

Oil Palm 1, 2, 30, 40, 45, 46
- By-product Utilization 4
- Demonstration Project (OPD) 54
- Development Project (OPDP) 54
- Diversification 2
- Helps to Balance out Carbon Emission 4
- Most Versatile and Useful Plant 5
- Non-conventional Energy 3
- Save the Environment 4
- Value Addition 3

Organic Recycling 196

P

Packing 121
Palm Oil 5
- Health and Nutritional Security 5
- Import Substitute 5

Palms 167
- Crown Damaged 167
- Performance 168
- Soil 199
  Soils Conditions 89
  Soil Type 90

Pests 231
- Adult 226
  Red Palm Weevil 229
  Rhinoceros Beetle 226
- Minor 231
  Aphids 231
  Bagworm 231
  Cockchafer Beetle 232
  Hopper 232
  Mealy Bugs 231
  Nettle Caterpillar 231
  Scales 231
  Termites 232
- Vertebrate Pests 233
  Birds 233
  Mammals 233

Plantation Establishment 151
- Age of Seedlings 153
- Field Preparation 151
- Growing Green Manure Crops 157
- Lay Out 152
- Pitting 153
- Planting Method 154
- Putting Wire Netting 157
- Spacing 152
- Special Planting Technique 157
- Time of Planting 154

Plantations 226
- Adult 226
  Disease 234
    Basal Stem Rot 236
    Bud Rot 235
    Bunch Failure 238
    Bunch Rot 239
    Crown Disease 239
    Spear Rot 238
    Stem Wet Rot 237
    Upper Stem Rot 238
    Vascular Wilt Disease 239
- Captive Plantations 174
- Commercial Plantations 46
  Karnataka 47
  Kerala 46
  Little Andaman 47

Planting 138
Pollen Collection 117
- Female Inflorescence 118
- Pisifera Pollen 117
- Pollen Quality 118

Pollinating Weevils 160
Pollination 117

Polyembryoid Cultures 124
Post-Harvest Technology 39
Potential Areas 50
Preparation of Seed 119
- Depericarping 119
- Seed Treatment 120
Processing 271
- Clarification 272
- Digestion 272
- Pressing 272
- Purification 272
- Sterilization 271
- Stripping 272
- Vacuum Drying 273
Production 7
- Global 7
  Africa 30
  France 31
  India 32
  Indonesia 31
  Malaysia 32
  Myanmar 31
  Papua New Guinea 31
  Phillipines 31
  South America 31
  Thailand 31
- India 74
  Andaman and Nicobar Islands 75
  Andhra Pradesh 77
  Arunachal Pradesh 77
  Assam 78
  Bihar 78
  Chhattisgarh 79
  Goa 79
  Gujarat 80
  Karnataka 80
  Kerala 81
  Maharashtra 81
  Meghalaya 81
  Mizoram 82
  Nagaland 82
  Odisha 82
  Tamil Nadu 83
  Tripura 83
  West Bengal 84
Products 281
- Chocolate 281
- Confectionery 281
- Ice Cream 281
- Margarine 280
- Shortening 281
- Soap Noodles 281
- Vegetable Ghee 281
Propagation 123
- Germination Techniques 121
- *In vitro* 123
Pruning 158

R

Ripeness Standards Minimum 252

S

Seed Gardens 116
- Selection of Dura Parents 116
- Selection of Pisifera Parents 117
Seedlings Selection 143
- Culling 143
- Over-aged 146
- Planting Doubletons 145
- Precautions 146
- Splitting 145
Seed Requirement 122
Shading 139
Shoot Regeneration 124
Slanted Palms 167
Spacing 138
State Committees 50
Steam Management 277
Super Cyclone Affected Palms 166

T

Transport 248
- Bunches 248
- Loose Fruits 249

V

Varieties 125
- Bamenda x Ekona 127
- Deli X AVROS 126
- Deli x Ekona 126
- Deli x Ghana 126
- Deli x La Mé 127
- Deli x Nigeria 127

- Deli x Yangambi 127
- Tanzania x Ekona 127
- Varieties Compact
  Compact x Ghana 128
  Compact x Nigeria 128
  Deli x Compact 128

Vegetable Oil 11
- Area 12
- Consumption 11, 14
- Future Projections 15
- Indian Scenario 11
- Production 12
- Productivity 12

W

Wastelands 169
- Cultivable 171
- Preparation 172
- Uncultivable 172

Wastes 287
- Empty Fruit Bunches 289
- Fibres 289
- Fronds 288
- Shells 289
- Trunks 288

Water Harvesting 187

Water Management 140, 141, 179
- Hose Pipe and Rose Can 140
- Sprinkler System 140

Weeding 139

Weed Management 162
- Methods 162
  Biological Control 165
  Chemical Weeding 164
  Integrated Weed Management 165
  Legume Covers 162
  Manual Weeding 164
  Natural Ground Covers 163
  Plastic Mulching 165

Weed Shifts 166

Z

Zero Discharge 298

Zero Waste 298

Figure 3.1: Power Operated Harvestor. (p. 40)

Figure 6.3: Leaflet Arrangement Along the Rachis in *E. guineensis* (a) *E. oleifera* and OxG Hybrids (b). (p. 97)

Figure 6.4: Flower Emergence to Ripe Bunch. (p. 100)

Figure 6.5: Inflorescences–a) Male and b) Female. (p. 101)

Figure 6.5: Inflorescences–a) Male and b) Female. (p. 101)

Figure 6.8: Various Segments of Fruit Bunches. (P. 103)

Figure 9.2: Double Stage Nursery. (P. 136)

Figure 9.2: Double Stage Nursery. (P. 136)

Figure 9.3: Nursery Bags Arranged Over Polysheet. (p. 139)

Figure 9.4: Shading Nursery Bags with Palmyrah Leaves during the Summer. (p. 140)

Figure 9.5: Sprinkler irrigation. (p. 141)

Figure 9.6a and b: Drip Irrigation. (p. 141)

a) Culling of Seedlings at Preliminary Nursery

Figure 9.7: Various Symptoms of Abnormal Seedlings. (p. 144)

b) Culling of Seedlings at Main Nursery

Figure 9.7: Various Symptoms of Abnormal Seedlings. (p. 145)

a) Tractor mounted Pot hole digger

Figure 10.2: Preparation of Planting Pits. (p. 154)

c) A view of the field with pits

Figure 10.2: Preparation of Planting Pits. (p. 154)

a) Preparing pit for planting

Figure 10.3: Method of Planting. (p. 155)

b) Filling pit and poly bag removal

Figure 10.3: Method of Planting. (p. 155)

c) Planting seedling and irrigation

Figure 10.3: Method of Planting. (p. 155)

d) Planting, pressing all around and irrigation

Figure 10.3: Method of Planting. (p. 155)

Figure 10.4: Making One Meter Wide Basins after Planting. (p. 156)

Figure 10.4: Making One Meter Wide Basins after Planting. (p. 156)

Figure 10.5: Forming Irrigation Channels-Correct Method. (p. 156)

Figure 10.6: Drip Irrigation. (p. 156)

a) Sowing *dhaincha* seeds in basins

Figure 10.8: Growing Green Manure Crop in the Basin of the Seedling. (p. 157)

b) Sowing sun hemp seeds in the basins

Figure 10.8: Growing Green Manure Crop in the Basin of the Seedling. (p. 157)

a) Power tiller for farming basins

Figure 10.9: Use of Power Tiller for Basin Weeding, Mixing Manure, and Forming Basins. (p. 159)

Figure 10.11: Cover Crops in Oil Palm Plantation. (p. 163)

a) Wasteland with little slope

b) Wasteland plain

c) Wasteland in the foothills

d) Wasteland with shrubs

Figure 10.12: Different Types of Wastelands (p. 170)

e) Wasteland with rocky patches

f. Clearing of the shrubs

g. Burning of the cut shrubs

Figure 10.12: Different Types of Wastelands (p. 170)

Figure 10.13: Planting Pits and Mixing Sand. (p. 172)

Figure 10.14: Planting Process and Mixing Soil. (p. 172)

Figure 10.14: Forming Wider Pits in the Stony Patches and Planting. (p. 174)

Figure 10.15: Forming Basin after Planting and Mulching in the Basin. (p. 174)

Figure 10.18: Oil Palm Growth at 26 Months after Planting. (p. 176)

Figure 10.19: Oil Palm in Rocky Patches of Birpradapur, Boudh District. (p. 176)

Figure 10.20: Five Year Old Palms in Yielding Stage. (p. 177)

Figure 10.21: River Water Pumping for Irrigation through Drip. (p. 177)

Figure 10.22: Water Harvesting Structures Developed for Irrigating Oil Palm. (p. 178)

Figure 10.22: Water Harvesting Structures Developed for Irrigating Oil Palm. (p. 178)

Figure 10.22: Water Harvesting Structures Developed for Irrigating Oil Palm. (p. 178)

Figure 10.22: Water Harvesting Structures Developed for Irrigating Oil Palm. (p. 178)

a) More unopened spindles and leaf yellowing and drying

b) Severe shortage of water-leaf drying

Figure 11.1: Water Deficiency Symptoms in Oil Palm Trees. (p. 181)

a) River Water Pumping

Figure 11.3: Water Harvesting Structures (p. 187)

b) Constructing Sunk Well in Rivers

Figure 11.3: Water Harvesting Structures (p. 187)

c) Small Farm Pond along *Nala* Side.

Figure 11.3: Water Harvesting Structures (p. 187)

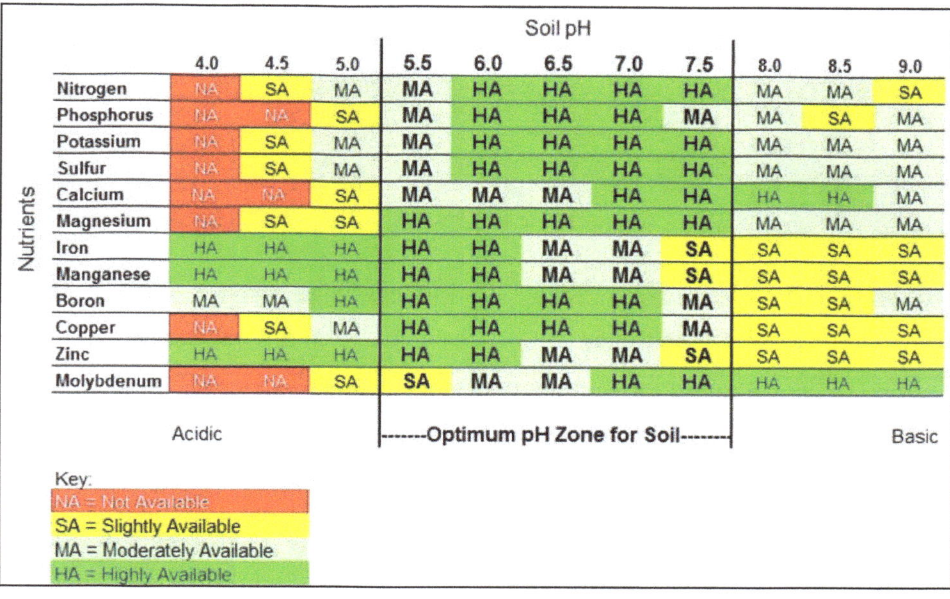

Soil pH

Nutrients	4.0	4.5	5.0	5.5	6.0	6.5	7.0	7.5	8.0	8.5	9.0
Nitrogen	NA	SA	MA	MA	HA	HA	HA	HA	MA	MA	SA
Phosphorus	NA	NA	SA	MA	HA	HA	HA	MA	MA	SA	MA
Potassium	NA	SA	MA	MA	HA	HA	HA	HA	MA	MA	MA
Sulfur	NA	SA	MA	MA	HA	HA	HA	HA	MA	MA	MA
Calcium	NA	NA	SA	MA	MA	MA	HA	HA	HA	HA	MA
Magnesium	NA	SA	SA	HA	HA	HA	HA	HA	MA	MA	MA
Iron	HA	HA	HA	HA	HA	MA	MA	SA	SA	SA	SA
Manganese	HA	HA	HA	HA	HA	MA	MA	SA	SA	SA	SA
Boron	MA	MA	HA	HA	HA	HA	HA	MA	SA	SA	MA
Copper	NA	SA	MA	HA	HA	HA	HA	MA	SA	SA	SA
Zinc	HA	HA	HA	HA	HA	MA	MA	SA	SA	SA	SA
Molybdenum	NA	NA	SA	SA	MA	MA	HA	HA	HA	HA	HA

Acidic -------Optimum pH Zone for Soil------- Basic

Key:
NA = Not Available
SA = Slightly Available
MA = Moderately Available
HA = Highly Available

Figure 12.1: Nutrient Availability Chart. (p. 198)

a) Nitrogen Deficiency

b) Magnesium Deficiency

c) Copper Deficiency

d) Potassium Deficiency

e) Boron Deficiency

f) N/K imbalance

Figure 12.3: Nutrient Deficiency Symptoms. (p. 205)

Figure 12.4: Boron Deficiency Symptoms. (p. 209)

Figure 12.6: Bunch Failure. (p. 211)

a) Lily in Oil palm

b) Marigold in Oil palm

c) Turmeric in Oil palm

d) Banana in Oil palm

c) Brinjal in Oil palm

Figure 13.1: Different Intercrops in Oil Palm. (p. 218)

a) Betelvine in Oil palm

Figure 13.2: Various Mixed Crops in Oil Palm. (p. 220)

b) Black pepper in Oil palm

c) Cocoa in Oil palm

Figure 13.2: Various Mixed Crops in Oil Palm. (p. 220)

d) Bush pepper in Oil palm **e) Vanilla in Oil palm**

Figure 13.2: Various Mixed Crops in Oil Palm. (p. 220)

Figure 13.3: Mixed Farming in Oil Palm with Sheep, Goat and Cattle. (p. 221)

Figure 14.1: Rhinocerous Beetles (Adults) and Symptoms on Young Palm. (p. 227)

Figure 15.4a: Harvesting Three Year Old Palm with Chisel. (p. 247)

Figure 15.4b: Harvesting in Young Oil Palm. (p. 247)

Figure 14.2: Symptoms, Life Cycle, Breeding Site and Bio-control. (p. 228)

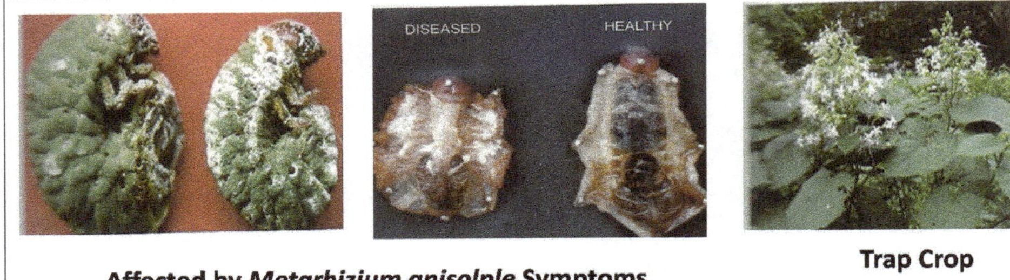

Figure 14.2: Symptoms, Life Cycle, Breeding Site and Bio-control. (p. 228)

Figure 14.3: Pheramone Trapping. (p. 230)

Figure 14.4: Red Palm Weevil Adult and Life Cycle. (p. 231)

Figure 15.5: (a-b) Harvesting of Matured Palms and (c) Harvest Pole. (p. 247)

Figure 15.6: Harvesting with Power Operated Harvester. (p. 248)

Figure 15.7: Lightest Harvesting Pole in Operation. (p. 248)

a) Harvest Using Ladder

Figure 15.8: Harvesting FFB Using Local Knowledge. (p. 249)

b) Using cycle tyre for climbing for harvest

Figure 15.8: Harvesting FFB Using Local Knowledge. (p. 249)

Figure 15.9: Gathering and Transport of FFB. (p. 249)

Figure 15.10: Collection and Gathering of Loose Fruits. (p. 250)

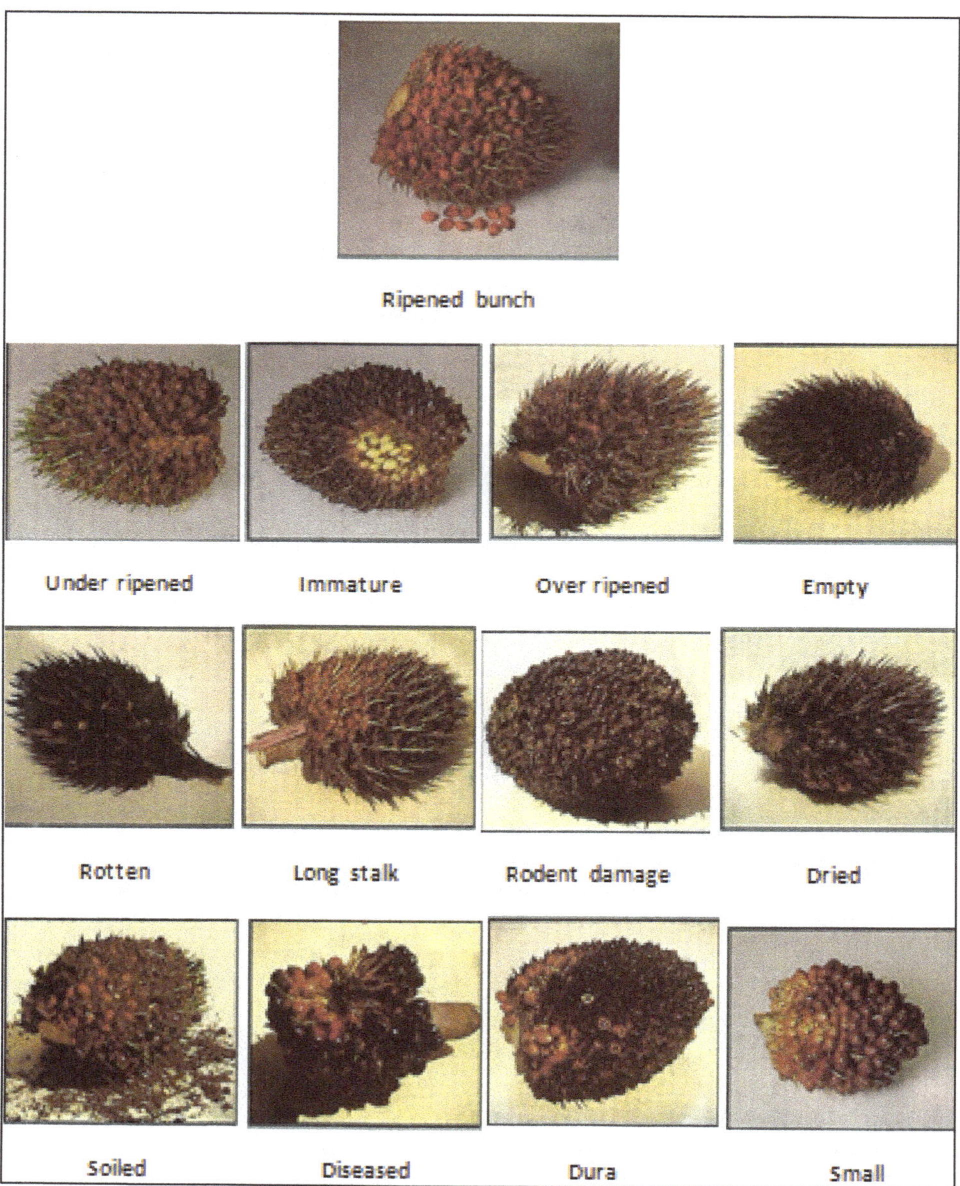

Figure 15.11: Types of Branches. (p. 255)

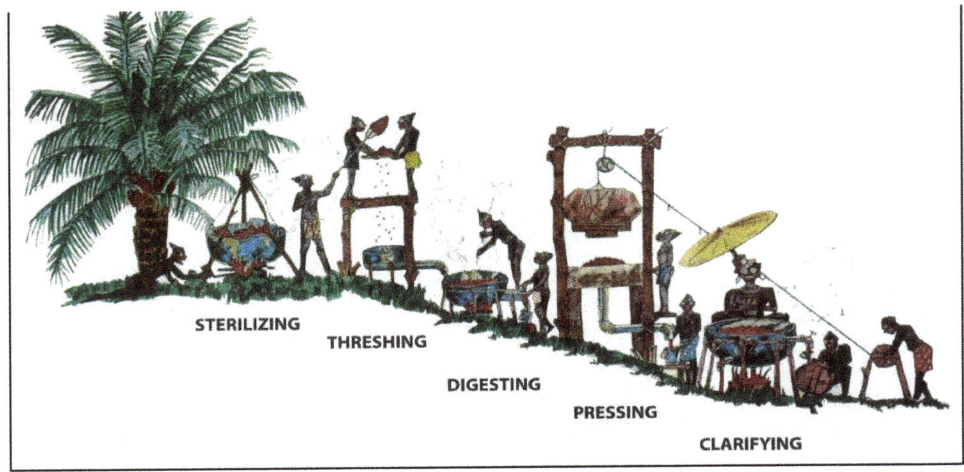

Figure 16.1: First Palm Oil Factory in Africa. (p. 270)

Figure 16.2: Modern Palm Oil Processing Mill. (p . 271)

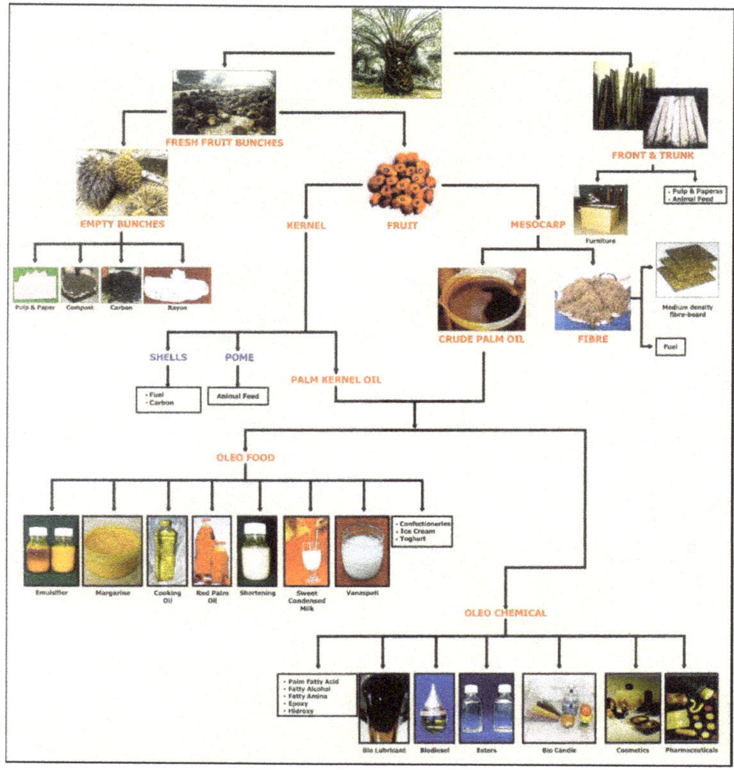

Figure 18.1: Oil Palm by-product Utilization. (p. 288)

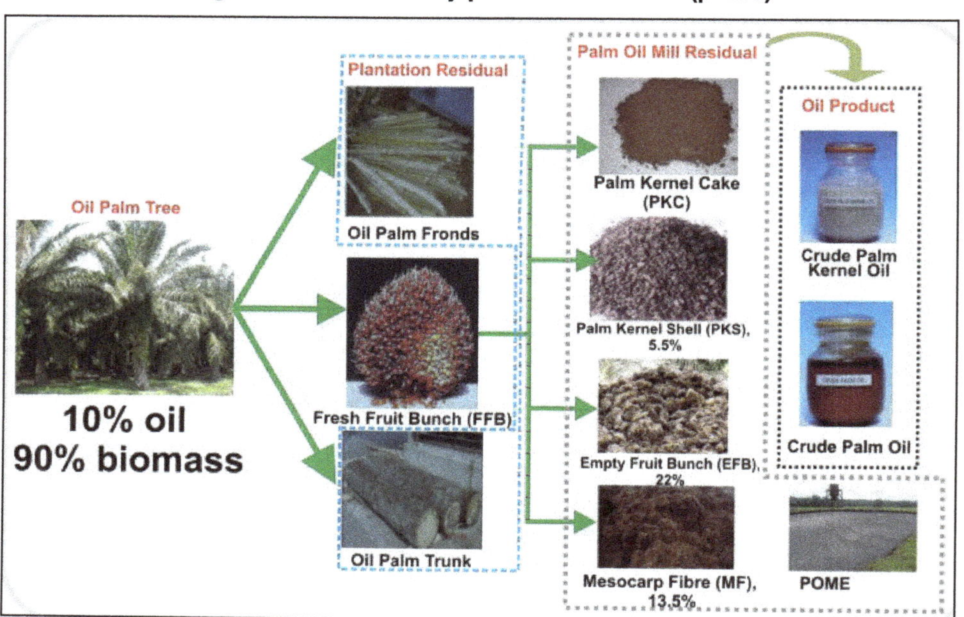

Figure 18.2: Distribution of Various Components in Oil Palm. (p. 289)

Figure 18.3: Empty Bunch Fibre Utilization. (p. 292)

Figure 18.4: Fibre Briquetes. (p. 292)

Figure 18.5: Compressed Laminated Products. (p. 292)

www.ingramcontent.com/pod-product-compliance
Ingram Content Group UK Ltd.
Pitfield, Milton Keynes, MK11 3LW, UK
UKHW021009290726
14059UKWH00001BA/39

9 789388 173858